THE DYNAMICS OF COASTAL MODELS

Coastal basins are defined as estuaries, lagoons, and embayments. This book deals with the science of coastal basins using simple models, many of which are presented in either analytical form or through numerical code in Microsoft Excel or MATLAB™. The book introduces simple hydrodynamics and its applications to mixing, flushing, roughness, coral reefs, sediment dynamics, and Stommel transitions. The topics covered extend from the use of simple box and one-dimensional models to flow over coral reefs, highlighting applications to biogeochemical processes. The book also emphasizes models as a scientific tool in our understanding of coasts, and introduces the value of the most modern flexible mesh combined wave–current models. The author has picked examples from shallow basins around the world to illustrate the wonders of the scientific method and the power of simple dynamics.

This book is ideal for use as an advanced textbook for students and as an introduction to the topic for researchers, especially those from other fields of science needing a basic understanding of the fundamental ideas of the dynamics of coastal embayments and the way that they can be modeled.

CLIFFORD J. HEARN is Director of the Tampa Bay Modeling Program for the United States Geological Survey.

THE DYNAMICS OF
COASTAL MODELS

Clifford J. Hearn

CAMBRIDGE
UNIVERSITY PRESS

CAMBRIDGE
UNIVERSITY PRESS

Shaftesbury Road, Cambridge CB2 8EA, United Kingdom

One Liberty Plaza, 20th Floor, New York, NY 10006, USA

477 Williamstown Road, Port Melbourne, VIC 3207, Australia

314–321, 3rd Floor, Plot 3, Splendor Forum, Jasola District Centre, New Delhi – 110025, India

103 Penang Road, #05–06/07, Visioncrest Commercial, Singapore 238467

Cambridge University Press is part of Cambridge University Press & Assessment,
a department of the University of Cambridge.

We share the University's mission to contribute to society through the pursuit of
education, learning and research at the highest international levels of excellence.

www.cambridge.org
Information on this title: www.cambridge.org/9780521807401

First published 2008

A catalogue record for this publication is available from the British Library

ISBN   978-0-521-80740-1   Hardback

Contents

Preface

My intention is to cover the material that would normally be relevant to an environment study but only in so far as this can be represented by comparatively simple models. It is not my intention to consider very specific models that describe particular systems in great detail. Indeed that would distract from the very purpose of the book which is to illustrate basic concepts in terms of simple models. Furthermore, modeling real coastal basins starts with the simple models described in this book. The models can then be extended in terms of detail and aggregated so that they become realistic simulations of actual coastal basins. In so doing, most of the fundamental processes represented by the models remain essentially unchanged. Simple models deal with individual processes and fortunately real systems can usually be simulated (to a first approximation) by the sum of these processes (although understanding of the interaction between these processes is often critical at a later stage). Modelers tend to assess the basic science of coastal basins by implementing such elementary models and their *modus operandi*, in the initial stages of investigating coastal basins, is to keep models fairly simple. Models are built slowly as more data become available, and the structure of the basic scientific ingredients changes very little as sophistication is added.

I have added a very extensive bibliography of books (as "Further reading") at the end of each chapter. This includes both recent and newer textbooks that deal with the basic science of coastal basins and models. I have not hesitated to include some texts which may be more difficult to obtain since they are often unique in terms of their presentations. The present book deals with a wide spectrum of science and data analysis and I have broadened the bibliographies appropriately. My intention is that the book should be read as a *textbook*, and so I have presented the modeling ideas in an educational context and not as an extensive review of recent research. For that reason, I have limited references within the text and have adopted a colloquial style in which I try to mention something of the contributions made by distinguished scientists as their work is mentioned. My apologies for so much important work that is omitted, and the reader will find much more detail in books devoted to particular aspects of coastal and ocean science some of which are included in the bibliography.

There are changes in meanings of terms in different branches of science and the dynamics of coastal basins certainly straddles several of these divides. The meaning of the word *model*, or *modeling*, is a classic case which I have explored in some detail in Chapter 1 but there are other, less obvious, cases that do cause considerable confusion. Good examples are the trio of terms *estuarine circulation*, *density driven flow*, and *stratified basin* which describe various states, and processes, in a coastal basin which are discussed in the later chapters of this book. Caution is needed in using these terms without the necessary qualification as to their meaning and they should often be avoided in a technical context.

My intention is that this book should present models as tools for understanding science and as a direct challenge to the notion that *models are for modelers*. Models are simply a way of articulating scientific ideas and are a simple application of mathematics which is the basis of all science.

Acknowledgements

The book reflects lectures given to a mixture of classes in several continents. I have had the advantage of discussions and research with many colleagues with deep understanding of the science of coastal basins and many of whom have become close personal friends. They are too numerous to name but I want to record my very special thanks to my late colleagues Cyril A. Hogarth and Paul N. Butcher, and to my advisor Peter T. Landsberg, from whom I learnt the joy of simple analytical models as a way of conceptualizing ideas about processes. My fundamental interests in the properties of non-linear processes in coastal basins comes from years at the University of Warwick with George Rowlands and these are discussed in Chapters 2, 8, and 10. My early interest in physical processes in the ocean came from conversations with Henry Stommel when I was at Harvard University and this is reflected in the text on *Stommel transitions* in Chapter 10. My interest in shallow coastal basins came originally from my work with John H. Simpson and then Jörg Imberger. My interest in biological processes in coastal basins, and especially coral reefs, stems from work with Bruce G. Hatcher, Stephen V. Smith, Arthur J. McComb, and Marlin Atkinson. The discussion of the coastal boundary zone in Chapter 11 was motivated by conversations with William M. Hamner during my time at the University of California at Los Angeles. Individual parts of the book use model technology and ideas for which a special gratitude is due to several individuals and companies who are acknowledged in the text. Kimberley K. Yates has been responsible for much of my appreciation of the value of modeling in the context of integrated systems and to her go very special thanks. I owe a special debt of gratitude to Dr. John R. Hunter of the University of Tasmania with whom I worked on many projects in modeling, and parts of Chapter 8 come from an idea that he and I pursued together. This book was completed during the time that I was engaged on the Integrated Science Study of Tampa Bay in Florida, and I acknowledge the support of the US Geological Survey and ETI Professionals Inc. I have included reference to models and data for Tampa Bay because of my personal involvement in that project and because they represent good example of the ideas presented in the book. The model development in Tampa Bay is the cumulative effort of many scientists at the US Geological Survey, University of South Florida, colleagues at DHI, and the Danish Technical University in Copenhagen. To all these

scientists and colleagues go my special thanks. I am grateful to my friends at Delft
Hydraulics and the Technical University of Delft (Netherlands) for making my time
working with them on models of coastal basins so very productive in that historic town.

The task of removing typographical errors and other inconsistencies has been
performed to the best of my ability but there are, no doubt, many remaining and I
appreciate the time of any readers in pointing them out to me. My thanks to the staff at
Cambridge University Press for so much help and all my colleagues in the US
Geological Survey for their support.

Note on mathematics and model codes

I will not generally present detailed computer codes for the simple models given in the book except where it is possible to use easily available software such as *MS Excel*, or the Mathworks *Matlab* programs. The codes that are presented here are sufficiently simple that they should provide valuable insight into the basic science which is the *raison d'être* for this book. I feel that the languages of both mathematics, and computer code, can assist our understanding of the basis processes of coastal basins, but should never be allowed to obscure the fundamental science. For this reason, I will very deliberately minimize mathematics and avoid any attempt at code beyond a little occasional *Excel* and some simple *Matlab* scripts. However, it is unfortunately impossible to discuss basic processes without some mathematics. I have successfully taught this material to university and college graduates, and undergraduates, with only very limited mathematical training. The basic mathematical requirement is a knowledge of simple differential and integral calculus and the methods by which we can represent derivatives by finite differences. I make no excuse for the inclusion of material based on this elementary calculus, and indeed mathematics is the basic language with which physical scientists establish and exchange ideas. The book is aimed at a wide spectrum of scientists interested in coastal basins with the caveat that they have a minimal mathematical knowledge. This includes biologists, chemists, geologists, and physical scientists starting to work in coastal basins, plus students of any undergraduate, or graduate, course in oceanography, marine science, or environmental science. I have tried to make the book one that delivers results and usability based on simple models. Each such model will be largely self-contained, and so it should be possible to consider each model separately although, there will be references made between the models in the sense that one model may be a progression of an earlier model (usually in the same chapter).

I have tried to avoid double use of mathematical symbols as far as possible but occasionally this would only be possible at the expense of producing mathematics that is quite difficult to read and my philosophy is that price is too high. As an undergraduate, I had classes from a professor who would just pull notation out of the air and said this was a good way of making sure one really understood the physical meaning of mathematics. With that I did not agree at the time, and still do not agree, but on the other hand we have cases like the symbol f used for the Coriolis parameter. To change that symbol makes a text on physical oceanography very obscure to the eye of most modelers. Yet f is used so naturally in applied mathematics to denote *some function* that we are bound to have double use of the symbol. Similarly ϕ is used for the fineness scale of sediments and to use any other symbol would be unthinkable and yet in other contexts, ϕ is used universally as an angle and especially for latitude on the Earth. The use of the same symbol for different quantities is usually safe provided that the usage is in very different aspects of a subject but danger does lurk in cross-disciplinary studies; there are few theoreticians who have not caused themselves hours of worry over a confusion of that type. But that is a difference between a textbook and a piece of personal analysis. So, I have been careful with symbols but occasionally one symbol is used for different entities but hopefully in different parts of the book. I have often (but not always) used a convention adopted in fluid mechanics of using a double symbol for dimensionless numbers such as Reynolds number Re, Ekman number *Ek*, Froude number *Fr*, and many others. My personal view is that these double symbols can cause confusion in equations and to avoid that mathematical confusion, I often revert to a single symbol sometimes with a subscript.

1

Prelude to modeling coastal basins

1.1 Coastal basins

This book explains the basic dynamics of bays, estuaries, and lagoons through the use of simple models. There is a focus on physically simple systems for which easy-to-understand models give good insight into basic processes. These models may be either analytical or numerical (or ideally both). The book uses these simple models to present our basic ideas of processes in coastal basins with a deliberate emphasis on box models and simple one-dimensional and (occasionally) higher-dimensional models. The book avoids, as far as possible, the complexities of three-dimensional models in favor of the simplicity of lower-dimensional models.

I use the term *coastal basins* to represent the myriad of different water bodies which we find in the region between the land and the open continental shelf. They carry local names such as *estuary, bay, sound, inlet, gulf* although those names are also used for systems which lie on the continental shelf or form a part of deep ocean basins. Finding a generally accepted name is not easy and the obvious choice of *estuaries* is not really correct in so far as estuaries normally have a very dominant circulation due to density differences; whereas much of the basic dynamics of many coastal basins is due to tides, winds, and waves. The term *lagoon* is also a possibility but carries a connotation of certain types of morphology. I have therefore settled on the term *coastal basins* although I must stress that I am not including aquifer or drainage basins and I have not included models of surface water flow or groundwater systems. So, a *coastal basin*, in the context of this book, is a shallow system, i.e., typically less than 10 m (although possibly some tens of meters deep) forced mainly by wind, tide, and river flow. A key requirement of a coastal basin (in the present context) is that the coast has a dominant influence.

Much of our understanding of coastal basins is directly relevant to lakes and rivers but that is not a primary focus of this book. Although I have included a chapter on sediment dynamics, and wave models, I have avoided, any detailed discussion of *nearshore processes*. My reason is that these processes act on much smaller spatial and time scales and usually involve highly phenomenological laws of mechanical interaction between currents and solids. The processes are extremely important but

1

do require more space than was available in the present book and are treated extremely well elsewhere.[1]

A coastal basin, in the context of this book is a shallow system, i.e., typically less than 10 m (although possibly some tens of meters deep) forced by wind, waves, tide, river flow, and other types of buoyancy flux such as surface heating, cooling, and evaporation. Topography and bathymetry play a major role in the dynamics of a coastal basin. Individual basins may carry names such as Bay, Estuary, Lagoon, Sound, Gulf, etc. We shall cover the material relevant to environment studies that can be represented by comparatively simple models, which illustrate basic concepts.

Coastal basins differ from the deep ocean and continental shelf in many respects. One of the most important differences comes through the effect of astronomical tides. It is usual to refer to the astronomical tides as simply *tides*, and we shall often use that abbreviation. We need to be aware, however, that a *tide* really refers to a change in water level which may come from a variety of causes, including winds and changes in atmospheric pressure. All of these tides are important in coastal basins, but astronomical tides are usually (but not exclusively) the most important because they are always present. Tidal ranges in most of the world's oceans are in the order of 1 or 2 m. Occasionally we find tidal resonances that produce *macroscopic* tides of ranges up to 10 m, or *microscopic* tides which range a few tens of centimeters.

1.2 Geomorphic classification of ocean basins

Coastal basins have historically attracted human populations and associated harbor construction, especially since they often provide very low energy wave environments. Construction of coastal outfalls from industrial and sewage plants have followed human development, and the *built environment* has markedly changed the physical and ecological dynamics of many of these basins.

There are many hundreds of thousands of coastal basins in the world, ranging in size from tiny inlets to huge estuaries and coastal seas. We try to develop ideas that are applicable to whole ranges of basins, because we cannot expect to develop new ideas for each particular basin. Science always tries to generalize the dynamic behavior of systems, so that ideas are portable between systems. So, an estuary in China may be very similar to another in south-west Australia, and we attempt to construct a conceptual model that applies to both of these basins and very many others involving only very minor changes in the value of some parameters. Clearly, we cannot expect one conceptual model to apply to all basins, but we can expect one model to apply to basins that are in some sense *very similar*. Hence, we try to develop a *classification system*, so that we can divide all basins into categories and develop conceptual models of each category.

---

[1] For example: Fredsøe and Diehards (1992).

Unfortunately, there is no single classification system because coastal basins have so many different facets. We might, for example, just classify basins in terms of area: small, medium, and large, but that would not be very helpful because there are so many factors other than area that control the behavior of basins. So, we try to devise classification systems that are based on what we consider to be the important controlling factors for some particular type of behavior. The two simple systems are *geomorphic* and *stratification*. The geomorphic classification system is based on topography, geology, and sediment dynamics, while stratification refers to the existence of strata, or layers, of water with different density (and often having dissimilar sources such as different rivers or the open ocean).

There are various types of geomorphology. All present-day estuaries date from the end of the last ice age (about 20 000 years BP) when sea level had its last significant change, and most have been severely modified by secondary changes since that date. All estuaries are in a constant state of geomorphic change, and so their classification may alter with time. There are basically seven types of geomorphically distinct basins, and these are discussed in the sections below. Most coastal basins have features that fall within several of the classes which we list here. So the geomorphic classification of a basin is not unique, and most basins have characteristics of several classes of basins. The names which have historically been given to basins do not usually help in the classification system. Just as botanists use strict Latin names for plants and ignore their common names, we need to ignore the historical names for basins. So, whether a basin appears on a map as a *lagoon, bay, sound, inlet, gulf, estuary, strait, sea, river, channel,* or *passage,* or any local variant, should not influence our attempt to classify the basin.

1.2.1 Coastal plain (or drowned river valley) basins

These basins are due to the invasion of an existing river valley by rising sea levels. A good example is Sydney Harbour shown as Figure 1.1.

The boundaries of this type of coastal basin are essentially one of the height contours of the terrain that was flooded. The terrain around Sydney is hilly, with complex and convoluted height contours. This produces a coastline with the same type of characteristics, making the area one with an abundant supply of *waterfront property* along its many small creeks and inlets. The name *coastal plain* is accepted but not very accurate, since the flooded terrain may be far from the coastline that existed prior to flooding, so we prefer the term *flooded valley.*

1.2.2 Fjords

Flooded valleys cut by a glacier are given the name *fjord.* These valleys have a distinct sill near the mouth of the basin, and are now filled with marine water with a river running into the basin at its head. Fjords are generally deeper than the coastal basins

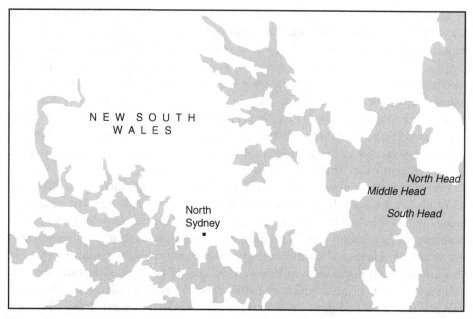

Figure 1.1 Sydney Harbour is a coastal plain estuary resulting from the rise of sea level at the end of the last ice age.

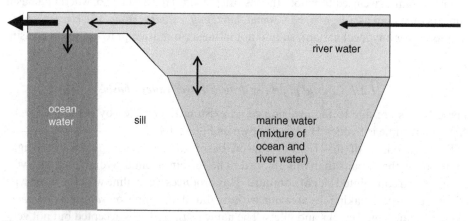

Figure 1.2 Schematic side view of a fjord.

which we consider in this book, and those depths have been created by the cutting action of the glacier (Figure 1.2).

The sill consists of debris carried by the glacier and dumped as the glacier melted into the ocean. Heavier marine water tends to be trapped at the bottom of the basin behind the sill. Similar sills are found acting as natural dams in mountain lakes. One occasionally finds sills in coastal basins that are not associated with glacial action, and these are termed *fjord-like* basins.

The sills again tend to trap ocean water at the bottom of the basin behind the sill. Usually the ocean water is brought into the basin by exceptionally high tides or by a storm event. The basin will then show a high degree of stratification, which inhibits vertical mixing. The usual sign of water that has been long trapped at the bottom of a basin is that it becomes hypoxic, i.e., the oxygen concentration is reduced well below the saturated concentration. This is a characteristic of stratified bottom water because the stratification greatly reduces the downward mixing of oxygen from the atmosphere, while the bottom water loses oxygen due to respiration by creatures and decaying bottom detritus. If the divide between the less dense surface water and denser deeper water (called the *pycnocline*) is close to the top of the sill, *internal waves* (waves along the pycnocline) can transport bottom water over the sill. Wind forcing can also be responsible for moving water over the sill. This is illustrated in Figure 1.2. It is basically an internal wind-driven tide that produces *upwelling* (upward movement of bottom water). If the basin is of uniform depth, the surface of the water is raised downwind by the action of the surface wind stress. This tends to produce a balance between wind stress and the pressure gradient produced by the surface elevation. Because the wind stress acts at the surface, it requires some mixing process to carry that stress down through the water column. In a rotating basin the penetration of the wind stress is limited by the Ekman layer. There is a water circulation downwind at the surface (wind driven) and a return upwind flow at the bottom driven by the pressure gradient.

1.2.3 Bar-built estuaries and inland lagoons

Shallow basins enclosed by a spit or bar are produced by littoral drift as in Figure 1.3. The basins are created by one or more rivers entering the coastal ocean and may have been wide open embayments at some time since the end of the last ice age.

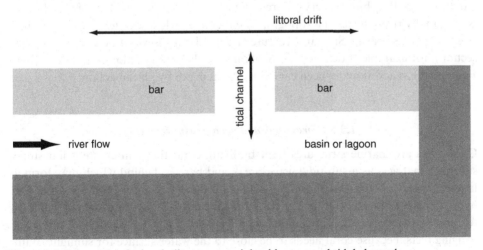

Figure 1.3 Schematic of a bar-built estuary or inland lagoon and tidal channel.

The size of the opening in the bar depends on flow of river (or other fresh water) or tidal flow, and in some cases the basin is closed except during floods. Bar-built basins are very common in regions having seasonal rainfall, so that there are months of limited precipitation in which the bar can form via *littoral* (alongshore) drift of sand along the coast. Most bar-built basins are very shallow (few meters) and have usually been formed by the retreat of sea level, as was common after its initial rise at the end of the last ice age. The opening in the bar usually carries strong tidal currents, with the size of this tidal channel contracting until tidal currents are strong enough to resuspend sand or larger-sized sediment ($0.15 \, \text{m s}^{-1}$ or higher). Other types of estuaries and lagoons may also have bars. The tidal channel usually restricts the magnitude of astronomical tides in the main part of the basin, because the channel itself requires a substantial pressure gradient to force the flow of water. This also creates a phase lag between the basin and the open ocean which can approach 90° in the limit of a very restrictive channel, so that the elevation of the water level in the basin is near zero at the times when the ocean elevation is maximal or minimal. Tidal channels can be very narrow and meandering, so that many *inland lagoons* are virtually hidden to an observer on the coastal ocean. Dutch explorers of the south-west coast of Australia in the late seventeenth century observed no river mouths.

1.2.4 Geological basins

Geological or tectonic basins are due to geological formations at the coast. The archetypal basin in this class is San Francisco Bay (Figure 1.4), which is associated with the San Andreas Fault. Such basins are often very elongated and also may have a bar and tidal channel, such as Tomales Bay (illustrated in Figure 2.3), and are the subject of many one-dimensional hydrodynamic models. Tomales Bay is just north of San Francisco Bay, but has very different dynamics. A common feature of such basins is the rapidly rising terrestrial topography either side, which can tend to direct wind stress along the basin. Similar structures exist inland (as we noted earlier in this section), for example, Loch Ness in Scotland. Loch Ness lies along the Great Glen Fault, and has additionally been carved to 200 m depth by glacial action.

1.2.5 Coral reefs and open-coast lagoons

Coral reefs are marine structures built by living coral that can create a limestone substrate that acts as a sub-tidal wave-break and barrier, behind which may form a lagoon. Such reefs may form around islands or along the continental margins. The hydrodynamics of water flow across reefs and through their lagoons has many unique features and properties that are familiar to other coastal basins. All coral reefs start as fringing reefs, because coral needs to be close to the water surface for sunlight. With changes in sea level these fringing reefs may move into deeper water and become

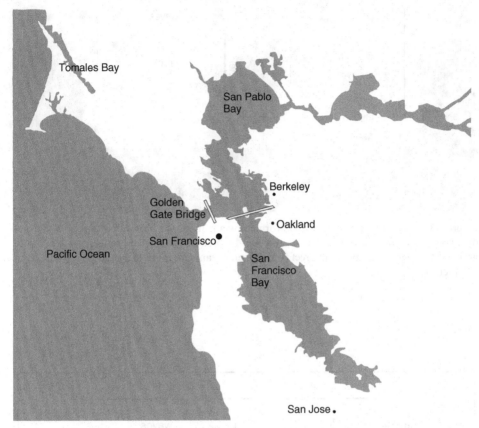

Figure 1.4 San Francisco Bay is a geological, or tectonic, basin with topography and bathymetry controlled by the San Andreas Fault. The city of San Francisco lies at the northern tip of the peninsula on the south-west side of the Bay. Notice Tomales Bay northward along the Californian coast.

barrier reefs, and possibly *atolls*. Most reefs have lagoons and most lagoons are very shallow, and either backed by land or other reefs. Water circulation in coral reef lagoons is often dominated by forcing due to breaking waves. We will consider some of these properties of coral reef lagoons and flow across coral reefs in Chapter 11. We have included the group *open-coastal lagoons* in the present class as distinct from lagoons in class 1.2.3. The intended distinction is that the lagoons in class 1.2.3 are separated from the open coast by a narrow channel. These are often called *inland lagoons*, despite the fact that they usually have a major marine component. Lagoons in the present class are on the open coast. The reef that forms the barrier around the lagoon is not necessarily a coral reef and may be, for example, a limestone reef formed by sea level rising over old sand dunes. Figures 1.5 and 1.6 show open-coast lagoons formed by a coral and limestone reef respectively.

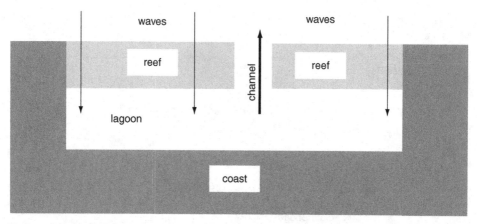

Figure 1.5 Schematic of a coastal basin formed by a fringing, or barrier, coral reef. The usual dynamic is that water is pumped across the reef by breaking waves and returns via a pressure gradient through a channel in the reef to the open ocean. Compare Kaneohe Bay discussed in Chapter 11.

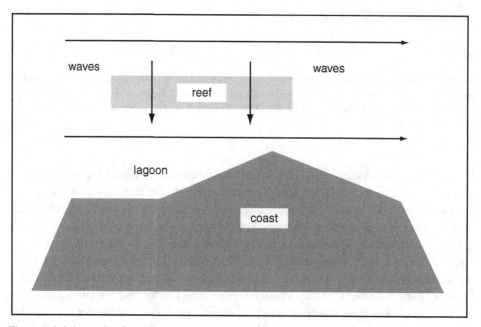

Figure 1.6 Schematic of the plan view of an open-coast lagoon. Such lagoons are often formed by limestone reefs. Dominant water flow is alongshore with some wave pumping across the reef. Compare Marmion Lagoon on the south-west coast of Australia. See Hatcher (1989).

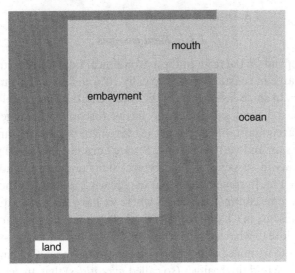

Figure 1.7 Schematic of a typical coastal embayment.

1.2.6 Coastal embayments

Open bays in the coastal ocean often form basins that are distinct from the coastal ocean. This is usually due to their topography (Figure 1.7). Water circulation inside embayments differs from that of the totally open coast, mainly through differences in tidal flow and regional coastal currents. As a result, coastal embayments usually show much more temperature and salinity variation than the open coast.

These embayments may be called Gulf, Sound, etc. Many embayments that have been produced by erosion since the rise in sea level at end of the last ice age can have many of the attributes of an estuary or lagoon, because fresh water flows into the bay and the exchange of water with the open shelf is slow.

1.2.7 Continental seas

These are essentially outside the scope of the present book but a part of our understanding of basin dynamics. They consist of enclosed, or partially enclosed, pieces of continental seas such as the Irish Sea, Baltic Sea, or Bass Strait. The distinction with most of the basins that we do consider in this book is mainly due to spatial scale, which is responsible for such processes as macrotides and high-energy wave conditions. In other ways they share many of the properties of coastal basins. For example, the Baltic Sea has lowered seasonal salinity due to freshwater runoff. The Baltic is on the outer limit of what we might reasonably call a coastal basin, but nevertheless shares some of their characteristics (albeit at much larger scale): it is strongly influenced by freshwater inflows, can freeze over, and also has comparatively limited entrances.

1.3 Distinctive features of coastal basins

1.3.1 Tidal currents

Tidal currents depend on the ratio of the astronomical tidal range to the water depth, which becomes large in coastal basins: typically 10% to 100%. This means that tidal currents are large. In some cases, parts of the continental shelf may well fall within our definition of a coastal basin, since tides may be macroscopic with ranges of up to 10 m. These currents have important effects on the movement of particles, and also on their horizontal dispersion and vertical mixing. Strong currents also erode the seabed and maintain sediments in suspension. The typical tidal range throughout the world's oceans is of order 1 m. In some oceans that may reach 2 to 3 m, but the range is rarely greater except on some continental shelves where we have *macrotides* due to resonance effect (which we discuss in Chapter 6). The tidal range may also drop to a fraction of a meter in regions close to what we call an *amphidrodal point*, but if we assume a range of 1 m that is correct to within a factor of perhaps five at the most. Only in very restricted coastal basins do we find tidal ranges (so called *microtides*) that are less than 20 cm. If there is a restriction in the width, currents are amplified accordingly. This is especially important in coastal basins where we find *tidal channels*, in which tidal currents can reach values only limited by the stability of the banks of the channel (typically 20 cm s^{-1} for sand, but reaching 1 to 2 m s^{-1} through rock or engineered surfaces such as armored walls). So, tides have a major influence on the dynamics of coastal basins, while in the deep ocean (and most continental shelves) they have minor effect. Most dynamical models of deep oceans are run quite independently of tides and indeed many are *rigid lid* models, i.e., the surface assumed fixed in time.

1.3.2 Wetting and drying

Wetting and drying of the shallow regions of coastal basins due to either tides or wind stress is a very important process for coastal models, and here lies a distinction from the deep ocean and most of the world's continental shelf; modelers talk of *fixed lid* models (for the deep ocean) and *variable surface* models for coastal basins. However, wetting and drying is not necessarily a major influence in all basins. This depends very much on the distribution of depth in the basin.

Figure 1.8 shows the depth distributions for some typical basins. The upper panel's depth (in meters) as a function of distance from the shore and the lower panels provide the corresponding histograms of basin area for each 1 m of depth. Basin (a) has steep sides and flat bottom over most of the basin. Basins of this type have often resulted from rivers cutting down through rock. Basin (b) has almost constant bottom slope and may have resulted from erosion through softer limestone whilst basin (c) has extensive shallows and a central deeper channel probably due to local sea level rise. The steep-sided basin has virtually no wetting and drying until water level drops below the bottom of the basin, while that with gently sloping sides may show a more gradual

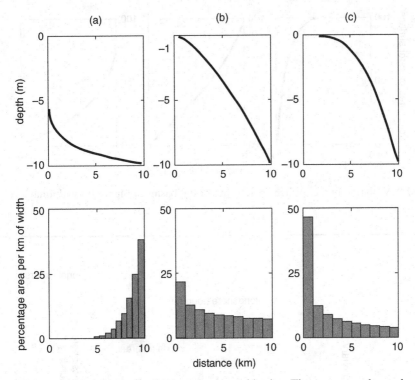

Figure 1.8 Some typical depth distributions for coastal basins. The upper panels are depth (in meters) as a function of distance from the shore and the lower panels provide the corresponding histograms of basin area.

change in wetted area with water level. The variation of wetted area with water level is displayed in Figure 1.9 for each of the depth distributions shown in Figure 1.8.

Wetting and drying in coastal basins creates the *intertidal* region, which has a unique influence on coastal ecology. It is probably true that *wetting–drying* is one of the most difficult processes to model numerically. It is a challenge absent from deep ocean and continental shelf models.

1.3.3 Modeling tidal influences

Modeling tidal influences in the deep ocean is a comparatively simple task, and essentially a solution of a wave equation. It has become much simpler since the 1990s due to the measurement of water elevation over the open ocean by satellite altimeters. These tidal models are useful for models of coastal basins, in that they supply the water level variations for the boundaries of coastal models. We call these *open boundary conditions,* and it is very important that we know the variation of tidal phase around these boundaries; especially if the current along the coast outside the

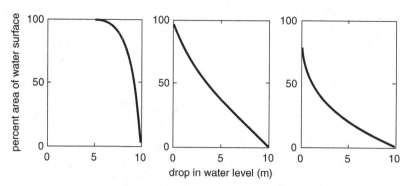

Figure 1.9 Variation in the surface area of the coastal basins in Figure 1.8 with tidal level.

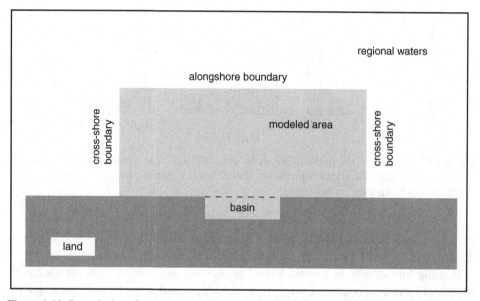

Figure 1.10 Boundaries of a coastal model. The basin which is of interest is indicated by the inner rectangle enclosed by land and open to the ocean at its mouth. The model area is greater than that of the basin in order that the boundaries with the open ocean are removed from the basin. The ocean boundaries are termed *cross-shore* and *alongshore*.

coastal basin is important. This is illustrated in Figure 1.10 which shows two types of open boundaries between a coastal model and the open ocean. One has cross-shore open boundaries and the other has only alongshore boundaries. For the case of alongshore boundaries, the current flowing along shore depends on small differences in the phase of the tidal elevation at the two across shelf boundaries.

The basin that is of prime interest in the model shown in Figure 1.10 is the small embayment surrounded by land on three sides and open to the ocean on its fourth side,

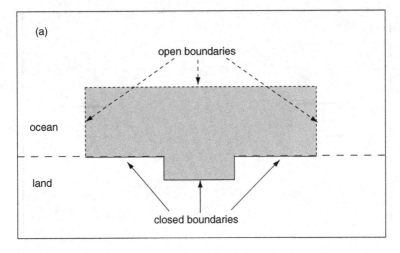

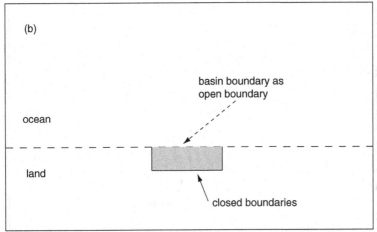

Figure 1.11 Panel (a) shows that there are two types of boundaries to the model in Figure 1.10: *closed* boundaries (which are boundaries with land) and *open* boundaries (which are boundaries with the open ocean). Panel (b) shows the *basin* boundary being used as an open boundary which has the disadvantage that it allows only one way transport of material originating in the basin; this is further illustrated in Figure 1.12(b).

which is the mouth of the basin. It is possible to put the open boundary straight across this mouth as shown in Figure 1.11(b), but we usually prefer to take the boundary further away from the basin as shown in Figure 1.11 (a). The reason is that boundaries involve conditions imposed on the model that are intrinsic restrictions, and our basic philosophy is that these we should therefore move the boundaries as far away as possible from the area of real interest to the model. Panels (a) and (b) of Figure 1.12 illustrate the two types of flow that can occur across open boundaries: residual flow (unidirectional arrows) and tidal flow (double-headed arrows).

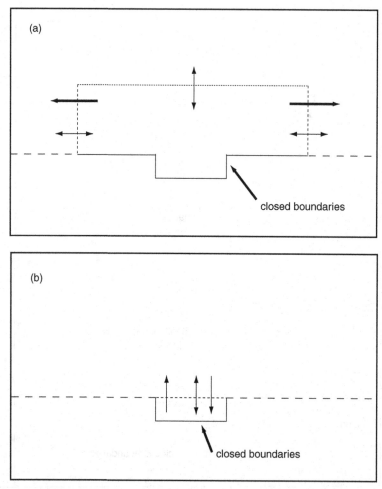

Figure 1.12 The types of water flow that can occur across the open boundaries of models shown in panels (a) and (b) of Figure 1.11. The unidirectional arrows are residual flows and the double-headed arrows are purely tidal flows which have zero residual current.

It might seem, at first sight, that we can always measure quantities along a boundary and so a boundary like that shown in panels (b) of Figures 1.11 and 1.12 is both economical and effective. This is found not to be true, and furthermore, models of contaminants (or tracers) introduced into the basin must use boundaries in regions far enough away from the source so as to be totally unaffected by the source. This is illustrated by panels (a) and (b) of Figure 1.13.

The coastline of the model produces *closed boundaries* and these are real boundaries with no ambiguity as to their nature, i.e., they block all transport. This simplicity is however a conceptual model in its own right since the coast is not a straight line and does not even conform to a geometric shape (as we shall discuss in Chapter 11). As

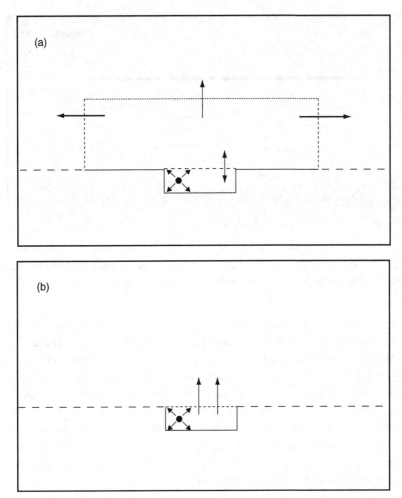

Figure 1.13 The paths of a tracer, or pollutant, introduced by a source in the basin (the dot with radial arrows) for panels (a) and (b) of Figures 1.11 and 1.12. The unidirectional arrows represent outward transport of tracer and the double-headed arrows indicate two-way movement of tracer. Because the tracer originates in the basin, it can only be transported *outward* across the open boundaries since the tracer concentration is necessarily set to zero on those boundaries. For panel (b), this means that the tracer swept out from the basin on the ebb tide does not return on the flood tide.

shown in Figure 1.14, modelers have a useful experimental tool at their disposal in transforming the alongshore open boundary into a closed boundary. This simplifies flow in the outer zone of the model and is perhaps justified as long as that boundary is far from the basin.

Figure 1.12 illustrates the types of flow which occur across open boundaries. The alongshore residual current may, or may not, be important to the dynamics of the

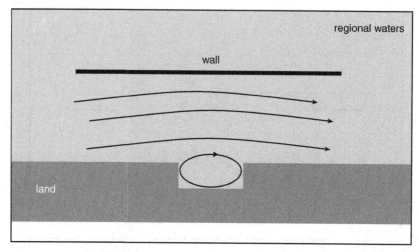

Figure 1.14 The artifact of closing an open alongshore boundary condition known colloquially as *building a wall* to simplify flow in the model zone outside of a basin.

coastal basin. This involves the interaction between the basin and the neighboring waters of the continental shelf. We often use the term *regional waters* (Figure 1.14) to describe the neighborhood of the open ocean outside the coastal basin. The interaction with regional waters is important to the tidal *flushing* of the coastal basin, because material that is swept out of the basin on an ebb tide may re-enter the basin on the next flood tide. This is affected by the transport of material along shore during the time between ebb and flood. If we simply draw an alongshore boundary across the mouth of the basin as shown in panels (b) of Figures 1.11 to 1.13, we are unable to properly model this re-entry process, and the flushing time is affected by the value assumed for the concentration of material on the boundary.

1.3.4 Bottom friction

Bottom friction becomes an important force in the dynamic balance of coastal basins as a consequence of limited water depth. This occurs for two reasons: the first is that bottom friction is a surface force, and therefore increases in relative importance as the depth decreases. So coastal models need to consider in great detail the form of bottom friction, and this force has a major effect on both the physical and ecological dynamics. This is in contrast to the deep ocean, and in contrast to the continental shelf (except for shelves with macroscopic tides), where friction is virtually negligible in much of the dynamics. The second reason that bottom friction is so important is that it usually (but again not exclusively) increases as the square of the current speed, while most other forces tend to be nearly linear in current speed. Coastal basins tend to show much larger currents than the deep ocean (although there are many counter examples to this statement) and therefore bottom friction is generally more important.

1.3.5 Wind forcing

The importance of wind forcing increases in coastal basins due to the limited water depth, in the sense that the whole of water column is affected by the wind rather than just a surface layer (as in the deep ocean).

1.3.6 Freshwater inflow

Fresh water enters coastal embayments from rivers, creeks, and as submarine ground-water discharge. It also flows in basins as simple surface runoff from the land. Models tend to subdivide these sources into *point* sources (rivers, creeks, and gullies) and *distributed* sources (groundwater and runoff). It also enters basins through direct precipitation (which is a distributed source). The relative magnitude of all of these sources depends on climate, coastal terrain, urbanization, diversion of water for irrigation, industrial and human use, and the depth of the coastal basin. Fresh water influences the dynamics of basins through changes which it produces in salinity and hence density and large volumes of water flow can physically force water out of the basin into the ocean. Evaporation also may affect coastal basins and in regions having arid periods, evaporation may be more important than freshwater inflow. An increase or decrease in salinity affects density, which produces what is called positive, or negative (respectively) changes in buoyancy. These effects are also evident over a part of some continental shelves, such as that off the Amazon River in South America. Special names have been coined for these regions, such as *region of freshwater input* and *runoff controlled buoyancy*; their existence illustrates the rather *fuzzy boundary* between a *coastal basin* and *continental shelf* or *continental sea* (the distinction between these last two categories is that the term *sea* implies some large measure of enclosure by land).

1.3.7 Rapid response

Rapid response of coastal basins to external forcing is a consequence of the limited water depth. For example, wind forced currents change quickly with changes in wind speed and direction, whereas the wind forced layer in the deep ocean (which may be many tens of meters deep) has much more *inertia*, and flows in a manner which reflects longer-term averages of wind speed and direction. This rapid response characteristic is also seen in the temperature of the basin due to changing solar radiation. In many basins the temperature is controlled primarily by the surface heat exchange, for which the characteristic time is proportional to depth, and is typically 1 or 2 days. The same effect is seen in the rapid response of a basin to river floods following storms.

1.3.8 Changes in environmental forcing

This ability of coastal basins to respond quickly to changes in environmental forcing is a very basic characteristic and one that is important not only to their dynamics, but also to their response to human influences. The classical view of the ocean is that it is a natural reserve of unchanging constancy. It is true that the deep ocean and continental shelf have slowly changing temperatures, whereas the land and lower atmosphere can fluctuate in temperature on time scales of hours. However, we are rapidly learning that the ocean is in no sense an infinite reservoir for all of the pollutants that come from humans, and in particular that coastal basins change quite quickly under such *stressors*. A term to describe the capacity of coastal basins to absorb human impact is *assimilative capacity*. This limited capacity of coastal basins is illustrated by their temperatures which, as already mentioned, can change quickly, on a time scale of many hours (as well as the seasonal and annual variability associated with the heat forcing from the sun, and regional oceans). In a sense, coastal basins are a transition region between land and sea. They are greatly affected by many factors associated with the coast and coastal terrain, and change quickly both in temperature, salinity, and other parameters.

1.3.9 Pressure gradient

The dominant force in perhaps all ocean systems is the horizontal gradient of pressure. It arises from the slope of the ocean surface (if the water is of constant density), and also from horizontal variations in density at a given height in the water. The pressure gradient is the most immediate responses of the ocean to any forcing, such as the gravitational force of the Sun and Moon, the effect of the rotation of the Earth (the Coriolis force), wind, density variations (due to changes in salinity and temperature), and breaking waves. The pressure force is a consequence of water starting to move under these external forces. In coastal basins, this movement of water produces a surface slope (or in modeling parlance the surface *sets up*) to produce a pressure force that opposes the force that created the current. We shall see that this balance of pressure gradient with other forces is very common in ocean systems. The total pressure gradient (summed over the water column) created by surface slope is proportional to the depth of the water column, while wind stress is independent of the depth of the water. Consequently, when a wind blows over a coastal basin the resultant *set up* of the surface is inversely proportional to depth, and so coastal basins tend to exhibit large surface set ups. For example, a typical strong wind (but not of storm strength) blowing over a typical coastal basin of depth (say) 2 m and length 20 km will produce a surface set up of some tens of centimeters. The same wind blowing over the continental shelf, alongshore, produces a surface set up of less than a millimeter.

One of the most important characteristics of the pressure force is that it has an entirely different depth dependence than most of the other forces that we consider for coastal basins. For a fluid of uniform density, the pressure force relies on a slope of the surface of the fluid. In this case the force is independent of depth. This means that the *total* force summed over the entire water column is greater in deeper water than in shallow water. This has some interesting effects, especially in coastal basins where depth can vary over much smaller horizontal spatial scales. If, for example, the total pressure force is balanced by bottom friction, we will obtain much higher average currents in deeper water leading to much high volume flux. We see the same effect in wind-driven gyres discussed later in this section. This independence of pressure force on depth in the water column remains true even if the density varies horizontally but is momentarily constant in the vertical. If the depth of the basin is constant, the total force on each water column can be equalized by a suitable surface slope. However, a steady state can only be achieved by the basin becoming at least partially stratified; water at the bottom of the denser portions of the basin has higher pressure than its neighbors, and flows outward from those regions to be compensated by lighter water moving inward at the surface. In essence, we can see the same effect if we pour water onto the surface of a table; the water is denser than air and moves outward to be replaced by air.

If the fluid is stratified and a surface slope is created by, for example, a wind blowing over the surface (which causes water to accumulate at the upwind end of the basin), the pycnocline tends to inhibit the downward transfer of the stress created by the wind blowing over the surface. This can cause the pycnocline to slope downward in the direction of the wind, i.e., opposite to the slope of the surface. This produces a compensation of the pressure gradient in the lower layer, and we determine a critical slope of the pycnocline such that the pressure force is zero in the lower layer. So, in the absence of stress, there is no motion. We say that the pycnocline *downwells* downwind and *upwells* upwind. The pycnocline can break through the surface and form a *density front*.

1.3.10 Time and space scales

Time and space scales are much smaller in coastal basins than on the continental shelf or the deep ocean. So, a detailed study of, say, the coastal basin between the Catalina Islands off the coast of California would resolve processes on spatial scales too small to be seen in detailed models of the California Current. For a regional model that describes the California Current, these islands are merely a minor obstruction. At most, the basin between them would contribute as an element of coastal roughness.

All oceanographic systems contain water movement with periods ranging from a fraction of a second to years. This is a consequence of a fluid having a virtually infinite number of degrees of motion. So, the ocean does not conform to our perception of a thick viscous fluid which moves smoothly around a coastal basin, characterized only by its mean current. We group ocean phenomena, or processes, according to these time periods or (as we say) *time scales*. We shall see various examples of these processes throughout this book. Time scales bring with them concomitant spatial scales connected to the time scales by the speed of propagation of the phenomena.

Processes occurring at very high frequency, or on small time or spatial scales, are classified as *turbulence*. However, the size of the system under study controls the minimum spatial scale resolved in the model. Modelers are always limited in their treatment of high frequency, small spatial scale phenomena. For small laboratory sized systems we can use methods known as *computational fluid dynamics* (CFD), which can solve for fluid motion at high frequencies. This is not presently practical for environmental systems, and we are only able to include the effects of turbulence through methods known as *turbulence closure* discussed in Chapter 7. These closure systems work with certain averages of the motion, and in so doing try to replicate the sort of processes that would be seen in CFD simulations at small spatial scale. The limiting upper time/space scale of turbulence is determined by the type of ocean system under study. For the open ocean this is much larger than for the small coastal basins considered in this book, and is very much affected by the rotation of the Earth; we refer to this as *geophysical turbulence*. As we move to higher frequencies, i.e., longer periods, we observe the effect of astronomical tides (periods from 12.4 hours to days), effects of the rotation of the Earth (also from half a day to a day or two, depending on latitude), and then effects of changing weather conditions (the so-called *weather band* with a typical period of 10 days). With care we can see much longer period processes up to a year of even longer.

1.3.11 Surface gravity waves

Surface gravity waves are a part of the spectrum of processes in the ocean. They are dominantly due to wind forcing and are one of the most obvious processes that we all observe, whether watching the ocean during a walk along the beach or experiencing the motion of a boat. In the deep ocean, the motion associated with surface gravity waves extends to only a depth roughly equivalent to a few multiples of their wavelengths, and their speed of propagation increases with wavelength. Gravity waves are produced by almost any disturbance to the ocean surface such as winds, ship traffic, etc. Waves from a storm cannot propagate much faster than the maximum wind speed, and this sets a limit to their wavelength. The longer waves move away from the source more rapidly, and at a distance they create a distinct wave front. For a major storm, this front can be mapped as it moves across the ocean. The shorter waves from these events are damped more rapidly and affected by interference from multiple sources, with the result that the waves from typical distant storms that arrive at our coasts normally have periods of 15 to 20 s, which we call *swell*. Swell waves arriving at a coastal basin have a preferred direction, and there is a seasonal variation to both this direction and the strength of these swell waves (due to the seasonal movement of large-scale weather systems). These swell waves have a major influence in sculpting the shape of beaches (which often vary seasonally) and the overall topography of a coast. However within the deep ocean itself, or on the continental shelf, surface gravity waves usually have little influence on longer period dynamics.

A difference between waves in the deep ocean, or continental shelf and coastal basins, is that waves in shallow water have a speed of propagation that increases with water depth, but not with wavelength. This is only true for waves with a wavelength greater than the water depth, the so-called *long waves*. This certainly applies to swell waves entering the basin which have periods greater than 10 s, and wavelengths of some tens of meters. The dependence of speed on water depth causes such waves to *refract* in very much the same way that light refracts, or bends, as it goes from air to glass. A rule of thumb is that refraction causes the wave to swing in the direction of slower propagation speed (into shallow water) so that, for example, waves tend to converge onto a headland, and there is a focusing effect rather like that produced by an optical lens. Refraction has a major influence on coastal erosion and sediment accumulation. However, waves that are sufficiently short continue to behave rather as waves in the deep ocean. Waves from local winds have not propagated over the long distances required to "filter out" the short period components, and result in much more disorderly motions that we call *wind chop*. Such higher frequency waves are greatly damped as they approach our beaches, and the regular breaking waves we see on our walk along an ocean beach are mainly due to swell. In many coastal basins, the characteristic wave period away from the mouth is only of order 1 to 2 s. In a basin of depth 1 m the speed of long waves is about $3\,\mathrm{m\,s^{-1}}$, and so the equivalent wavelength for 1 s waves is only just in the *long wave* category; we call such waves *transitional*.

Another major difference between waves on the continental shelf and in coastal basins is that *long waves* in coastal basins penetrate to the ocean bed, i.e., the associated motion of water is felt all the way down through the water to the bottom so that waves impose a stress on the bed of the basin. This can suspend sediments and causes an increased bottom friction for currents in the basin. We refer to this as the *wave–current* interaction. The stress produced by the waves passing over the rough seabed only penetrates a finite distance into the water from the bottom, to produce the *wave boundary layer*. The reason is that the direction of the bottom stress is periodic in time, and so it can only penetrate into the water column a distance equivalent to that which it can travel upwards in about half a wave period.

1.3.12 Stratification

Stratification occurs in coastal basins due to surface heating and freshwater runoff. The major differences from the continental shelf arise through the increased fluxes of fresh water, limited depth, and stronger tidal and wind-driven currents. As has been noted above, some continental shelves do have macroscopic tides and major fresh-water inflows, and so the distinction is not sharp. Or, we might say that some shelves have some characteristics of coastal basins. Generally speaking, coastal basins have much more variable stratification than do continental shelves, both in a spatial and temporal sense. This is a consequence of the much greater spatial and temporal variability of both buoyancy input (due to fresh water and heating) and mixing (due to tides and wind). It is exacerbated by the limited depth, which reduces the time of response and accentuates the mixing processes. Care is needed in shallow basins not to assume that stratification will be absent simply because of the reduced water depth. Also note that horizontal gradients in density can drive currents in the basin. This is always associated with some vertical variation in density, although the water column may not be actually *stratified* in the sense of a well defined pycnocline, which divides the water column into two strata or *layers*.

1.3.13 Topographic effects

Topographic effects are a special feature of coastal basins and not usually found in the same form on the continental shelf. Topography is effective through both the limited depth of the basin and constrictions of the coastline, including reefs and embankments.
   Two of the most common effects are:

Classic basin circulation

This is counterclockwise *around the basin* in the northern hemisphere, and clockwise in the southern hemisphere. This is more exactly described as horizontal separation (across the lateral axis of the basin) of inflowing and outflowing water due to the Coriolis force. In the northern hemisphere the water moves to its right towards the nearest coastline and to its left in the southern hemisphere. This type of *gyre* can be seen in density forced flow due to fresh water from rivers or localized cooling events. It is also evident in *tidal gyres* with water running around the basin from flood to ebb. Associated with these gyres is a pressure gradient which involves the surface setting up or down against the coast. The water column may be stratified, in which case the pycnocline is tilted away from the horizontal to partially balance the pressure gradient in the lower water column; in extreme cases the pycnocline can break the surface to produce a surface discontinuity in density or *density front*.

Wind-driven gyres

Such gyres are a feature of bottom topography. Basically, the net flow in the shallow regions tends to be downwind and the net flow in the deeper water tends to return the water upwind. These gyres are best seen in the net flow averaged from top to bottom of the water column. Usually the surface flow in all parts of the basin travels downwind (or at least at an angle less than 90° to the wind) with the net upwind flow in deeper water coming from the lower part of the water column. *Wind-driven gyres* require depth variation, and in a basin with no variation in depth the wind-driven flow (in the absence of stratification) would be simply downwind at the surface everywhere and upwind in the lower water column. This flow is altered by the Coriolis force and acquires something of the *classic basin circulation* with the water moving to the left or right across the basin to give net inflow on one side and net outflow on the other. There is a distinction between the major gyres of the deep ocean basins, which are primarily a consequence of latitudinal changes in wind direction. The continental shelf can contain close cousins of the gyres seen in basins, especially when there is partial enclosure by the coast.

1.3.14 Variable roughness

Variable roughness is a feature of coastal basins and a product of the coastal complexity created by reefs, *submerged aquatic vegetation* (SAV) and engineered structures such as pipelines, bridge supports, and dredged channels. Although these may exist on continental shelves, the reduced spatial scale of basins gives them much greater relative importance.

1.3.15 Biogeochemical processes

Coastal basins are often regions of high biological productivity and intense activity in terms of biological and chemical processes. Estuaries represent a unique ecological environment as river water meets the coastal ocean and much of human development has been based on the abundance of all the trophic levels that prosper in these waters. The complexity and diversity of the species and their interactions cannot be treated in the present book, but we do attempt to show the way through some examples. It is also true that much of this modeling is appropriate to parts of the continental shelf and conceptually the models are similar, but are mounted on different hydrodynamic realizations.

1.4 Types of model

1.4.1 The changing etymology of modeling

The words *model*, and *modeling*, have reached almost universal vogue in the early 2000s to a point where they have lost much of the specific meaning of earlier decades, and the scientist needs to take care as to what the community may understand by a *model*. This is part of a trend that has been under way for several decades. In earlier times the word *model* was used much more selectively, and perhaps it is unfortunate that there is now some degree of misunderstanding as what different groups of people mean by the word *modeling*.

In the middle of the twentieth century, physicists used the word *model* to mean a theoretical construct that was well defined, intended to be capable of exact solution, and considered to have some reasonable relevance to much more complex systems in nature. The construction of a model was always based on sets of observations from nature combined with the *modeler's* basic understanding of the laws of nature. The term *model* carried with it the connation of simplification and idealization. With the advent of readily available analog computers in the last one or two decades of the twentieth century, models were conceived that were technically very simple and yet only "soluble" using computers. Computers had been in existence for many decades earlier, but they were sufficiently slow and unavailable that scientists needed great dedication to use them. Suddenly, around 1980 *desktop* computers appeared that provided easy access to computation power, and in the first decade of the twenty-first century scientists have the power on their own desk to quickly compute, analyze, and display information that was beyond the limits of any machine on Earth just decades earlier. According to *Moore's law*, computing power doubled every 2 years in the late 1990s. So, not surprisingly, the meaning of the word *model* has evolved to suit this revolution in computation.

The word for *measure* or *standard* in Latin was *modulus* and this had associated with it the diminutive form *modellus*. Thence came the old Italian *modello*, meaning a *mould* for producing things, or the *standard*, archived form of an object. The word drifted into French as *modèle* in the sixteenth century, and then into English in the sense of a small representation of some object and with an increasing connotation as *ideal*. Hence, for example, we had model trains, model homes, meaning displays of new building styles. The word moved also from mould to mean a version of a product, and so we had the *Model T* Ford motor car. People who showed new versions of a product were said to *model* that version.

1.4.2 Physical models

The nineteenth century saw small-scale representations of bridges and other planned civil engineering product and eventually *models*,[2] or *miniatures*, of prototype aircraft, dams, etc., and then *physical models* of coastal systems. These are truly scientific models because they were based on scaling laws derived from our understanding of fluid mechanics, the laws of aerodynamics, and the mechanics of rigid bodies. There are a wealth of field stations, testing stations, and experimental tanks and *flumes* throughout the world in which such physical models have been built and used. They are still quite vital to many aspects of the study of processes in coastal basins for which we have limited understanding, or as a means of providing calibration and verification of theoretical studies. Examples are wave tanks in which studies are made of breaking and non-linear waves, and effects of waves on structures. Many of these physical models provide basic empirical laws upon which simulation models are based, and we refer to them especially in our chapter on sediment transport, which makes extensive use of such laws. Otherwise this book does not generally treat the theory of physical models. It is important not to confuse such models with models of physical processes.

---

[2] Note the distinction between *physical* models, i.e., miniature constructions in a test tank, and models of physical processes which are either analytic or numerical (or just conceptual).

1.4.3 Theory or model

For scientists the word *model* has always meant a *representation* of a system, whether that representation is conceptual, mathematical, numerical, or physical. The word *model* has nevertheless increasingly replaced *theory* in some branches of science, and so we often now hear "Darwin's *model* of evolution" rather than the earlier "Darwin's *theory* of evolution." To most scientists the latter form is superior because the word *theory* still maintains a perception of a general and fundamental idea, whereas *model* implies a basic simplification, or acceptable approximation. Darwin's *theory* of evolution is a very fundamental statement about the mechanism by which living forms adapt to their environment, and contains no leeway for approximation. In contrast, we can have a *model* of evolution which attempts to define processes by which the Darwin mechanism functions. For example, there are many *life games* available for simple computers which illustrate the way that species can evolve through *bottom–up* laws of interaction.

1.4.4 Simulation models

Traditional modeling of ocean and estuarine systems has its origins in the physical sciences, and is built around *simulation of processes*. We therefore refer to such models as *simulation models*; they are the heart of the modeling science that scientists have developed for the ocean and the atmosphere. So, for example, a hydrodynamic model of an estuary starts with some approximations to the Navier–Stokes equations for a fluid, i.e., we basically assume Newton's law of motion and the continuity equation for mass. It is true that any hydrodynamic model will contain parameters whose values have some degree of uncertainty and for which a process of *calibration* against data is needed. So, in a sense the model is *tuned* against data sets and is therefore not entirely based on simulation of well-understood processes, but it does remain true that the heart of the model is *simulation*. All physical models of (for example) coastal systems also require specification of conditions at their open ocean boundaries and at their surfaces (often called *data assimilation*). This is necessary because our physical models are *open* rather than *closed* in the sense that they only simulate processes within a finite physical domain. Thus, for example, a coastal model of a nearshore coral reef will have at least one boundary with the open ocean, with its other boundaries closed by the land boundaries; the model does not simulate processes in the open ocean, although these processes may have major influences on the coral reef.

So, our coastal models only simulate processes within their spatial and temporal domains and rely on data from outside those domains as forcing functions. Nevertheless the heart of any model lies in the *simulation* of processes. Hydrodynamic models have become increasingly sophisticated over the last few decades because of their improved simulation algorithms. In so doing, they have placed increasing demands on data collection as a means of verification and understanding of processes, but *not* because they are merely processing data in a more intensive manner. It is also true that improved instrument methodology has produced high-quality data and this technological advance has spurred modelers to produce more sophisticated models, but this is always based on fundamental understanding of *processes*. For example the *acoustic Doppler velocimeter* (ADV) allows us to measure the spectrum of high-frequency oscillations above 1 Hz, i.e., one cycle per second. Existing instruments can measure at frequencies above (at least) 50 Hz. These fluctuations in coastal basins represent what we refer to as *turbulence*. At the present time we cannot model such turbulence in a detailed sense, but we do have *turbulent closure schemes* (or *closure models*) that fit inside our hydrodynamic models and simulate the basic processes by which turbulent energy is dissipated within the system.

Observation and experiment are an important part of the scientific process and integral to our *theories* and *models*, but the heart of science lies in the understanding of processes either at a local or system level. So, simulation models do much more than assimilate data. It possible for a modeler to create a model of a coastal system for which there is almost no data just by knowing its basic locality and topography. The model is simulating the known processes that work in all coastal systems and making a judgement as to forcing functions, such as wind and tide and perhaps river flow. The same would not be true of a model based primarily on data. So a simulation model is not merely processing data. There is a very important difference between *simulation models* and what we might call *data assimilation models*, or *data synthesis models* (I prefer the latter) as will be discussed next. Of the latter, a well-known adage is "rubbish in: rubbish out." That is not equally true of simulation models, and for them, modelers selectively reject bad forcing data (or the numerical system goes into convulsions) but a simulation model will still produce acceptable simulations. Most of this book is about the ideas that have lead to simulation modeling of coastal basins. This is now augmented by data synthesis modeling at a practical level, and I find it a very valuable part of the modelers' toolbox. The marriage of simulation and synthesis modeling produces the new art of so-called *integrated* modeling, which will be discussed below.

1.4.5 Synthesis models

Modeling is also increasingly used in science to include the collection and processing of data. As emphasized earlier, the understanding and intelligent interpretation of data is a vital part of the modeling process. However, there is much that needs to be done with most field data before any sort of analysis can commence. The business of instrument deployment, *downloading* of data, and the processing of those data lie in an area most scientists would describe as *data acquisition and processing*. Nevertheless, the processing and understanding of data sets is a vital part of the scientific process and is one in which modelers have a very active, perhaps even key, role. The dominant place of computers in much of science dating from the late 1980s certainly tends to smear the boundaries between the, once distinct, activities of *theorizing* or *modeling* processes in a system and analyzing data sets. Despite that merging of modeling with data collection and processing, there is an essential task for data collectors in digesting their data sets. Data from modern instruments is collected in digital form and usually stored in a binary format specific to the manufacturers of the particular instrument. The user then downloads the data into an ASCII format (*American Standard Code for Information Interchange*) is a character encoding based on the English alphabet; ASCII codes represent text in computers, communications equipment, and other devices that work with text using software supplied by the manufacturers). The user then needs to examine the data and edit them so as to remove spurious records such as those obtained prior to the instrument being deployed, or after recovery. Most files taken from field instruments require considerable "massage" and editing before they can be passed into the next stage of scientific analysis. This is often a task requiring significant scientific skills. Many instruments used in the coastal environment require maintenance in the field, and the technical task of editing data includes corrections and calibrations based on a knowledge of those maintenance routines.

The power provided by computers in the analysis of data has given birth to a new type of modeling which is here called *synthesis modeling*. Models that involve data processing and relationships between data sets (or "layers" as they are termed in geospatial analysis) are called *synthesis models* and are different in a fundamental sense from models that have been traditionally used by scientists. There is a sense in which this modeling is the traditional precursor of simulation modeling methods, but there is also an important caveat that it needs careful differentiation from simulation modeling methods. One of the most familiar branches of synthesis modeling is that of *modeling* within Geographic Information Systems (GIS) in which *models* look for links between spatial data sets. This is basically a method of storing, processing, mapping, and displaying spatial data and has found immense application in the geographical, mapping, and surveying sciences. It does nothing outside of the scope of the mapping techniques available in programming languages already familiar to most modelers, but it does so in a very compact and "user friendly" way with especially convenient spatial grids, so that programming skills are built around simple menu choices. It is possible to use GIS for synthesis modeling of spatial data in a very effective manner.

An example of a synthesis model is the change in seagrass coverage in an estuary over the last 100 years in relation to urbanization of the surrounding coastline. The model looks for a statistical trend between the area of seagrass in a particular part of the estuary and the number of homes within a radius of 10 km. That *model* is amenable to a great deal of modification as more *layers* of data are included, such as the use of the estuary for recreational boating, commercial fishing, etc. Synthesis modeling, like all branches of modeling, involves intelligent and inspired choices of layers of data for use in the model. If the synthesis of seagrass and depth data suggests a dominant process in a particular coastal basin, then we might try to use it in a predictive sense for that basin; for example to model the effect on seagrass coverage of major dredging of the shallow regions of the basin. The veracity of such predictive modeling is usually dependent on the construction of some sort of theoretical relationship from the synthesis model. Otherwise, the model might be considered to be only a correlation between sets of data and we must always be aware that *correlation does not equate to causality*.

The distinction between simulation and synthesis models can become blurred in some areas of the science of coastal systems. This is especially true in what has become known as *ecological* sub-models and models of higher trophic systems. We should mention that these are not truly *ecological* models, but simulations of chemical and biological processes in a coastal system. They can include human influences through such factors as nutrient input and fisheries. The number of components in the system can be as large as the modeler is prepared to accept. Each component is called a *species* and examples are: nitrogen, phosphorus, oxygen, phytoplankton, zooplankton, seagrass, and so on. Pressure from users of the model is always to increase the number of species, since they often have interest in some particular component. For example, they may wish to introduce subspecies of phytoplankton. However, more species means more interactions between components of the system, and the number of interaction parameter values increases approximately as the square of the number of components. For example, a system with 16 components might have over 100 parameters, and with 25 components it could rise to 250 parameters. Many of these *ecological* models are available in the public domain, with active exchange of information between users, or can be licensed from central suppliers. Fortunately, many of the parameters are probably largely system independent, and so the wide use of these models could allow a library of parameters to be assembled with individual users making only minor modifications for their own systems. However, the complexity of natural systems and the potential for spatial and temporal variability means that this approach is fraught with uncertainty.

1.4.6 Integrated models

The real power of modeling is seen in the synthesis modeling of observational data and output from simulation models of physical and biogeochemical processes. Modern trends are to refer to the output information as *model data*, although the output of a simulation model is not *real* data in the sense that it has not been collected from nature, it is nevertheless a data set. So, we will conform to the trend to refer to the model output as *data*. In the example of synthesis modeling that we discussed earlier, and which involved the comparison of data sets for seagrass coverage and water depth, it would clearly be advantageous if that comparison were assisted by output from a simulation model. The simulation model might, for example, predict wave fields in the estuary. We could then give more substance to a conjecture that the dependence of seagrass coverage on water depth is a function of wave activity. Wave fields can be measured, but the advantage of a reliable simulation model is that we can produce wave fields everywhere inside the estuary and under any weather situations of our choice.

The combination of synthesis and a range of simulation models is called *integrated modeling* and in the context of this book, an *integrated coastal model* is one that combines all such tools. It can be seen as a tool for researchers in the many diverse aspects of coastal sciences and for those responsible for management policies. Integrated modeling can lead to prediction of *behavior*. These system aspects of integrated models are especially important when we consider the way that management policies affect the complex environment of a coastal system, and the use of that system by the community. System models have their origins in strategic planning for defense and business operations and are the basis for much economic planning. In a sense, they are simulation models built on the top of synthesis models, and there is a wealth of literature on the basic linear programming techniques on which these models are founded.

1.5 Terminology in the sciences of water flow

Coastal basins encompass many different sorts of fluid environments ranging from creeks and rivers, through wetlands to shallow nearshore waters and waves breaking on coral reefs, to open embayments, outward onto the continental shelf. Scientists and engineers working in different parts of the subject, or within different types of flow conditions, or environments, have traditionally used different names for the science of water flow. Coastal basins include many of those environments.

The science of *hydraulics* is mainly concerned with the study of flow through channels and pipes. It is characterized by the dominance of momentum advection which we will study in Chapter 8. These processes are occasionally important in coastal basins and lead to the phenomena of (aptly named) *hydraulic jumps*. By way of contrast, the science of *hydrology* is concerned with the movement of water through the various compartments of the Earth's *hydrosphere*, i.e., from ocean, lake, and river to atmosphere, and back as rainfall to surface flow, groundwater, and all of the forms in which water occurs on our planet. However, the same term *hydrology* is also used for smaller-scale studies such as individual wetland systems. We also find another term – *hydrography* – in use for coastal basins, which comes from the measuring of the bathymetry and currents by *hydrographers*. The term *fluid mechanics* is often reserved for much smaller spatial scales than occur in coastal basins but nevertheless small scale motions do have a fundamental role and we later mention *computational fluid mechanics or computational fluid dynamics* (CFD). An exception is *geophysical fluid dynamics* which refers to the dynamics of large-scale motions of the oceans and atmosphere.

The science of *nearshore processes* deals with the surf zone and wave driven sculpting of the geomorphology of the coastline. This is also part of the traditional field of *coastal science*. The engineering of coastal structures is part of *coastal engineering*. All of these nearshore sciences work on much smaller scales than we consider in this book and use models that are specific to features on scales of meters to tens of meters. In many cases such models are highly phenomenological or even empirical. We border these fields in Chapter 12 when we discuss sediments. Empirical models are a powerful part of synthesis modeling that we have mentioned in this chapter and necessarily require individual treatment as they are less easily translated between systems.

Ocean sciences have traditionally referred to the science of water movement at large spatial scales in the oceans as *hydrodynamics*. The inclusion of the term *dynamics* is an essential ingredient of the simulation modeler's approach to fluid flow because it implies that we identify (or model) the forces that cause water to flow and then use the basic principles of Newtonian dynamics to find the resulting motion. For that reason we use the term *hydrodynamics* in this book for all circulation modeling independently of spatial scale. Some of the examples in this book do consider processes on scales of tens of meters and this takes our models into an area bordering on *nearshore processes*. However, there is a difference of emphasis between models of coastal basins and models of individual nearshore processes. Temporal scales are generally linked to spatial scales, and so, whilst nearshore modelers may be interested in individual waves, modelers working at large scales are usually more interested in waves averaged over many hours which give *wave climates*. In the same way our studies of biological and chemical processes are based on whole *communities* rather than individual entities within that community and over periods of weeks and months, rather than some specific series of events. This is in marked contrast to studies of biological processes in laboratory tanks in which a specific sets of organisms are studied in conditions of varying flow and other environmental parameters.

Integrated models will gradually merged these fields as research advances our understanding of processes at all scales.

I often use the term *velocity* for the components of velocity in a particular direction and I do that interchangeably with the term *current*. There is a strong doctrine in physics that velocity is a *vector* and the term *speed* should be used for the magnitude of the velocity which is a *scalar*, i.e., a scalar has magnitude but not direction. The somewhat debatable issue in applied mathematics is whether velocity better describes a quantity that may be either positive or negative. If I say that the speed of water flow across a coral reef is 1 m s^{-1}, I may be asked "in what direction" and I reply "from the ocean to the lagoon" whereas I prefer to say that the water velocity, or current, is 1 m s^{-1} from ocean to lagoon. However, I always use *current speed*, or *particle speed* to describe motions which may be in any direction such as a tidal, or wave, motion. I use angular brackets to denote averages of various kinds which I always define locally in a model or piece of analysis. The use of such parenthesis is common in applied mathematics and is often equivalent to the over-bar, i.e., $\langle u \rangle$ is equivalent to $\bar{u}$ but the angular brackets are easier for long and more complex expressions. I also occasionally use the *tilde*, e.g., $\tilde{u}$, for some special form of a variable. Primes are used occasionally and especially for the departure of a variable from its mean value such as in turbulence.

1.6 Further reading

Alongi, D. M. (1997). *Coastal Ecosystem Processes*. Boca Raton, FL: CRC Press.

Arctur, D. and M. Zeiler (2004). *Designing Geodatabases: Case Studies in GIS Data Modeling*. Redlands, CA: ESRI Press.

Beer, T. (1997). *Environmental Oceanography*. Boca Raton, FL: CRC Press.

Bowden, K. F. (1984). *The Physical Oceanography of Coastal Waters*. Chichester, UK: Ellis Horwood.

Davis, D. E. (2003). *GIS for Everyone*. Redlands, CA: ESRI Press.

Emery, W. J. and R. E. Thomson (2001). *Data Analysis Methods in Physical Oceanography*. New York: Elsevier Science.

Ferziger, J. H. and M. Peric (2001). *Computational Methods for Fluid Dynamics*. New York: Springer.

Flannery, T. (2006). *The Weather Makers: How Man Is Changing the Climate and What It Means for Life on Earth*. Melbourne, Australia: Atlantic Monthly Press.

Goodchild, M. F., B. O. Parks, and L. T. Steyaert (1993). *Environmental Modeling with GIS (Spatial Information Systems)*. Oxford, UK: Oxford University Press.

Hanselman, D. C. and B. L. Littlefield (2004). *Mastering MATLAB 7*. New York: Prentice Hall.

Hilbor, R. and M. Mangel (1997). *The Ecological Detective*. Princeton, NJ: Princeton University Press.

Kjerfve, B. (1988). *Hydrodynamics of Estuaries*. Boca Raton, FL: CRC Press.

Mitchell, A. (1999). *The ESRI Guide to GIS Analysis,* vol. 1, *Geographic Patterns and Relationships*. Redlands, CA: ESRI Press.

Nixon, S., B. Buckley, S. Granger, and J. Bintz (2001). Responses of very shallow marine ecosystems to nutrient enrichment. *Hum. Ecol. Risk Assessment* **7**(5) 1457–1481.

Oppel, A. (2004). *Databases Demystified*. New York: McGraw-Hill Osborne Media.

Pedlosky, J. (2004). *Ocean Circulation Theory*. New York: Springer.

Pickard, G. L. and W. J. Emery (1990). *Descriptive Physical Oceanography*. Oxford, UK: Butterworth-Heinemann.

Ramaswami, A., J. B. Milford, and M. J. Small (2005). *Integrated Environmental Modeling: Pollutant Transport, Fate, and Risk in the Environment*. New York: John Wiley.

Ross, D. A. (1982). *Introduction to Oceanography*. New York: Prentice Hall.

Silberschatz, A., H. F. Korth, and S. Sudarshan (2005). *Database Systems Concepts*. New York: McGraw-Hill.

Sinclair, M. (1988). *Marine Populations: An Essay on Population Regulation and Speciation (Recruitment Fisheries Oceanography)*. Seattle, WA: University of Washington Press.

Stommel, H. M. (1991). *A View of the Sea*. Princeton, NJ: Princeton University Press.

Thurman, H. V. and A. P. Trujillo (2003). *Introductory Oceanography*, 10th edn. New York: Prentice Hall.

von Arx, W. S. (1967). *Introduction to Physical Oceanography*. Reading, MA: Addison-Wesley.

Weyl, P. K. (1970). *Oceanography*. New York: John Wiley.

2

Currents and continuity

Modeling in coastal basins is based on conservation, or continuity, laws for a series of quantities that are input at boundaries and point sources into the modeled basin. These laws include the conservation of mass (or in the simplest case, volume). They provide the fundamental ingredients of all models of coastal basins. Conservation of momentum provides the dynamic, or hydrodynamic, part of a model and corresponds to the use of Newton's law of motion. Although we delay consideration of the details of these dynamics until Chapter 4, it is important to emphasize that most models contain a dynamical component (although that component may be largely implicit). In this chapter and Chapter 3, we consider continuity for a wide spectrum of substances such as heat, salt, sediments, nutrients, plankton, etc. Not all these substances are necessarily conserved in the truest sense, but we are able trace their movement between various compartments of the model. Continuity, or conservation of these quantities, with a limited hydrodynamic component does provide some very useful modeling techniques, and also lays the basis of our understanding of coastal basins.

2.1 Position of a point

Models of ocean dynamics are usually based on a set of *Cartesian coordinates* that are universally labeled x, y in the horizontal, called the *horizontal distances*, and z measured *upwards* in the vertical, called the *height*. The coordinates are universally assumed to be *right-handed*: placing a somewhat closed right hand on the plane with the thumb pointing up, the fingers point from the x axis to the y axis, and the thumb along the z axis. Occasionally, Cartesian coordinates may be replaced by *polar* coordinates or *spherical* coordinates in order to help solve problems with special geometries, but it remains easier to formulate our general ideas in Cartesian coordinates. We try to avoid the use of the word *depth* for z and call z *height in the water column* (or just *height*), but *depth* is nevertheless acceptable with some caution. The word depth is properly reserved for the distance down to the seabed.

We need a horizontal plane from which to measure z. At first sight, we might suggest the bottom of the coastal basin as a convenient surface, especially since all values of z will then be positive. Unfortunately, the bottom of the ocean is not generally a horizontal plane. As a next attempt, we might consider using the surface of the ocean as the plane from which to measure z, but like the ocean bottom, the surface is not usually horizontal (and furthermore its height varies in time due to tide and winds, etc). Hence, we have to invent a horizontal plane from which to measure z. This plane is called the *height datum*. It is usually chosen as somewhere in the vicinity of the actual surface, but since that surface moves with time that is not necessarily always true. We can pick any height datum, and provided that we specify where that datum is located, this is perfectly satisfactory. The height datum for a model may not correspond with the *standard* height datum used in maps and charts and is chosen for convenience of the model, but the difference between the model and the standard datum must always be known.

Depth is the distance of the seabed (the bottom of the coastal basin) below the datum, i.e., the value of z at the bottom is $z = -$depth. We usually denote depth by the symbol h, so at the bottom $z = -h$. Remember that h is usually positive and that usually h varies with position, i.e., is a function of x and y, $h(x, y)$. The origin chosen for x and y is essentially arbitrary within a model, but that origin must be related to standard map coordinates. The interface between the seabed and the water column may not be sharply defined, especially in regions in which we have loose sediments or a very open structure (such as a coral reef or seagrass meadow). The model will adopt a well-defined boundary but include a layer of sediments and other material. A map of h across a coastal basin is called a *bathymetric chart*. This chart is produced to some specified spatial resolution. Most coastal basins were traditionally mapped to resolutions consistent with their overall size, and typically some hundreds of meters, or perhaps several kilometers. Electronic surveying techniques now allow this resolution to be reduced to order 10 m or less, and over very rough surfaces (such as coral reefs) it may be necessary to attempt to work at even smaller spatial scales.

The value of z at the surface, i.e., the height of the surface above the height datum at a particular value of (x, y), is called the surface height or *surface elevation* (Figure 2.1). It is usually denoted by the Greek letter corresponding to either h, i.e., η (eta), or to z, i.e., ξ (xi). We use the former. Remember this elevation varies with time t and with x, y so that $\eta(x, y, t)$. The *total height of the water column* is denoted by H:

$$H \equiv h + \eta. \qquad (2.1)$$

Surveyors picking a chart height datum traditionally adopted the convention that the elevation should *almost never* be negative, so that if mariners interpreted the chart depth h as the total height of the water column below a boat (which is really H), they would almost always err on the side of extreme caution, insofar as they would never overestimate the value of H. For that reason, the chart datum was usually chosen as the *lowest low water* ever observed at a particular port. So charts tended to give the appearance that water was shallower than is usually the case. Making the chart datum lower would, of course, give the appearance of a port that is less navigable than is really the case. So, in ports with large tides, the elevation η can be as large as 10 m relative to chart datum.

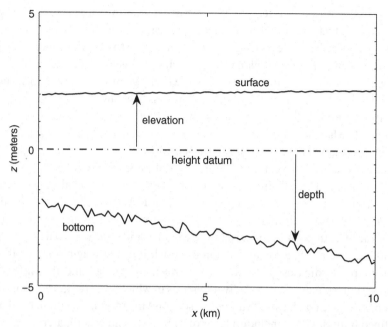

Figure 2.1 Schematic of the way that heights are measured in a coastal basin. This is a snapshot of the water surface at some time (*t*) showing the height datum, depth (*h*), and surface elevation (*η*). The vertical axis (*z*) is measured positive upwards with $z = 0$ at the height datum.

When a bathymetric chart is produced by using an echo sounder from a boat (which measures *H*), it is necessary to combine the depth data with other data which give the height of the boat *η*. Traditionally this involved the use of a tide gauge (or *water level recorder*) which had been *leveled*, i.e., its height known relative to the chart datum. Since the water level recorder was at a fixed point, and the water elevation does (in general) vary across a basin, this required a system of corrections based on either a number of recorders (at different locations) or (much better) a numerical tidal model of the basin. The latter was not totally reliable because the water level varies due to many influences, which are additional to astronomical tides such as wind induced fluctuations and many *seiching* motions (seiching involves a long-wavelength water wave reflecting back and forth across the coastal basin and is similar to the sloshing of water in a bath tub). Since the 1990s it has been possible to use Global Positioning System (GPS) to find the height of the boat directly, and provided that proper time averaging is used this also effectively gives the surface elevation. So by either method, one can find *h* via $h = H - \eta$. In shallow water (depth just a few meters) with good clarity, i.e., low suspended solids and color, the use of boats has also been superseded by remote sensing from either airplane or satellite, which look for reflections of some form of electromagnetic radiation from both the surface and bottom. The method has its problems because of penetration of the light signal through the water column.

Whatever method is used, it is important for models to use actual bathymetric data. How accurate the data need to be depends on the spatial resolution of the model and the types of predictions required from the model. It is possible to understand basic physics by the simplest approximations to basin bathymetry, but often we need precise bathymetry in order to understand important processes in coastal basins.

2.2 Height datum and map projections

The use of the original chart datum for a port is now largely superseded by adoption of a national, or global, height datum. Over the size of most coastal basins the difference between a truly horizontal plane and the idealized shape of the Earth is small, but nevertheless these differences can be important at fine spatial scales as, for example, on reefs. The Earth was traditionally assumed to be approximately spherical, but we now understand that it can be best approximated as an ellipsoid. Due to rotational effects, the ideal equilibrium shape of the oceans, called the *geoid*, differs from an ellipsoid. So there is a choice as to how we reference heights. For example, in North America, heights are presently commonly referred to NAV88 (North America Vertical Datum of 1988). Map projections are also required to approximate the curved Earth onto a horizontal plane, especially when dealing with global or regional scale models. These issues are important for applied models of coastal basins but are not treated in this book, since our emphasis remains with the basic science of modeling coastal basins.

2.3 Velocities

Velocities in the water column are the velocities of water particles *while those particles are at a particular point* (x, y, z). Their components along the x, y, and z axes are denoted by u, v, and w. This is a fairly universal notation in oceanography. The components u and v in the horizontal plane are also called the *currents* along the x and y axes (Figure 2.2). Because of the many degrees of freedom of a fluid, these velocities tend to fluctuate in both space and time around a mean value and we are here discussing only those mean values.

Notice that current in oceanography is a vector which shows where the water particle is *headed*, whereas in meteorology the *wind vector* indicates the direction from which the wind *originated*. This *origin* convention is also true of wave direction, so a wave direction quoted as *due south* means the wave are coming from the south. To avoid confusion between oceanographic and weather maps, we therefore tend to talk about a *southward* or *eastward* current to emphasize that it is *headed* to the south or east. If the x, y axes are orientated west-to-east and south-to-north and $u = 0.1 \, \mathrm{m\,s^{-1}}$ and $v = 0$, we would say that the current is eastward at $0.1 \, \mathrm{m\,s^{-1}}$, i.e., about 2 knots (1 knot $= 1$ nautical mile per hour $\sim 0.5 \, \mathrm{m\,s^{-1}}$). These velocities are referred to by oceanographers as *advective* velocities, and we say that the ocean is *advecting* at the speeds u, v in the two directions. The velocities represent the *advection* of water.

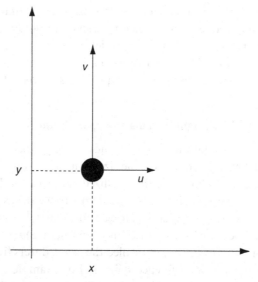

Figure 2.2 Notation for horizontal spatial coordinates (x, y) and horizontal velocities (u, v). The diagram shows a horizontal plane in the coastal basin looking downwards from above. The vertical z axis shown in Figure 2.1 would point upwards out of the paper towards the reader.

The speeds traditionally denoted by u, v, and w are assumed to be averaged in time over periods of minutes and over small volumes of the water column. The reason is that the water in a coastal basin is *turbulent*, and we shall deal with this in more detail in Chapter 7. Hydrodynamic models of coastal basins only determine these averages over the turbulent field, i.e., they do not solve for the details of the turbulent motion although they can determine certain *other average* properties of the turbulent motion, as we shall discuss in Chapter 7. Similarly, our models only resolve an average over a small volume of a coastal basin. The size of that volume depends on the spatial resolution of the model. This marks a major difference between models of the ocean and coastal basins, and models of small systems such as tanks and pipes. The latter models are referred to as computational fluid dynamics (CFD).

In the ocean, the mean vertical velocity w is *usually* (but not exclusively) small compared to u, v. This is mainly because the vertical distances over which particles may move in the ocean are usually very much smaller than the horizontal distances over which we consider model ocean dynamics. This is one of the ways that the dynamics of water in the oceans differs from that of the water in a small tank in the laboratory. Thus a *scaling analysis* of the comparative values of u, v, and w indicates that they scale as the characteristic horizontal and vertical distances (respectively) in the basin. Since the height of the water column in an ocean model is *typically* 1000 times less than the horizontal distances in a model, we might expect the same to be true of w relative to u and v. If we consider modeling basins at very small horizontal scales, i.e., tens of meters, then the scaling analysis would suggest that we *do* need to consider the vertical velocity w. We need to emphasize that when we work at our usual horizontal spatial scales of (say) hundreds of meters, those vertical currents (which we would resolve at much smaller scales) are still present in the basin (although not in our model). The difference is that when we view the *average* over a horizontal distance of (say) 100 m, the average of those vertical motions must be small compared to the corresponding horizontal motion. If that average w were not small the water surface would be rising, or falling, at a corresponding rate. The surface of a coastal basin does, of course, change with time. The best example is a surface wave in which the surface may rise and fall by several meters in 10 s. However, such motions only usually occur on small spatial scales; although we can build models of such waves (but we do not usually work at such small spatial scales).

As an aside, we should mention that we do incorporate wave modeling into our studies of coastal basins but we do not usually consider *individual* waves, but rather consider the *spectrum* of such waves and how that spectrum varies on much larger spatial scales. An example of larger-scale motion is the rise of the surface of a coastal basin during a flood tide. A maximum tidal range is probably 10 m and astronomical tides rise in a minimum period of 6 hours: this gives a vertical velocity of $5 * 10^{-4} \, \mathrm{m \, s^{-1}}$. This is indeed very much smaller than the horizontal velocities that would be associated with such a tide in a coastal basin, which are of order a fraction of a meter per second. This discussion also helps us understand the linking of spatial and temporal scales in coastal basins, which is a connection that dominates much of our thinking about their dynamics. So, small spatial scales usually involve small time scales, and it is quite common to merely use the term *small scale* without specifying *spatial* or *temporal*. The reason is that spatial and temporal scales in coastal basins are connected by the speed of gravity waves. Most fields of physics have such linking, and for example, spatial and temporal scales in cosmology are linked by the speed of light so that as we look further into the universe we go back further in time.

Clearly, we cannot neglect the vertical velocity in a model of a coastal basin because otherwise the surface could not move; although motion of the water surface is neglected in many models of the deep ocean (called *rigid lid* models). But we do accept that the *mean* vertical motion is *small* in relation to the horizontal velocities unless we work at small spatial scales. However, there is a vitally important caveat to this discussion. The vertical velocities that we saw on smaller scales play a *very important role* in the dynamics of coastal basins and are treated in our *theory of fluctuations*. Vertical movements transport momentum and other quantities, such as salt and heat, through a process that we call *diffusion*. We shall see that fluctuations are also important in the horizontal and lead to horizontal diffusion or *dispersion*. This illustrates a very important point for us: the behavior of coastal basins is not controlled entirely by their currents, i.e., by the *circulation*. Instead, many of the properties of coastal basins have a vital dependence on processes of diffusion. For example, the movement of salt in a tidal basin is strongly affected by diffusion with the actual periodic tidal circulation playing a fundamental, but not solitary, part in the process.

All of the fluctuations that are responsible for diffusion of materials in a coastal basin occur on scales (spatial and temporal) that are beyond the direct resolution of our models are treated only in a *phenomenological* manner.

The argument which we used for proving that the mean of the vertical velocity w, as a fraction of the mean horizontal vertical u, scales with the horizontal spatial scales, can be used for comparisons of the means of u and v. So, for example, in a long thin basin we might assume that the lateral speed (say v) across the basin is small relative to the longitudinal speed along the basin (u). An example is provided by Tomales Bay in California shown in Figure 2.3., which has a small aspect ratio of width to length. Generally such assumptions are valid, but need to be treated with care. In particular, it is important to consider role of *lateral fluctuations* in longitudinal transport of materials such as salt.

We generally need to consider that u and v are functions of space and time, i.e., $u(x, y, z, t)$, $v(x, y, z, t)$. We often consider ocean systems in which there is negligible change of u, v, and η with time, and these are called *steady state* systems. In steady state systems u, v, and η are independent of time. An important aspect of the spatial variation of a current is its dependence on height z. So, we are frequently interested in the form of $u(z)$ and $v(z)$. These are called the *height profiles of current*. We draw vertical profiles of currents as shown in Figure 2.4; these are the curves of $u(z)$ and $v(z)$ with the height axis z vertical. The value of the currents at the surface are simply $u(\eta)$ and $v(\eta)$; called the *surface currents*. The *bottom currents* are $u(-h)$ and $v(-h)$.

The slope of the curves in Figure 2.4 is the derivative of u or v with respect to z, and this is known as the *vertical shear* in the horizontal currents, i.e., $\partial u/\partial z$, $\partial v/\partial z$. It is very important to understand the meaning of the symbols $u(x, y, z, t)$ and $v(x, y, z, t)$. These are the components of velocity of a particle as it passes the point (x, y, z) at time t. They do *not* refer to the velocity of a particle in the sense that x, y, and z are functions of t and it is important to understand that x, y, z, and t are *independent* variables. If u and v are constant in time (we have a steady state), this does not mean that the velocities of particles are constant, because at a later time each particle will have moved to a different point where u and v will be different. So the time rate of change of the x component of velocity of a *particle* is not $\partial u/\partial t$; this is, in fact, the rate of change of the component of velocity that particles have while passing through (x, y, z). In most steady states, the velocities of particles *do* change with time.

2.4 Fluxes

Modelers use the term *total flux* to describe the time average of the *net* quantity of a substance passing through a particular surface per unit time. Thus the total flux of water from a tap might be $0.1 \, \mathrm{m^3 \, s^{-1}}$ and the total flux of water from a river to the sea might be $1000 \, \mathrm{m^3 \, s^{-1}}$. The total flux of cars along a one-way road might be 230 cars per hour averaged over one morning period. If the road runs north–south and carries traffic in both directions, the flux might be 200 per hour to the north and 100 to the

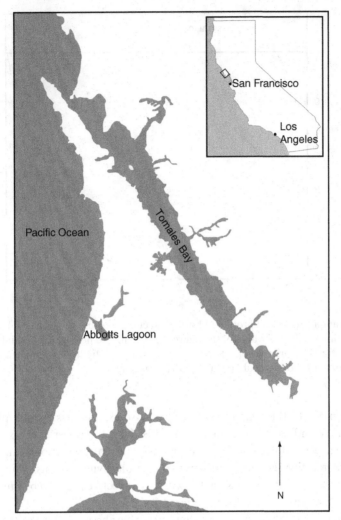

Figure 2.3 Tomales Bay on the west coast of California (compare Figure 1.4) illustrating a coastal basin, whose width is small relative to its length and for which a one-dimensional model along the axis of the basin might be justified. The basin lies along the line of the San Andreas Fault just north of San Francisco Bay. A one-dimensional model would align the longitudinal axis (x) along the center line of the basin. The ratio of width to length is about 5%. Tomales Bay is discussed in relation to Stommel transitions in Chapter 10.

south giving a *total flux* of 100 per hour northwards or -100 per hour southwards. In the ocean we often measure net flux across unit area of a plane and talk about *volume flux density per unit cross-sectional area*. This has units of volume per unit time, per unit area, or m^{-1} s, which is the same as velocity. We also determine flux densities *per unit width* of a river by dividing the total flux by the width of the river. We do the same thing for flow in the ocean, i.e., just consider the flow through a plane of unit width.

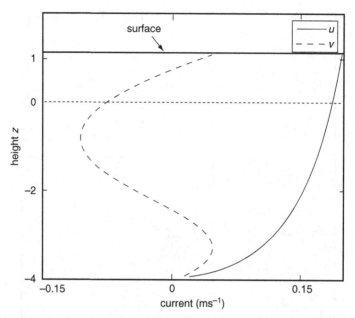

Figure 2.4 Example of vertical profiles of horizontal currents u and v in a coastal basin of depth 4 m and elevation $\eta = 1.1$. The axes of the figure are current and height z, i.e., u and v are shown as functions of z. The orientation of the axes shown in this figure is a conventional way of representing functions of vertical height z in oceanography.

We can determine the flux, and respective flux densities, for mass by multiplying speed by mass density and for any quantity associated with the water such as salt, or any dissolved or suspended material by multiplying by the concentration of that quantity. We can also determine the flux of momentum by multiplying by density times velocity, and the flux of kinetic energy (associated with the advective motion) by multiplying by density times half advective velocity squared.

These fluxes are used in equations which describe the way that the total quantity changes with time, or remains constant, i.e., is conserved, called *continuity equations*. Here are some examples of fluxes:

(1) The total flux of water (density 1000 kg m^{-3}) flowing through a pipe of cross-sectional area 0.001 m^2 is 0.0001 m^3 s^{-1}. The average velocity of the water is $0.0001/0.001 = 0.1$ m s^{-1} which is identical to the volume flux density. The mass flux is $1000 * 0.0001 = 0.1$ kg s^{-1} and the mass flux density is $0.1/0.001 = 100$ kg m^{-2} s^{-1}.

(2) A channel of width 10 km and depth h of 9 m (relative to the height datum) with a surface elevation η of 1 m (relative to that same datum) contains sea water of density 1026 kg m^{-3} and has a current $u = 0.1$ m s^{-1}. The total height of the water column is $H = h + \eta = 9 + 1 = 10$ m. The cross-sectional area of the flow is 10^5 m^2. The flux of volume is area * speed $= 10^4$ m^3 s^{-1}. The volume flux density (per unit width) is 10^4 m^3 s^{-1} $/10\,000$ m $= 1$ m^2 s^{-1}. The volume flux density per unit area is 10^4 m^3 s$^{-1}/10^5$ m$^2 = 0.1$ m s^{-1}. To find the corresponding mass

fluxes multiply by 1026 kg m^{-3} to obtain a total mass flux of 1.026 *10^7 kg s^{-1}. The mass flux density per unit width is 1026 kg m^{-1} s^{-1}. The mass flux density per unit area is 102.6 kg m^{-2} s^{-1}.

The *total horizontal volume flux* in the x direction is denoted by Qx and obtained by considering the water column as composed of a large number of thin layers of thickness δz. If the layers are numbered $i = 1, N$ and the value of u is written as u_i in layer i, the flux from layer i is simply $u_i \delta z$, so adding all the values of u down through the water column,

$$Q_x = u_1 \delta z + u_2 \delta z + \cdots + u_N \delta z \rightarrow \int_{-h}^{\eta} u dz. \tag{2.2}$$

We like to work also with the total volume flux *density* Qx/H which is simply the volume of water per unit time per unit width *per unit height* of the water column. Note that the dimensions of this flux density are the same as velocity and it is simply the average of u over the water column, called the *depth average velocity*; we denote it by an over-bar (although in many models in which we only consider the depth average motion we drop this over-bar):

$$\bar{u} \equiv \frac{Q_x}{H} = \frac{1}{H} \int_{-h}^{\eta} u dz. \tag{2.3}$$

Similarly for the component in the y direction:

$$\bar{v} \equiv \frac{Q_y}{H} = \frac{1}{H} \int_{-h}^{\eta} v dz. \tag{2.4}$$

The volume flux densities at individual heights in the water column are defined as the volume of water per unit time, per unit height per unit width passing through a plane perpendicular to each axis and if denoted by $q_x(z)$ and $q_y(z)$:

$$\begin{aligned} q_x(z) &= u(z) \\ q_y(z) &= v(z). \end{aligned} \tag{2.5}$$

2.5 Two-dimensional models

In a two-dimensional model, we work only with the depth average velocities defined above (a two-dimensional model is *not* usually a model of a coastal basin at a particular height). This means that we do not interest ourselves in the height (z) dependence of currents, i.e., we do not obtain $u(z)$ and $v(z)$. A two-dimensional model of a coastal basin which has constant density is a perfectly valid way of analyzing circulation (provided certain non-linear terms can be avoided), and simply lacks information on the height dependence of current. Mean currents and surface elevation due to tides and wind stress (for example) are very well simulated by two-dimensional models. In two-dimensional models we often drop the use of the over-bar on u and v; we accept that u and v are actually depth average velocities.

As a caveat, we should mention that two-dimensional constant density models do often contain minor errors which arise from their lack of information on the z dependence of current, of which the most often quoted is bottom stress, i.e., bottom friction (and as mentioned they may also fail in certain aspects of non-linear advection). Processes due to density variation cannot usually be modeled in two dimensions, although there are some exceptions when the basin has almost constant density in the vertical. Models which work with the full height dependence are called *three-dimensional models*. Such models require greater computational effort than two-dimensional models. The use of two-dimensional models was quite standard in coastal basins until the late 1980s, when computation power improved to the point in which three-dimensional models could be easily run. As we shall see later, there is a very considerable body of detail as to how two- and three-dimensional models are developed into numerical form.

2.6 Volume continuity equation

Mass conservation is one of the most important rules used in the development of models of coastal basins. In shallow water simple models, density is often constant and so volume may be considered to be conserved; we discuss this issue in more detail in Chapter 3. The volume continuity equation is a perfect approximation in conditions where the density of a fluid is constant. In other cases we do need care to work with mass and not volume, but volume continuity is a good starting point for our study of the hydrodynamics of coastal basins.

Consider water flow through a channel into a closed coastal basin. Although surface water may flow into the basin and deeper water may flow out of the basin (or vice versa), we can determine the *total* volume flux into, or out of the basin. Suppose that we find that the depth average velocity into the basin is $u_{in} = 0.1 \, \mathrm{ms}^{-1}$, and the channel has a total water height $H_{in} = 60 \, \mathrm{m}$: this gives a flux density of $Q_{in} = 6 \, \mathrm{m}^3 \, \mathrm{s}^{-1}$ per meter width. If the channel is 3000 m wide, the total flux is $18\,000 \, \mathrm{m}^3 \, \mathrm{s}^{-1}$. This inward flux has the effect of increasing the surface elevation in the closed basin. If the surface area of the basin is $50 \, \mathrm{km}^2$, adding $18\,000 \, \mathrm{m}^3 \, \mathrm{s}^{-1}$ would increase the elevation at a rate $18\,000/(50*10^6) \, \mathrm{m \, s}^{-1}$, or $1.3 \, \mathrm{m \, h}^{-1}$. The principle that we have used here is called the *conservation of volume*. It assumes, of course, that the density is unchanged from channel to basin. We can express this result as

$$\frac{\partial \eta}{\partial t} = \frac{bQ_{in}}{A} \tag{2.6}$$

where b is the width of the channel and A the surface area of the basin. Suppose now that the basin has the same width as the channel and its length is L:

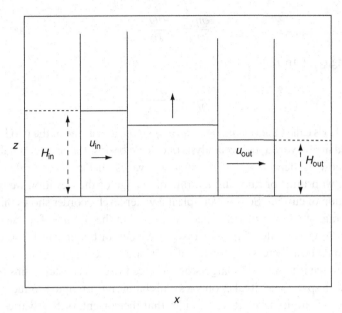

Figure 2.5 Illustration of two channels running in and out of a basin, which is used in the text to establish the basic ideas of volume continuity. The incoming channel is on the left and outgoing channel on the right. The two channels have different total heights H_{in} and H_{out} and different currents u_{in} and u_{out}. The vertical upward arrow shows the surface of the basin rising due to the inward flux exceeding the outward flux of water.

$$\frac{\partial \eta}{\partial t} = \frac{H_{in}\bar{u}_{in}}{L}. \tag{2.7}$$

Let us now suppose that there is another channel that flows out of the basin, parallel to the incoming channel but on the other side of the basin (Figure 2.5). Assume it has the same width b but a different total water height H_{out} and depth average velocity u_{out}, so the equation for volume conservation becomes

$$\frac{\partial \eta}{\partial t} = \frac{H_{in}\bar{u}_{in} - H_{out}\bar{u}_{out}}{L}. \tag{2.8}$$

What we now have is a continuous channel, so the notation should include the x coordinate to measure position along the channel and this can replace the old notation of *in* and *out*. If the point of entry of the channel is located at x, then the point of exit of the exit channel is located at $x + L$. Hence

$$\frac{\partial \eta}{\partial t} = \frac{H(x)\bar{u}(x) - H(x+L)\bar{u}(x+L)}{L}. \tag{2.9}$$

If we now let $L \to 0$, we see that the right-hand side is just minus the derivative of Hu with respect to x,

$$\frac{\partial \eta}{\partial t} = -\frac{\partial Q_x}{\partial x},$$ (2.10)

or identifying Q_x with $H\bar{u}$,

$$\frac{\partial \eta}{\partial t} = -\frac{\partial H\bar{u}}{\partial x}.$$ (2.11)

Equation (2.11) is called the *volume continuity equation,* and is one of the two fundamental relations used in all hydrodynamic analysis of coastal basins. It is a very valuable relation, since if we know the dependence of H and u on x we can determine the rate of change of the elevation with time; or knowing the rate of change of the elevation we can find the spatial variation of current. So, for example if a water level recorder shows that the water is rising at a rate of $1\,\mathrm{m\,h^{-1}}$ in a channel, we know that the volume of water passing that point is decreasing at a rate of $1/3600\,\mathrm{m^2\,s^{-1}}$ per meter of travel, which for a channel of $10\,\mathrm{m}$ deep would be a decrease of current of $0.027\,\mathrm{m\,s^{-1}}$ every $1\,\mathrm{km}$. A worthwhile lesson from this example is to note that large-scale volume fluxes in coastal basins tend to vary only slowly with space, since it takes only a small change in current to produce a significant rate of change of surface level. If we consider that the current, or flux, varies in both the x and y directions we get the more general form of the continuity equation for volume,

$$\frac{\partial \eta}{\partial t} = -\left[\frac{\partial Q_x}{\partial x} + \frac{\partial Q_y}{\partial y}\right] \rightarrow -\left[\frac{\partial H\bar{u}}{\partial x} + \frac{\partial H\bar{v}}{\partial y}\right].$$ (2.12)

The term on the right is called the *divergence* of the flux and often written as $\nabla.Q$ where Q is the vector flux, i.e., a vector with the components Qx and Qy. If we use u as vector notation for current then $Q = Hu$.

In three dimensions the volume continuity equation is based on a three-dimensional elemental volume with sides dx, dy, dz (Figure 2.6).

The total volume flux into the element must equal the total flux out, since the volume cannot change (and we are assuming the water is incompressible):

$$dxdy(w + dw - w) + dwdy(u + du - u) + dwdx(v + dv - v) = 0$$ (2.13)

or dividing through by the volume of the element $dxdydz$,

$$\frac{\partial u}{\partial x} + \frac{\partial v}{\partial x} + \frac{\partial w}{\partial x} = 0$$ (2.14)

and so the current (considered in three dimensions) is *non-divergent*, i.e., the vector form of (2.14) is $\nabla.u = 0$.

If we integrate (2.14) with respect to z from the bottom ($z = -h$) to the surface ($z = \eta$),

$$\int_{-h}^{\eta} \left(\frac{\partial u}{\partial x} + \frac{\partial v}{\partial y} + \frac{\partial w}{\partial z}\right) dz = 0$$ (2.15)

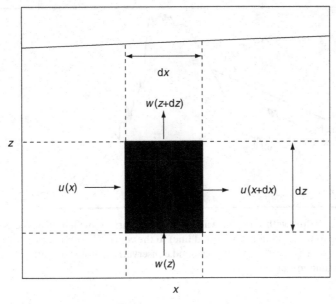

Figure 2.6 Illustration of the elemental volume used to establish the continuity equation in three dimensions; viewed through the vertical (x, z) plane.

and since x, y, and z are independent variables,

$$\frac{\partial}{\partial x} \int_{-h}^{\eta} u\, dz + \frac{\partial}{\partial y} \int_{-h}^{\eta} v\, dz + w|_{-h}^{\eta} = 0. \tag{2.16}$$

Noting that w must vanish at $z=-h$ and that the rate of increase of the surface elevation is exactly equal to the vertical velocity at the surface, i.e.,

$$\frac{\partial \eta}{\partial t} = w(\eta) \tag{2.17}$$

and from our definition of total flux (2.2), (2.16) gives

$$\frac{\partial \eta}{\partial t} = -\left(\frac{\partial Q_x}{\partial x} + \frac{\partial Q_y}{\partial y}\right) \tag{2.18}$$

which recaptures the two-dimensional volume continuity equation (2.12).

The volume continuity equation readily generalizes to continuity of any quantity with a flux density P, and generally the time derivative is replaced by the rate of change of the concentration p of that quantity with time (in the same sense that $\partial \eta / \partial t$ is the rate of change of volume per unit area),

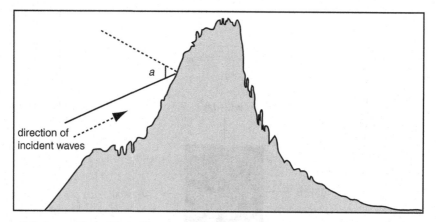

Figure 2.7 A coastline with incident waves (shown by arrow) which meet the coastline at an angle a relative to the local normal (dotted line) to the coast. These waves produce littoral drift along the coast which varies with a so that sand conservation implies accretion or erosion of the coast line as a function of a.

$$\frac{\partial p}{\partial t} = -\left(\frac{\partial P_x}{\partial x} + \frac{\partial P_y}{\partial y}\right). \tag{2.19}$$

The continuity equation applies, for example, to the rate of change of salt concentration due to the spatial change of the flux of salt. In that case p becomes the concentration of salt per unit volume, and $\boldsymbol{P}$ is the flux of salt. In that application, (2.19) provides the basic method of determining salt distribution in a coastal basin. However, the continuity equation for salt cannot be used alone, since we need a method of finding the salt flux and this requires a dynamical equation, i.e., a relationship between the state of the basin and $\boldsymbol{P}$. Salt is very important in models of coastal basins both through its effect on buoyancy and as a conservative tracer of marine water as it mixes with fresh water of terrestrial origin. This is modeled through the salt continuity equation and discussed in detail in later chapters.

The applications of continuity equations are very diverse and very powerful in all of modeling studies. To appreciate this efficacy of the continuity equations, it is important to understand the meaning of the terms in (2.19). For example, we can consider the flux of sediment parallel to the shore of a coastal basin driven by waves hitting the shore at an angle less than 90°. This flux is called *littoral drift* (*littoral* means shore and drift is another word for current) and is responsible for much of the dynamics of beaches (Figure 2.7).

The strength of littoral drift depends on the angle at which the waves hit the coast and is about maximal when that angle is near 45°. So when waves with constant direction hit a coast that is itself varying in direction, the strength of the littoral flux changes with position along the coast. This means that the flux must deposit sediment if the flux is decreasing, or entrap sediment if the flux is increasing. Consequently, the beach must either grow in width (*accretion*) or decrease in width (*erosion*), respectively. We can determine these rates from the continuity equation for sediment, once we know the dependence of the flux on the angle of waves relative to the normal to the coastline. Application of the continuity equation provides us with a simple model of coastal dynamics based on littoral drift, and indeed this model explains the formation of sand spits and the undulations in beach width that we observe as we walk along the beach. This model depends on our knowing the dependence of the littoral drift on wave angle, which we would describe as the *dynamic* part of the model. Most models of coastal basins tend to have a part based on continuity, i.e., conservation, and a part based on dynamics.

2.7 Sources and sinks

Equation (2.12) assumes that the flow has no *sources* or *sinks* of fluid, i.e., points in the basin where fluid enters or leaves the basin. Suppose now that we have the possibility of an inflow at any point in the basin, so that in general a volume $r \mathrm{d}x \mathrm{d}y$ of fluid flows in unit time into the element of area $\mathrm{d}x\mathrm{d}y$ located at position (x, y). Here r is a function of x, and y has the dimensions of a velocity. The continuity equation (2.12) now becomes

$$\frac{\partial \eta}{\partial t} + \frac{\partial Hu}{\partial x} + \frac{\partial Hv}{\partial y} = r(x, y) \tag{2.20}$$

or in vector notation where $\boldsymbol{u}$ represents the vector velocity,

$$\frac{\partial \eta}{\partial t} + \nabla.H\boldsymbol{u} = r(x, y). \tag{2.21}$$

If the basin were entirely closed (no external boundaries), then the total volume of water in the basin would change with time at a rate determined by the integral of the source/sink term in (2.21) over the volume of the basin, i.e.,

$$\frac{\partial \langle \eta \rangle}{\partial t} = \iint_{\text{basin}} r(x, y)\mathrm{d}x\mathrm{d}y \tag{2.22}$$

where we have used the angular brackets to denote a spatial average over the basin, i.e.,

$$\langle \eta \rangle \equiv \iint_{\text{basin}} \eta(x, y)\mathrm{d}x\mathrm{d}y \bigg/ \iint_{\text{basin}} \mathrm{d}x\mathrm{d}y \tag{2.23}$$

and note that the right-hand side of (2.22) is the total rate of inflow into the basin. Equation (2.22) requires us to prove that for a totally closed basin the integral of the surface volume flux terms in (2.20) is zero. This is straightforward for a one-dimensional basin, since

$$\int_{\text{basin}} \frac{\partial Hu}{\partial x} dx = Hu|_{\text{boundaries}},$$ (2.24)

and the right-hand side of (2.24) is evidently zero for a closed basin (no volume flux across ends). For a two-dimensional basin we need to use the *Gauss divergence theorem*, i.e.,

$$\iint_{\text{basin}} \nabla.H\boldsymbol{u}\mathrm{d}x\mathrm{d}y = \int_{\text{boundaries}} H\boldsymbol{u}.\mathrm{d}\boldsymbol{a}$$ (2.25)

where the right-hand side is the surface integral of the volume flux normal to the surface (d$\boldsymbol{a}$ is a vector unit of surface area which has magnitude da and direction parallel to the normal to the surface at the particular point). The right-hand side of (2.25) is obviously the total surface flux and therefore vanishes.

If more generally we assume that there is a finite volume flux through the surface, then (2.20) gives

$$\frac{\partial \langle \eta \rangle}{\partial t} = \iint_{\text{basin}} r(x,y)\mathrm{d}x\mathrm{d}y - \int_{\text{boundaries}} H\boldsymbol{u}.\mathrm{d}\boldsymbol{a}.$$ (2.26)

Equation (2.26) describes in a general form most states of a coastal basin with water entering, or leaving, through the open boundaries and sources and sinks such as rivers and evaporation. If the volume of the basin is fixed, as may occur in models of long term averages of coastal basins, the right-hand side of (2.26) vanishes with boundary flow exactly compensating sources and sinks. If there is inflow into one small area of the basin as occurs, for example, with a river entrance, then the function $r(x, y)$ is very highly peaked at one particular point in the basin. Let us reduce this to a one-dimensional problem, in which case r is a function of x only. As an example, consider $r(x) = s(x)$ defined as

$$s(x - x_1) = \frac{1}{w\sqrt{\pi}} e^{-(x-x_1)^2/w^2}.$$ (2.27)

This function is illustrated in Figure 2.8 for $x_1 = 0$ and half width $w = 1$. The function has a total width of $2w$ in the sense that s decreases by a factor of $1/e$ from its value at $x = 0$ on either side of the origin at $x = \pm w$, and the total area under the function is unity, i.e., the integral of s with respect to x over the entire domain of x is unity. As w decreases so the function becomes increasingly sharply peaked at x_1, but keeps the total area under the curve fixed at unity. In the extreme limit of infinitesimally small width, the function s has a special name: the Dirac delta function (after Paul Dirac [1902–84], a British mathematical and theoretical physicist), i.e.,

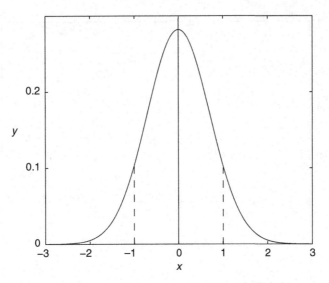

Figure 2.8 The function $s(x)$ drawn with its center at $x = 0$ and half width $w = 1$. It can be used as a source function in one-dimensional models of a coastal basin. For two- and three-dimensional models, source functions can be constructed from products of similar functions in y and z.

$$\delta(x - x_1) \equiv \lim_{w \to 0} \frac{1}{w\sqrt{\pi}} e^{-(x-x_1)^2/w^2}. \qquad (2.28)$$

The δ *function* is one of a class of *generalized functions* that are defined as the limits of continuous functions.

The other generalized function that we meet in our studies of coastal basins is the *unit step function* denoted by $U(x - x_1)$, or sometimes by $H(x - x_1)$, also known as the Heaviside step function (after Oliver Heaviside [1850–1925], a renowned British mathematical scientist), and defined as

$$U(x - x_1) \equiv \begin{cases} 0, & x < x_1 \\ 1/2, & x = x_1 \\ 1, & x > x_1 \end{cases} \qquad (2.29)$$

where U can be considered as the integral of the delta function, i.e.,

$$U(x - x_1) = \int_{-\infty}^{\infty} \delta(x - x_1) dx. \qquad (2.30)$$

The step function is illustrated in Figure 2.9 as the limit of the integral of (2.28) for decreasingly small values of w from $w = 1.5$ to $w = 0.5$. It follows from (2.30) that the derivative of the unit step function is the delta function, i.e.,

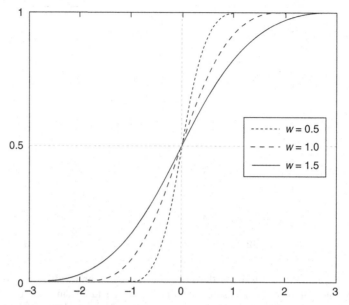

Figure 2.9 The generalized function which leads to the step function plotted with its center at $x = 0$ and decreasing half width w from $w = 1.5$ to $w = 0.5$.

$$\frac{\mathrm{d}}{\mathrm{d}} U(x - x_1) = \delta(x - x_1). \tag{2.31}$$

The delta function is used as a point source input in analytical models. So in two dimensions, inflow at one single point (x_1, y_1) of total magnitude R is

$$r \rightarrow R\delta(x - x_1)\delta(y - y_1). \tag{2.32}$$

The volume continuity equation without any source/sink terms (2.18) is said to be a *homogeneous* equation, basically because the terms in the equation all involve the dependent variables u and v. The source terms in (2.20) are *inhomogeneous* because they are usually independent of the dependent variables, i.e., u, v, and η. The sink terms are almost certainly inhomogeneous but, for example, might depend on η.

   When using delta functions and unit step functions in applied mathematical analysis, we should be careful to note that their arguments should be non-dimensional. So we cannot write a delta function of dimensional lengths, and for example, $\delta(1 \text{ m})$ is totally meaningless. So when using generalized functions in analytic models, we must make sure that we first non-dimensionalize the space coordinates. We can also use delta functions of time $\delta(t)$ but t must be non-dimensionalized. We are very familiar with such techniques since we use them for trigonometrical functions, such as sine and cosine, which can only be functions of dimensional quantities called angles.

   The set of model differential equations that represent the continuity and dynamical equation of a coastal basin are only homogeneous if there are no external forcing functions, such as a surface wind stress. This set of equations cannot be solved without boundary conditions, which also have the form of a set of differential conditions either equated to zero or to some constants. In the former case, we say that the boundary conditions are homogeneous (contain no external forcing) and in the latter case inhomogeneous (do contain *external forcing*). In the solution of the model equations, the steady state values of the dependent variables are entirely controlled by the values of the inhomogeneous terms (in either the model equations or boundary conditions), i.e., the solution is *driven* by the externally applied forcing. If there are no inhomogeneous terms, the model solution will simply be zero, or some constant value, called the *trivial solution* by applied mathematicians. If we retain only terms in the model equation which are linear in the dependent variables, it is easy to show that the steady state model solution is linear in the external forcing functions. In such a case, the response of the system to the sum of a number of external forcing functions is the sum of the responses to the individual forcings.

An example of external forcing is provided by the source–sink terms in the continuity equation. For a simple one-dimensional linear model we can determine the solution $G(x - x_1)$ for a delta function source, i.e., $\delta(x - x_1)$. If the model is otherwise homogeneous in η, the solution for a general source function $r(x)$ is then

$$\eta = \int_{\text{basin}} r(x_1)G(x - x_1)dx_1 \tag{2.33}$$

where we have replaced the sum over a large number of forcing functions by its integral. Equation (2.33) is seen physically to be the sum of responses to delta functions placed at every value of x of strength $r(x)$, i.e., we have written the distributed source as an infinite sum of point sources,

$$r(x) = \int_{\text{basin}} r(x_1)\delta(x - x_1)dx_1. \tag{2.34}$$

The term G is called the Green's function of the continuity equation; it can be equally well defined in two or three dimensions. The form of G does depend on the dynamical equation which will be considered later, but for the moment let us consider a very simple situation in which there is a constant flow velocity u along the basin, i.e., the dynamical equation is simply $u = constant$. This might, for example, describe the flow along a channel of constant cross section. We can now solve for G using (2.20). We first need to non-dimensionalize the variables so that h is measured as a fraction of some standard depth h_0, time is measured relative to a standard time t_0, and x is taken as a fraction of a standard length x_0. The velocity u is then measured relative to x_0/t_0 and the source strength is non-dimensionalized to h_0/t_0. We shall assume that all variables have been non-dimensionalized accordingly. Hence, (2.20) with the source term replaced by a delta function we get

$$\frac{\partial \eta}{\partial t} + u \frac{\partial (h + \eta)}{\partial x} = \delta(x - x_1) \qquad (2.35)$$

since u is independent of x. For a steady state, $\partial \eta / \partial t = 0$ and so we can integrate (2.35) to give

$$u(h + \eta) = \int_{-\infty}^{x} \delta(x - x_1) \mathrm{d}x \qquad (2.36)$$

where we have assumed that the model domain extends over the whole x axis. Using (2.30) and (2.36) gives

$$h + \eta = \frac{1}{u} U(x - x_1) \qquad (2.37)$$

so that the input of fluid at $x = x_1$, simply causes the height of the water column to jump discontinuously by an amount $1/u$ (note that u is non-dimensional), so that the volume flux downstream of the source is appropriately greater than the volume flux upstream. For a distributed set of sources of (relative) strength $r(x)$ the total height of the water column is simply

$$h + \eta = \frac{1}{u} \int_{\text{basin}} r(x_1) U(x - x_1) \mathrm{d}x_1. \qquad (2.38)$$

2.8 Linearized continuity equation

It is often possible, and usually advantageous in analytical models, to linearize the continuity equation (2.18), by which we mean remove products of dependent variables such as η and u. As an example, consider a coastal basin of length 18 km which has a uniform depth of $h = 5$ m, in which the water surface moves up and down with the astronomical tide in such a way that the surface is *almost horizontal* over the basin (the change in elevation over the basin is in some sense small). This type of model is applicable to many short shallow coastal basins, in which the time for a long gravitation wave to travel the length of the basin is small compared with the tidal period. Suppose that at a particular time the surface elevation η is measured to be zero (relative to the height datum) and is increasing at a rate of 1 m h^{-1}, i.e., $\partial \eta / \partial t = 1/3600$. The change in the depth average current from one end of the basin to the other is given by the continuity equation (2.10) which, neglecting η relative to h, gives

$$5 \frac{\partial u}{\partial x} = -\frac{1}{3600} \qquad (2.39)$$

where u is the depth average current (we have dropped the over-bar). So (2.39) gives the change in current from one end of the basin to the other as $-1.0 \, \mathrm{m \, s^{-1}}$ (where the minus sign means that the current decreases with x); water is there being diverted from the flow to elevate the surface. If the basin is closed at its far end, i.e., at $x = L = 18 \, \mathrm{km}$ so that $u(L)$ is zero, we would find the current at $x = 0$ is $u(0) = 1 \, \mathrm{m \, s^{-1}}$; note that the volume continuity equation only establishes *differences* in current and requires boundary conditions at the edge of the model to establish actual values of currents. To generalize this example, if the depth average current in a long channel has the form

$$u = A + Bx \tag{2.40}$$

where A and B are constants and the water depth is constant at h, we can find the variation of surface elevation with time, assuming that the elevation $\eta \ll h$, as follows:

$$\frac{\partial \eta}{\partial t} = -\frac{\partial Hu}{\partial x} = \frac{\partial (h + \eta) u}{\partial x}$$

$$\approx (h + \eta) \frac{\partial A + Bx}{\partial x} \sim hB \tag{2.41}$$

which illustrates the important result that if u varies linearly with x, the surface moves as a horizontal plane. Equation (2.41) illustrates two very common approximations made with the volume continuity equation.

(1) If we can choose the datum to be near the water surface, *at all times*, we can neglect the elevation in comparison with the depth, i.e.,

$$\frac{\partial \eta}{\partial t} \rightarrow -\frac{\partial hu}{\partial x}. \tag{2.42}$$

Equation (2.42) is said to be *linearized* because we have removed the product of two *independent variables*, namely elevation η and current u. Note that h may also vary with x but that does not affect the linearity of (2.42): linearity only requires non-linear terms in the *dependent* variables (for which the model is being solved) which in this case are u and η (h is known as a function of x).

We can always use the trick of putting the height datum near the surface, provided that the surface elevation does not vary too much across the basin in relation to the water depth. This approximation works very well in deep water, where the elevation typically changes by very small amounts relative to depth. For example, the surface elevation in the South Pacific Ocean varies by about 1 m from the coast of Australia to South America, while the depth is thousands of meters. So we could use the linearized equation. In Jervis Bay, on the eastern coast of Australia, the depth is 10 to 60 m and the elevation varies by (at most) 0.1 m across the bay, so we could still linearize the volume continuity equation. In Tampa Bay, on the west coast of Florida, the tidal range is about 1 m and the water depth is a few meters, and so linearity is of marginal validity. Across a coral reef the depth might be less than 1 m, and the surface elevation varies by at least 0.1 m across the reef. So, linearization needs to be treated with extreme care (although it is always a good first approximation); there are many cautionary tales about the effect of removing the non-linear terms at too early a stage in the development of a model.

Non-linear terms which cannot be removed by a suitable change of height datum are not just a mathematical inconvenience, but are instead a very integral part of volume continuity. This is especially true where the bottom is close to the surface and the bottom may become dry, and then wet again, as elevation changes with time. There are a lot of important effects in shallow water waves which are associated with these *non-linear* terms in the continuity equation, i.e., terms of form $\partial \eta u / \partial x$. In numerical models there is usually no special advantage in linearizing the continuity equation, but in analytical models the non-linear term often makes a solution intractable. We shall see that non-linear terms also arise in the dynamical equation.

(2) If h is constant the linearized continuity equation (2.42) becomes

$$\frac{\partial \eta}{\partial t} = -h \frac{\partial u}{\partial x} \tag{2.43}$$

or in two dimensions

$$\frac{\partial \eta}{\partial t} = -h \left(\frac{\partial u}{\partial x} + \frac{\partial v}{\partial y} \right). \tag{2.44}$$

This much simpler form of the continuity equation can also be justified if the variation of h with x is less important than the corresponding variation of u with x, i.e., writing (2.42) as

$$\frac{\partial \eta}{\partial t} \rightarrow -hu \left(\frac{\partial \log(u)}{\partial x} + \frac{\partial \log(h)}{\partial x} \right), \tag{2.45}$$

then

$$\frac{\partial \log(u)}{\partial x} \gg \frac{\partial \log(h)}{\partial x}. \tag{2.46}$$

This type of simplification may be useful in numerical models. If (2.45) is valid, the *divergence* of the depth average current is zero in the steady state, i.e.,

$$\frac{\partial u}{\partial x} + \frac{\partial v}{\partial y} = 0 \tag{2.47}$$

It is important to note that (2.47) only applies to flows with constant density, i.e., no salinity and temperature differences and under incompressible conditions.

2.9 Potential flow

The curl of the velocity (called the *vorticity*) is defined as

$$vorticity \equiv \frac{\partial v}{\partial x} - \frac{\partial u}{\partial y}. \tag{2.48}$$

Suppose that additionally to (2.47), the velocities can be derivable from a *potential* which we shall denote by Ω, i.e.,

$$u = -\frac{\partial \Omega}{\partial x} \quad v = -\frac{\partial \Omega}{\partial y}. \tag{2.49}$$

In coastal basins (2.49) is equivalent to saying that we ignore the Earth's rotation and assume a steady state balance between pressure gradient and linear friction. This potential is the pressure or effectively the surface elevation divided by a linear coefficient of friction. The flow described by (2.49) is referred to as *potential flow* because we can integrate (2.49) to find the potential Ω which is a unique function of x and y (to within an arbitrary additive constant). If we substitute (2.49) in (2.47),

$$\frac{\partial v}{\partial x} - \frac{\partial u}{\partial y} \rightarrow -\frac{\partial^2 \Omega}{\partial x \partial y} + \frac{\partial^2 \Omega}{\partial y \partial x} = 0 \tag{2.50}$$

and so the vorticity vanishes for potential flow which means that the flow is *irrotational*. Such problems are usually derived from the Euler equations, which are a special case of the full Navier–Stokes equations in the absence of viscosity (so called *inviscid* flow).

If we differentiate (2.47) with respect to x and then use (2.50) we obtain

$$\left(\frac{\partial^2}{\partial x^2} + \frac{\partial^2}{\partial y^2} \right) u = 0. \tag{2.51}$$

The operator on the left-hand side of (2.51) is known as the Laplacian ∇^2:

$$\nabla^2 \equiv \frac{\partial^2}{\partial x^2} + \frac{\partial^2}{\partial y^2}, \tag{2.52}$$

i.e.,

$$\nabla^2 u = 0 \tag{2.53}$$

or differentiating (2.47) with respect to y and using (2.53),

$$\nabla^2 v = 0. \tag{2.54}$$

Hence u and v satisfy the *Laplace equation* $\nabla^2 = 0$. Such functions are said to be *harmonic*. It is easy to show that u and $-v$ (note the minus sign) satisfy

$$\frac{\partial u}{\partial x} = \frac{\partial(-v)}{\partial y}, \quad \frac{\partial u}{\partial y} = -\frac{\partial(-v)}{\partial x} \tag{2.55}$$

as we can prove by differentiating the first term of (2.55) by x and the second by y and adding to give $\nabla^2 u = 0$; repeating the same process by differentiating the first by y, the second by x and subtracting gives $\nabla^2 v = 0$.

Many problems in physics can be effectively treated using the theory of functions of a complex variable. Vital ingredients of that theory are functions that are both continuous and differentiable for all directions in the complex plane, over a finite domain of that plane. Such functions are said to be *analytic* in such domains. The

Cartesian components of the position of a point (x, y) are treated as the real and imaginary parts of a complex variable, usually denoted by z. Since we use that notation for height, we here take an upper case and write

$$Z = x + iy. \tag{2.56}$$

The value of Z denotes position on the (x, y) plane. Hence, the components of velocity (u, v) are functions of Z. We can also write Z in polar coordinates as

$$Z = Re^{i\theta} \tag{2.57}$$

with R the length of the line from the origin to the point (called the modulus of Z) and θ is the angle, measured positive counterclockwise from the x-axis to the line from the origin to the point.

The theory of complex variables considers a complex function of this complex variable and in our case we are interested in the velocity. Accordingly, we can write the velocity as a complex number with real and imaginary parts corresponding to these components. For convenience we use a complex function V, known as the conjugate complex velocity,

$$V = u - iv \tag{2.58}$$

which has u as its real part and $-v$ as its imaginary part. The complex conjugate of V would have the same components u, v as the velocity. The relations in (2.55) are the so-called *Cauchy–Riemann* conditions, which means that V is an *analytic function* of the complex variable Z over all regions of the plane for which (2.55) applies, i.e., regions in which the divergence and curl of u vanish. The Cauchy–Riemann conditions are a necessary and sufficient condition for a function to be analytic. So the solution of the flow problem simply reduces to finding an analytic function which satisfies the boundary conditions.

It is convenient to work with a complex form of the potential Ω defined in (2.49), i.e.,

$$W = \Omega + i\Gamma \tag{2.59}$$

and in order to make W analytic we choose Ω and Γ to satisfy the Cauchy–Riemann conditions, i.e.,

$$\frac{\partial \Omega}{\partial x} = \frac{\partial(\Gamma)}{\partial y}, \frac{\partial \Omega}{\partial y} = -\frac{\partial(\Gamma)}{\partial x}, \tag{2.60}$$

and using (2.55) this implies

$$u = -\frac{\partial(\Gamma)}{\partial y}, v = \frac{\partial(\Gamma)}{\partial x}. \tag{2.61}$$

It follows that we can derive V from the complex derivative of W which we write as dW/dZ. The meaning of this derivative is basically identical to a derivative with respect to a real variable such as x or y, but has the major difference that dz is a small change in the *position* of a point in the z plane. The property of an analytic function which we referred to as *differentiability* means that the value of this derivative does not depend on the *direction* of dz. Let us illustrate this with an example. Suppose that $W = Z$. It happens that this is an example of an analytic function; the only place in the Z plane at which it is not analytic is the so-called *point at infinity*, i.e., very large Z. We need to be very careful in talking about large values of a complex variable. A real variable x can have two limits at either end of the x axis which we call $\pm\infty$. But a complex variable can approach large values of Z along any direction in the Z plane. We therefore introduce a device in which we transform from Z to a variable Z' in which $Z' = 1/Z$ and hence, W becomes $W = 1/Z'$ which has a singularity at $Z' = 0$. We say that this point maps to the point at infinity in the Z plane. Notice that $1/Z'$ is clearly discontinuous at $Z' = 0$ (and therefore non-analytic and certainly non-differentiable) no matter what direction we approach the origin of the Z' plane since we can write

$$\frac{1}{Z'} = \frac{1}{R'}e^{-i\theta} \tag{2.62}$$

where

$$Z' = R'e^{i\theta} \tag{2.63}$$

and so clearly $1/Z'$ diverges as $1/R'$. However, excluding the point at infinity the function $W = Z$ is analytic. In fact, the only function which is analytic everywhere in the Z plane (including the point at infinity) is $W = \text{constant}$. To continue with our example of determining dW/dz, we first consider a path for dz which is along the x axis:

$$\frac{dW}{dz} = \frac{dZ}{dz} = \frac{d(x+iy)}{dz} \rightarrow \frac{\partial(x+iy)}{\partial x} = 1, \tag{2.64}$$

since x and y are independent variables. If we choose a path along the y axis

$$\frac{dW}{dz} = \frac{dZ}{dz} = \frac{d(x+iy)}{dz} \rightarrow \frac{\partial(x+iy)}{\partial iy} = 1, \tag{2.65}$$

and very clearly any path will give the result $dW/dZ = 1$. We can provide this result for any analytic function simply by using the Cauchy–Riemann equations. Consequently, the rules for differentiating functions of a complex variable are the same as familiar rules for differentiating functions of a real variable.

Hence, having established the unique meaning of a derivative of an analytic function we finally write

$$\frac{dW}{dZ} = \frac{\Omega + i\Gamma}{dZ},$$

(2.66)

which can be written as either a derivative along the x axis, i.e.,

$$\frac{dW}{dZ} = \frac{\Omega + i\Gamma}{dx} = \frac{\partial\Omega}{\partial x} + i\frac{\partial\Gamma}{\partial x}$$

(2.67)

which reduces via (2.49) and (2.55) to

$$V = -\frac{dW}{dZ},$$

(2.68)

or by a derivative along the y axis (or in any other direction) and this simply reproduces (2.68).

The complex potential W is a useful function, and our problem is merely to find an analytic function that satisfies any externally applied conditions and from W, we can determine V using (2.68). The complex potential has the interesting property, that its imaginary part Γ is constant along a streamline of the flow. A streamline is simply the path traced out by a particle as it moves in the velocity field V. So as the particle moves through (dx, dy) the change $\Delta\Gamma$ in Γ is

$$\Delta\Gamma = \frac{\partial\Gamma}{\partial x}dx + \frac{\partial\Gamma}{\partial y}dy$$

(2.69)

and replacing dx by udt and dy by vdt, (2.61) gives

$$\Delta\Gamma = \left(-\frac{\partial\Gamma}{\partial x}\frac{\partial\Gamma}{\partial y} + \frac{\partial\Gamma}{\partial y}\frac{\partial\Gamma}{\partial x}\right)dt = 0$$

(2.70)

so that Γ is constant along a streamline. This means that if we draw contours of Γ, they will exactly coincide with the streamlines of the flow field. For an analytic function it is easy to show that the contours of the real and imaginary parts are orthogonal, i.e., they cut at right angles. The proof is very easy, and for example, if we consider a change in position of a particle (dx, dy) then for a path that would be at right angles to the true path, i.e., $dx \to -vdt$ and $dy \to udt$, we see from (2.49) that Ω (the real part of W) is constant. So the streamlines run at right angles to the contours of the real part of the potential. Modelers often prefer to work with the potential, rather than the velocity itself, because boundary conditions are more easily satisfied by using the potential (as, for example, with streamlines that must run parallel to a coastline).

A simple solution of the flow can be obtained in an infinite domain of (x, y), i.e., over the entire Z plane. That solution is $V = A$ or equivalently $W = AZ + B$ where A and B are constants. This corresponds to u and v in (2.58) being constant over the plane, which is a flow of constant speed and direction everywhere over the plane. The contours of the real and imaginary parts of W are a set of orthogonal straight lines and for example if A and B are real,

$$\Omega = Ax + B$$
$$\Gamma = Ay$$
(2.71)

so the Ω contours involve constant x and are therefore parallel to the y axis and the Γ contours parallel to the x axis, i.e., we have flow parallel to the x axis. By adjusting the value of A (which is in general complex), we can produce a potential which describes parallel flow in any direction and of any magnitude. The constant B in the potential is always arbitrary but useful, since it allows us to adjust the potential along a boundary. The solution we have described, while very simple, is the only solution over an infinite plane, since $V = $ constant is the only function that is analytic everywhere. This is physically reasonable since we are considering an infinite domain with potential flow. This means that there are no sources or sinks.

It is also possible to introduce isolated points in the existing Z plane at which V is non-analytic and these represent either sources or sinks. These are called *isolated singularities*, because in the region around each such singularity (no matter how small) the function is analytic. In three dimensions these would appear as vertical line sources. Sources and sinks are represented by the simplest type of isolated singularity which is a *simple pole* for which V has the form

$$V = \frac{S_1/2\pi}{Z - Z_1}$$
(2.72)

where Z_1 is the position of the singularity and S_1 is the strength of the source (if $S_1 > 0$) or sink (if $S_1 < 0$). We can express Z in polar coordinates, i.e.,

$$Z = Re^{i\theta}$$
(2.73)

where R is the modulus of Z which corresponds to the distance from the origin of the plane to the point (x, y) and θ is the angle between the x axis and the line from the origin to the point (x, y). If the source is at the origin, i.e., $Z_1 = 0$, then (2.72) can be written as

$$V = \frac{S_1/2\pi}{R} e^{-i\theta}$$
(2.74)

so that the velocity field associated with a point source decreases as the inverse of the distance from the source. The flow is radial since (2.74) is equivalent to

$$u - iv = \frac{S_1}{2\pi R} (\cos\theta - i\sin\theta).$$
(2.75)

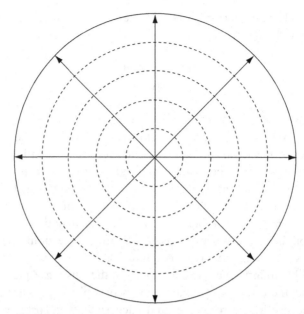

Figure 2.10 Streamlines (arrows) and contours (circles) of potential for a point source.

The complex potential corresponding to a single point source as in (2.72) is

$$W = -\frac{S_1}{2\pi}\log(Z - Z_1) + A \tag{2.76}$$

which is singular at $Z = Z_1$. Using the polar representation of Z, and for convenience taking the source to be at $Z_1 = 0$, (2.76) becomes for a source of strength S,

$$W = -\frac{S}{2\pi}\log(Re^{i\theta}) = -\frac{S}{2\pi}(\log R + i\theta) \tag{2.77}$$

or

$$\begin{aligned}\Omega &= -\frac{S}{2\pi}\log R \\ \Gamma &= -\frac{S}{2\pi}\theta.\end{aligned} \tag{2.78}$$

Equation (2.78) shows that the streamlines correspond to constant θ and the contours of potential are circles. Figure 2.10 illustrates a point source.

The total flux of fluid from the source can be obtained by integrating the flux around a circle, and (2.74) immediately verifies that this flux is S_1. More generally the flux across any closed path C can be written as

$$flux = \oint_C W\mathrm{d}Z. \tag{2.79}$$

If S is imaginary and $S=-iP$ (2.78) is replaced by

$$\Omega = -\frac{P}{2\pi}\theta$$

$$\Gamma = \frac{P}{2\pi}\log R$$ (2.80)

so that the streamlines are circles and the contours of constant potential are radial lines. This corresponds to a point source of vorticity, with W being singular only at the origin. In this case, (2.79) becomes the total circulation around the closed path.

2.10 Conformal mapping

Complex variable analysis provides a powerful model for solving potential flow through a method called *conformal mapping*. Potential flow in an infinite (x, y) plane without sources or sinks just consists of a uniform flow, i.e., u and v are independent of x and y, and it is easily seen that this is a trivial solution of the continuity and irrotational conditions. This also follows from the simple result that the only function that is analytic over the whole Z plane is a constant. Conformal mapping is based on finding an analytic mapping function

$$Z' = f(Z)$$ (2.81)

that maps our problem from our original Z plane (which corresponds to the coastal basin) to another plane called Z'. This new plane is one in which the geometry looks much simpler.

As an example, suppose that we wish to model the potential flow around a thin *semi-infinite* coastline that extends from $x=0$ to very large (infinite) x (see Figure 2.11). For the sake of simplicity, take the x axis to run from west to east, so that the y axis is south to north. We envision a westward flow north of the coastline and eastward south of the coastline, with a direction change around the west of the headland. Flow at the coastline must be parallel to that boundary and so $W=0$ would be a suitable boundary condition on the positive x axis. Finding an analytic function that satisfies that condition is not immediately obvious, but suppose that we map to another plane (the Z' plane) in which this boundary occupies the whole of the x' axis. This means that we need to imagine the positive x axis as a sheet folded about $x=0$; with its northern side one part of the sheet and its southern side the other part of the sheet. If we *unfold* the sheet, the southern side now fills the negative x axis. So, in the Z' plane, the flow is much simpler and goes from large positive x' to large negative x'. The solution in the mapped plane is $W=AZ'$ with A being a real constant, i.e., $\Omega=Ax'$, $\Gamma=Ay'$ so that the x' axis is a streamline with $\Gamma=0$. We need to find a mapping function $Z'=f(Z)$ which unfolds the coast in the way that we have described. Let us suppose that Z' can be written as

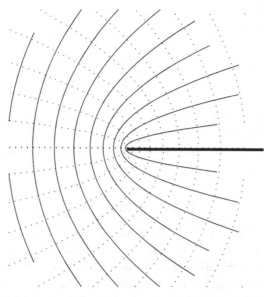

Figure 2.11 Potential flow around an infinitesimally thin headland. Full lines are streamlines and broken lines are contours of potential.

$$Z' = Z^n \tag{2.82}$$

where the exponent n is a positive constant. This is easiest to understand in polar coordinates (2.57), i.e.,

$$Z = Re^{i\theta} \quad Z' = R'e^{i\theta'}$$
$$R' = R^n \quad \theta' = n\theta. \tag{2.83}$$

Let us suppose in particular that the exponent $n = \frac{1}{2}$, and so as θ changes from 0 to 2π in the physical Z plane, θ' changes from 0 to π in the mapped Z' plane. This achieves the required unfolding of the x axis. Hence

$$W = AZ'^{1/2} \tag{2.84}$$

or in polar form

$$W = R^{1/2}e^{i\theta/2} \tag{2.85}$$

with real and imaginary parts

$$\Omega = R^{1/2}\cos\frac{\theta}{2} \quad \Gamma = R^{1/2}\sin\frac{\theta}{2} \tag{2.86}$$

and so the streamlines have the form

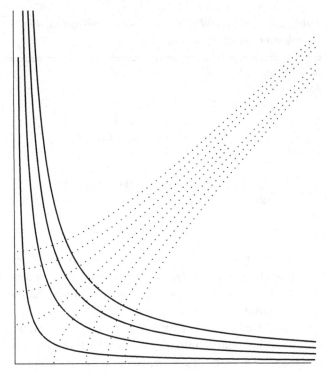

Figure 2.12 Potential flow meeting a sharp (90°) turn in the coastline. Full lines are streamlines and broken lines are contours of constant potential.

$$R = \left(\frac{\Gamma}{\sin \theta/2} \right)^2 \qquad (2.87)$$

which is shown in Figure 2.11, along with the contours of constant Ω. The resultant flow potential is only analytic on what we call a single *Riemann sheet*, and the flow pattern that is shown in the figure represents one of these sheets. The sheets are joined along the physical boundary, and this represents a *cut* in the doubled valued nature of the square root function; the origin $Z = 0$ is a *branch point*, as is the point at infinity in the Z plane (joined by the cut). This means that the flow at the origin is irrotational. The departure from potential flow is due to the boundaries, and in the present case is associated with the sharp headland. This is necessarily captured by the mapping function and incorporated into the complex potential. However, in problems that can be modeled by conformal mapping techniques the potential flow only usually fails at isolated points.

Figure 2.12 shows a potential flow meeting a sharp (90°) turn in the coastline. It uses the same model as in Figure 2.11, except we have replaced the exponent in n in (2.83) which previously had the value $n = \frac{1}{2}$ by $n = 2$. This has the effect of unfolding the sharp turn into a straight boundary along the x' axis.

Table 2.1. *Matlab code for conformal mappings used in this chapter (plotting routines not shown)*

```
M = 100; n= 2; N= 1/n;
for G = 1:2.5:10;
        for m = 1:M-1;
                theta(m) = (pi/4)*m/M;
                R(m) = (G/sin(n*theta(m)))^N;
                x(m) = R(m)*cos(theta(m));
                y(m) = R(m)*sin(theta(m));
                R1(m) = (G/cos(n*theta(m)))^N;
        end;
end;
```

The program used in our two examples of conformal mapping is shown in Table 2.1. Any value can be chosen for n, and our examples have used $n = \frac{1}{2}$ and $n = 2$. The table shows only the structure of the code, and individual cases may require attention to details; the plotting routines are omitted.

We see that the art of conformal mapping is finding an analytic function that transforms real physical boundaries into a much simpler form, for which an analytic potential is easily found. This can, of course, be done in steps changing the variable from Z to Z' and Z' to Z'', etc., i.e., we can map through a series of planes until we find suitable boundaries. This is aided by dictionaries of conformal mappings that perform particular changes in geometry. For example, the mapping $Z' = 1/Z$, which we discussed earlier, maps the inside of a circle to the outside of the same circle. A useful map is

$$Z' = \frac{a}{2}\left(Z + \frac{1}{Z}\right) \tag{2.88}$$

and this maps the outside of the unit circle (a circle in the Z plane of unit radius centered on the origin $Z = 0$) onto the upper half of the Z' plane. Consider three points on the upper circumference of the unit circle:

$$\begin{aligned}
A &: x = -1, y = 0 & x' = -a, y' = 0 \\
B &: x = 0, y = 1 & x' = 0, y' = 0 \\
C &: x = 1, y = 1 & x' = a, y' = 0,
\end{aligned} \tag{2.89}$$

i.e., the unit circle is mapped to the x' axis from $x' = -a$ to a. Hence the flow region around a circular island corresponds to the complex potential

$$W = \frac{1}{2}\left(Z + \frac{1}{Z}\right) \quad |Z| \geq 1 \tag{2.90}$$

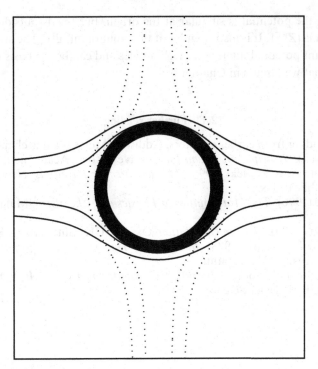

Figure 2.13 Potential flow around a circular island. Full lines are streamlines and broken lines are contours of potential.

to within a multiplicity and additive constant, i.e.,

$$\Omega = \frac{\cos\theta}{2}\left(R + \frac{1}{R}\right)$$
$$\Gamma = \frac{\sin\theta}{2}\left(R - \frac{1}{R}\right)$$

(2.91)

so that the streamlines correspond to

$$R = \frac{\Gamma}{\sin\theta} + \sqrt{\left(\frac{\Gamma}{\sin\theta}\right)^2 + 1}$$

(2.92)

and the contours of potential are

$$R = \frac{\Omega}{\cos\theta} + \sqrt{\left(\frac{\Omega}{\cos\theta}\right)^2 + 1}$$

(2.93)

both of which are shown in Figure 2.13.

Notice that the potential is singular at the origin, but this is outside the physical domain which is $|Z| \geq 1$. It is also singular at the point of infinity. The streamlines and lines of constant potential intersect at right angles and can be the basis of curvilinear coordinates that we discuss in Chapter 6.

2.11 Further reading

See further reading from earlier chapters. Additional texts for this chapter are:

Apel, J. R. (1987). *Principles of Ocean Physics*. New York: Academic Press.

Dennery, P. and A. Krzywicki (1996). *Mathematics for Physicists*. New York: Dover Publications.

Margenau, H. (1976). *The Mathematics of Physics and Chemistry*. Malabar, FL: Krieger.

Open University (2001). *Ocean Circulation*. Oxford, UK: Butterworth-Heinemann.

Pond, S. and G. L. Pickard (1983). *Introductory Dynamic Oceanography*. Oxford, UK: Butterworth-Heinemann.

Tomczak, M. and J. S. Godfrey (1994). *Regional Oceanography: An Introduction*. Oxford, UK: Elsevier Science.

3

Box and one-dimensional models

3.1 The value of box models

Conservation of mass and momentum are the two basic laws which govern the way that coastal basins respond to multiple environmental forcings. Box models capture much of the basic physics of mass conservation, and in this chapter we explore the use of box models. This is part of our general aim of using simple models to illustrate the basic science of coastal basins. Box models are also of practical value in determining some of the bulk, or overall properties, of coastal basins, and are arguably the first type of modeling tool that should be applied to any system before more detailed models are used. Momentum conservation is the basis of the dynamics which we will consider in Chapter 4, and is included in box models mainly through various calibration coefficients.

Apart from mass and momentum, many other quantities in coastal basins are conserved. The most important of these conserved quantities is salt. Coastal basins usually contain a mixture of marine and fresh water, so that their mean salinity provides a measure of the relative fluxes of marine and fresh water through the basin. By knowing long-term freshwater influx, and net precipitation/evaporation (from the surface of the basin), it is possible to determine marine fluxes, i.e., the rate of exchange of basin water with the open ocean (or continental shelf) outside the coastal basin. This is basically a *salt balance* or salt *budget* for the basin. This type of model inevitably gives information on timescales that are at least seasonal and spatial scales of the size of the basin. For coastal basins, freshwater inflow is one of the most difficult of all forcing functions to estimate accurately, because it is usually distributed amongst several gauged rivers with an appreciable amount of runoff in ungauged regions. The accuracy of gauging is often influenced by tidal flows in river mouths and many basins are also affected by submarine groundwater discharge, which is difficult to quantify. This affects all models of coastal basins, but is particularly important for models based on a salt balance. Salt balance also involves spatial averages of salinity and there can be some ambiguity about how such spatial averages should be taken. But these possible difficulties usually only affect the accuracy of the model to accuracies that are acceptable given the other major advantages of salt balance box models.

The most important advantage of the salt balance, or other box models, is that they are very powerful sources of information about the exchange of water between a coastal basin and the ocean. The strength of the methodology is that salt balance does not involve any details of the operation of specific exchange processes. Box models also have a major advantage in terms of scale, which means that they provide *system information*. Salt is basically just a conservative tracer for marine water (there are other possible tracers). It is also possible to compare seasonal salinities, and salinities of similar basins with other average biogeochemical properties, as a way of providing information on system behavior for nutrients and carbon. The advantage of the larger spatial scales is that the model is independent of assumptions about particular processes or biological communities.

3.2 Multi box models and one-dimensional models

The success of simple box models may lead to the idea of representing a coastal basin by two interconnected boxes or a series of boxes. This is only really feasible if the boxes represent systems which are physically distinct and whose internal interactions are much stronger than the interactions between the systems. A good example is two separate basins that are interconnected through a narrow channel, with perhaps one connected to the ocean. It is not generally good modeling practice merely to attempt to split a well-defined single basin into two boxes, because this introduces some ambiguity as to the relative exchange between the boxes (compared to the exchange between one of the boxes and the ocean). The reason is simply that the use of two boxes in this case contradicts the very philosophy of a box model, which involves ignoring the details of all *internal* processes and using only external inputs to, and outputs from, the basin.

If a box model does not provide sufficient information, then a probable next step is a one-dimensional model. Although working in one dimension does involve some significant conceptual limitation, the physical basis of the model is clearly stated. A box model is said to have zero dimension, and in a sense, a one-dimensional model is an infinity of zero-dimensional models connected together in a line. One-dimensional models are built around the advection diffusion equation, and are usually *calibrated* by adjusting the magnitude of the coefficient of diffusivity. One-dimensional models are able to provide good representations of some aspects of coastal basins and have the major advantage of ease of implementation and coding to simple computer spreadsheets. Several examples of one-dimensional models are given in this chapter.

3.3 Examples of box models

Box models represent a coastal basin as a box, i.e., they do not consider any spatial variations within the basin (but they do include time variation).

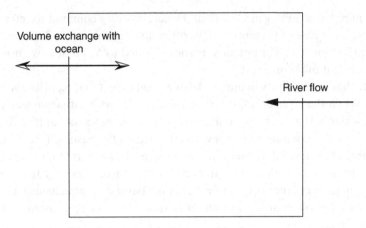

Figure 3.1 Schematic of simple box model of a coastal basin.

3.3.1 River flow

An example is a basin that is connected to the ocean at one end and has input from a river at the other end (Figure 3.1). Let us construct a box model for the salinity of a coastal basin. In reality, salinity varies from the ocean end of the basin (the mouth) to the other end (the head) but we choose to consider only the mean or average salinity of the basin, i.e., we consider the basin to be a box.

There are two connections to this box, one from the ocean (which flows both inward and outward to the open ocean) and the other from the river.

In physical oceanography salinity is defined as mass of salt per mass of water and expressed salinity as *parts per thousand (ppt)* or as *psu*, which means that a salinity of (say) 0.035 would be written as 35 ppt; however, remember that this number 35 really means 0.035. Models of salt and other dissolved (or particulate) materials in coastal basins can be framed either in terms of mass of solute, or particles, per unit mass of water, or the *concentration* of particles, i.e., mass of particles per unit volume of water. In this book we use the term *salinity* for the salt *concentration* in kilograms per cubic meter, and adopt a symbol S for this concentration. So a typical value of S is about 35 kg m^{-3}. Conventional salinity is denoted by s (mass of salt per unit mass of water which is a number). To convert S to salinity s we simply use the equation

$$S = s\rho. \tag{3.1}$$

As an example, take a salinity s of 35 ppt and ρ of 1026 kg m^{-3} (a typical value for the ocean and corresponding approximately to a salinity of 35 ppt and temperature of 20 °C at atmospheric pressure) the salt concentration S is 0.035 * 1026 = 35.91 kg m^{-3} and this differs numerically by only 2.6% from s expressed in ppt. The density of marine water changes by a maximum of only about 30 kg m^{-3} over the range from almost fresh, and warm, to very saline and cold. Hence the numerical difference between s in ppt and S in kg m^{-3} is small. Hence, it is very common to refer to S also as *salinity* and for ease of presentation we often do so in this book. To convert S to s, simply use (3.1), or take s in ppt as numerically equal to S. Finally, we usually omit explicit statement of the units of S.

We write the ocean salinity in our model as S_o and expect that S_o will have a value of about 35. When the river flows, the river water mixes with estuarine water and the volume of water in the box tends to increase, which decreases the salinity. We assume that the river water completely mixes with the water in the basin before it exits to the ocean. After a short time, the water level will become the same as that of the ocean, and we assume in our model that the water level in the basin is exactly the same as the ocean. This means that the volume of water in the basin does not change. This is not a necessary assumption, and we will see later that it is easy to construct alternative versions of the box model.

Let us write the volume of the basin as V and represent the salinity of the basin (the box) by S so that, for example, if the basin is fresh $S = 0$, and if it has ocean salinity we would expect S to be about 35. The quantity SV is the mass of salt in the basin. Suppose we move forward in time by some small finite time step Δt. In that time we add a small volume ΔV (which is small compared to V) of river water to the basin with zero salinity (the model is easily adapted to the input of water with non-zero salinity as, for example, from another basin). Initially we get a volume of water V', and we assume $V' = V + \Delta V$, but note that the volumes V and ΔV do change because volume is not a conserved quantity. Instead, it is the mass of water that is conserved and volume changes due to changes in density following mixing. We will discuss this in more detail later.

After river flow, the mass of salt is modeled as uniformly distributed through the new water mass by mixing (we have no other choice but to assume such mixing because in a box model we cannot consider vertical or horizontal variations in salinity). Thus the new salinity is $S + \Delta S$, where

$$S + \Delta S = \frac{SV}{V + \Delta V}. \tag{3.2}$$

Hence, multiplying through by $V + \Delta V$,

$$SV + V\Delta S + S\Delta V + \Delta V\Delta S = SV. \tag{3.3}$$

Equation (3.3) is based on the *salt continuity*, or the *salt balance*, equation. It merely states that the mass of salt is conserved during the initial mixing event as river water flows into the basin. It also assumes that the volume of water is conserved. In fact, volume is not an exactly conservative quantity and we shall discuss this later.

We ignore the *second order* term $\Delta V\Delta S$ in (3.3), because it is the product of two small quantities (remember that $\Delta V \ll V$ and so $\Delta S \ll S$, and thus $\Delta V\Delta S \ll V\Delta S$ or $S\Delta V$). By contrast $V\Delta S$ and $S\Delta V$ are *first order* terms (contain only one *small* quantity) and S and V are zero order terms (contain no *small* quantities). Strictly, we should say that these terms are zero, first, and second order in the relative volume $\Delta V/V$, which is infinitesimally small (and produces an infinitesimally small relative change in salinity $\Delta S/S$). Hence, (3.3) becomes

$$\Delta S = -S\Delta V/V. \tag{3.4}$$

Note that ΔS is negative indicating, as expected, that the salinity decreases in response to the river flow.

This technique of working with only the first order terms in ΔV and ΔS works increasingly well as we allow ΔV to become infinitesimally small, i.e., consider infinitesimally small time steps Δt. However, when used with finite time steps, a model that uses only the first order terms is said to be *explicit*, whereas a model with higher order terms is *implicit*. Implicit models require more advanced numerical schemes but allow longer time steps.

Let us use R to denote the volume flow of the river per unit time, e.g. $R = 10\,\mathrm{m}^3\,\mathrm{s}^{-1}$. Then in a small time Δt,

$$\Delta V = R\Delta t \tag{3.5}$$

so (3.4) becomes

$$\Delta S/\Delta t = -S\ R/V. \tag{3.6}$$

Notice that R/V is a characteristic time that controls the rate of decrease of S, and is physically the time for river flow to fill the basin. It is often referred to as the *river flushing time* of the basin, and implicitly assumes total mixing of the river water with that of the basin. If we intend to solve the model numerically, the above equation is sufficient. However, it is also useful to take the mathematical limit of infinitesimally small Δt.

$$\frac{\mathrm{d}S}{\mathrm{d}t} = -S\frac{R}{V} \tag{3.7}$$

which says that the rate of change of salinity with time is minus the river flow rate divided by volume of the basin. This model ignores the entry of salt from the ocean, i.e., the flushing of the basin by the ocean. We measure R in $\mathrm{m}^3\mathrm{s}^{-1}$ and we often call those units *cumecs*. Because S appears on the right-hand side of (3.7), as S decreases so $\mathrm{d}S/\mathrm{d}t$ decreases and in the limit that $S = 0$, river flow makes no further difference to S.

Let us finally return to the question of finding the value of $\Delta V'$. After mixing, the density of water in the basin changes from $\rho(S)$ to $\rho(S+\Delta S)$. The rate of increase of density with salinity is usually written in terms of a coefficient denoted here by β, and defined as $\beta \equiv \mathrm{d}\rho/\mathrm{d}S = (1/\rho)\mathrm{d}\rho/\mathrm{d}S$. If the addition of salt to water could be made without change of volume, the value of β would be unity since ρ and S are both masses

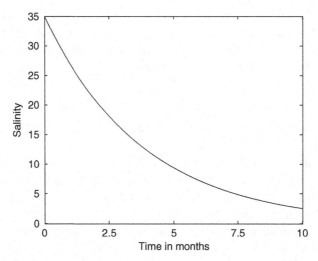

Figure 3.2 Modeled variation of salinity for a constant river flow of $6 \, m^3 \, s^{-1}$ into a basin of volume $60 \, 10^6 \, m^3$, ignoring salt entry from the ocean.

per unit volume. Instead, water needs to expand in volume to accommodate salt, and this reduces the increase in density so that typically $\beta \sim 0.7$ (depending slightly on the salinity and temperature). Therefore, the increase in volume δV of a water mass of volume V, and density ρ, due to addition of a salt mass δm is $\delta V = (1 - \beta)\delta m / \rho$. When river water mixes with basin water there is a transfer of salt from the basin water to river water of (say) mass δm. So, there is an exactly equal decrease in volume of the basin water and increase in volume of the river water provided that β is the same for both water masses; that is not quite true, but to a good approximation the total volume of water remains constant during mixing. This justifies our taking $V' = V + \Delta V$.

The essence of the present box model is that mixing precedes discharge to the ocean. It is possible to assume that these processes occur in reverse order. This modifies (3.4) to

$$S + \Delta S = \frac{S(V - \Delta V)}{V}, \tag{3.8}$$

i.e., the volume of basin water first reduces to $V - \Delta V$ and salt is then distributed over the volume $V - \Delta V + \Delta V$. This exactly reproduces (3.4), showing that the two box models are identical to *first order*.

If R is constant, (3.7) gives the variation of S with time assuming $S = S_o$ at time $t = 0$ as $S = S_o \exp(-Rt/V)$. This variation is sketched in Figure 3.2 for a coastal basin of volume $60 \, 10^6 \, m^3$, river flow of $6 \, m^3 \, s^{-1}$ assuming that the basin starts at salinity $S_o = 35 \, kg \, m^{-3}$. Note that we are ignoring the entry of salt from the ocean. The flushing time due to river flow, V/R, is 10^7 s or about 4 months.

If R varies with time, we may need to write a numerical version of this simple model which essentially just solves the simple difference equation (3.6) numerically. To do that we simply write the equation as

$$\Delta S = -S\frac{R}{V}\Delta t. \tag{3.9}$$

If we know S at one time t we can find its value at a slightly later time $t + \Delta t$ based on the value of R at that time. This is called *numerical integration* of the difference equation; because we are performing an operation that is the opposite of differentiation. Notice that this model needs a starting value of S (which we will call S_o). This is known as an *initial condition*. Most time-dependent models of coastal basins require initial conditions, and we have to start them off by assigning initial values to all the modeled quantities such as salinity, temperature, etc. We usually assume that if we run the model for long enough, the model will *forget* about these initial conditions. The time for this to happen is colloquially called the *spin-up time* of the model; it is the time for initial transients to decay away. For small systems like most coastal basins, the spin-up time is controlled by the characteristic time constants of the system but is usually considered to be short; the time depends on the properties under investigation and can vary from hours to a few weeks or months. For large basins such as the Southern Ocean, the spin-up time is several centuries because the system is so large. We discussed this property of coastal basins in Chapter 1.

Let us make up a spreadsheet in Microsoft Excel that illustrates our model. We are going to model a pulse of river water following a storm. Time is measured in days and R in $m^3 s^{-1}$. The variation of R is shown in the upper panel of Figure 3.3, and has been chosen to rise steadily to a peak of $20\,m^3\,s^{-1}$ after 6 days, and then to decay away to zero after 10 days. The lower panel of Figure 3.3 is derived numerically by the Excel program and shows that during this flood period, the predicted salinity drops from 35.0 to $31.1\,kg\,m^{-3}$, i.e., a decrease of $3.9\,kg\,m^{-3}$. We could have calculated the salinity drop approximately by finding the total volume of river water and using our original simple formula $\Delta S = -S\Delta V/V$, with $\Delta V =$ total flow and taking S as roughly constant at $35\,kg\,m^{-3}$. Adding the entries in the R column we get 81, and multiplying by seconds in a day we get total river flow volume as $\Delta V = 81 \cdot 3600 \cdot 24 = 7.0\ 10^6\,m^3$. So $R/V = 0.11$ (which is small compared to unity). Hence, $\Delta S = 35.\ 0.11 = 4.1\,kg\,m^{-3}$, which is slightly higher than the real decrease due to the assumption of S constant at 35. The Excel[1] code for Figure 3.3 that is shown in Table 3.1 gives the resultant spreadsheet as illustrated in Table 3.2. Table 3.3 also shows a Matlab[2] code for the same model. As with all such codes in this book, the graphical routines are not included (for brevity). The code is written in a form applicable to any array-based language (such as Fortran[3]) and does not utilize the full matrix capabilities of Matlab.

---

[1] See, for example, Bloch (2003).
[2] MATLAB® is proprietary computer software available from http://www.mathworks.com
[3] Fortran is free software and readily available for use on laptop computers.

Table 3.1. *Box model of salinity change due to river flow into a coastal basin shown in Figure 3.3*

Excel column	Quantity	Units	Excel formula[a]
A	Time	days	A2 = A1 + 1
B	Riverflow R	$m^3 s^{-1}$	Column entered by user
C	Average R	$m^3 s^{-1}$	C2 = (B1 + B1)/2
D	Time interval Δt	days	D2 = A2 − A1
E	Time interval Δt	seconds	E2 = Tday*D2
F	ΔS change in salinity	$kg\ m^{-3}$	F2 = − C3*E3*G2/Volume
G	S (new value of salinity)	$kg\ m^{-3}$	G2 = G1 + F2

[a] Microsoft Excel code for box model of basin with river flow only based on the model $\Delta S = - (R/V) \cdot S \cdot \Delta t$. Tday = 86400, Volume = 60 10^6 m^3.

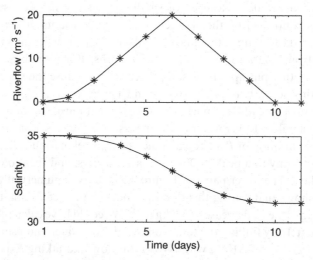

Figure 3.3 Box model of change in salinity of a coastal basin of volume 60 10^6 m^3 due to a pulse of river flow. The upper panel shows the time series of river flow and the lower panel the resultant time series of salinity. The crosses on the curves are the computed points using either the Excel spreadsheet (Tables 3.1 and 3.2) or Matlab code (Table 3.3).

3.3.2 Evaporation

Evaporation removes estuarine water from the basin, and in the absence of river flow this is replaced by ocean water at the salinity of the ocean (denoted by S_o). Evaporation does not remove salt (or any other chemical). The rate of evaporation is written as E and has the dimensions of velocity (volume of water lost by evaporation per unit surface area per unit time), i.e., $m s^{-1}$. Thus, in 1 s the surface would drop by E m (before water inflows from the ocean). There are a variety of methods of measuring rates of evaporation. The

Table 3.2. *Excel spreadsheet output for the model shown in Table 3.1*

A	B	C	D	E	F	G
Time t (days)	R $(m^3 s^{-1})$	Average R $(m^3 s^{-1})$	Δt (days)	Δt (s)	ΔS kg m^{-3}	S kg m^{-3}
1	0					35
2	1	0.5	1	86400	−0.025	34.97
3	5	3	1	86400	−0.151	34.82
4	10	7.5	1	86400	−0.376	34.48
5	15	12.5	1	86400	−0.620	33.83
6	20	17.5	1	86400	−0.852	32.98
7	15	17.5	1	86400	−0.831	32.14
8	10	12.5	1	86400	−0.579	31.57
9	5	7.5	1	86400	−0.341	31.22
10	0	2.5	1	86400	−0.112	31.11
11	0	0	1	86400	0	31.11

Table 3.3. *Matlab code for box model of salinity change
due to river inflow to a coastal basin as shown in Figure 3.3*

```
V = 60e6; t = 1:11; R = [0 1 5 10 15 20 15 10 5 0 0]; S(1) = 35;
for n = 2:11;
    R_average(n) = (R(n) + R(n − 1))/2;
    dt(n) = 3600*24*(t(n) − t(n − 1));
    dS = − S(n − 1)*R_average(n)*dt(n)/V;
    S(n) = S(n − 1) + dS;
end;
```

most common is the pan evaporometer, and for most coastal regions there are published mean monthly evaporometer rates. However, E is really a sensitive function of the temperatures of water and air, the relative humidity, and wind speed and so may be better determined by the model. A formula for E used commonly in the literature is

$$E = CW\frac{\rho_a}{\rho}[H_T - H_{T_{dew}}], \qquad (3.10)$$

ρ_a being the density of air, $C \sim 1.45 \ 10^{-3}$, H_T is the specific humidity of saturated water at temperature T (the surface temperature of the basin), $H_{T_{dew}}$ the specific humidity at the dew point T_{dew}, and W the wind speed. Specific humidity is the ratio of the density of the water vapor to the density of the air, i.e., the mixture of dry air and water vapor; standard formulae are available. Parameters of the atmosphere are to be measured at 10 m above sea level. The height of 10 m is a standard convention in these types of formula and is based on the assumption that 10 m is sufficient for the air flow to be above the roughness layer provided by surface objects. In practice measurements are usually made on land, although they should (and increasingly are now) made over the coastal basin itself. Evaporation is usually important in arid regions where there is limited precipitation and river flow, strong insolation, and low cloud cover and humidity. In such basins, evaporation tends to produce hypersalinity (salinity in excess of that of the adjoining open ocean). Evaporation also removes latent heat and has a major role in heat exchange between the ocean and atmosphere.

If the salinity (salt concentration) in the basin is originally S, then the mass of salt in the basin is SV. In time Δt, the volume of water lost by evaporation is $EA\mathrm{d}t$, where A is the surface area of the basin. This volume is made up by the inflow of ocean water. Thus the final mass of salt is $SV + S_0 EA\Delta t$ and so the final salinity, $S + \Delta S$, is given by $V(S + \Delta S)$ where

$$\Delta S = S_0 E \Delta t / h \tag{3.11}$$

and A/V has been written as the reciprocal of h defined as the mean depth of the basin. Hence,

$$\frac{\mathrm{d}S}{\mathrm{d}t} = S_0 \frac{E}{h} \tag{3.12}$$

and we note that the salinity *always* increases with evaporation. For example, if a coastal basin experiences summer evaporation of 10 mm per day, and the ocean salinity is 35 kg m^{-3}, mean water depth of the basin 1.5 m, the rate of change of salinity from 3.3.12 is 0.23 kg m^{-3} per day or 7.1 kg m^{-3} per month. If the mouth of the basin were suddenly blocked off at the start of summer, it would take evaporation $1.5/0.01$ days $= 150$ days to empty the basin. If there is no replenishment of the water in the basin by inflow, the salinity increases due to the original amount of salt being dissolved in an ever decreasing amount of water. Hence, $(V - EA\Delta t)(S + \Delta S) = VS$, and so to first order $\Delta S = SE\Delta t / h$ which has a differential form $\partial S / \partial t = SE/h$ and solution $S = S_0 \exp(tE/h)$, which diverges with increasing time.

Combining the processes of river flow and evaporation, using (3.6) and (3.12) requires some care. Let us rewrite (3.6) as

$$\frac{\Delta S}{\Delta t} = \frac{F_S}{h} \tag{3.13}$$

where $F_S = hSR/V$ is the flux of salt into the basin (negative for river flow) per unit surface area, i.e., kilogram per unit surface area per unit time. To simulate evaporation (3.12) shows that $F_S = ES_0$. In either case, the salt flux is composed of an advective velocity multiplied by the salt density at the point from which the flux originates. For water expelled by river flow, this is the salt density in the basin. For inflow created by evaporation, it is the salt density in the ocean. Clearly, this is an artifact of the model and occurs in all models where there is a boundary between the ocean and basin. We could, of course, take the flux as the velocity times the mean salinity of basin and ocean. These various forms for the advective salt flux are referred to as *up-winding* (meaning that the current carries the salt concentration from the *up-wind,* or in our case *up-current* direction) and *central differencing*, respectively. Hence, we can combine river flow and evaporation in a model of the form

$$\frac{\Delta S}{\Delta t} = -S_e \frac{P_e}{h}$$

$$P_e \equiv \left(\frac{R}{V/h} - E \right)$$ (3.14)

$$S_e \equiv \begin{cases} S, & P_e > 0 \\ S_o, & P_e \leq 0 \end{cases}$$

We will now construct a box model for a coastal basin in a so-called *mediterranean climate*. This type of climate is named after that famous region, from antiquity, of the Mediterranean Sea. It is found on many western coasts at latitudes of about 30°. It consists of long dry summers and winters dominated by frequent storms, interspersed with fine, cold periods. Good examples are the Mediterranean Sea itself, the coasts of California, and south-west Australia. The basin to be considered in our example has a mean depth of 1.5 m (very shallow basins are typical of coasts which have long periods of arid conditions) and volume 60 10^6 m^3, and is subject to winter river flow and evaporation in summer and autumn. We will use an Excel worksheet to find the salinity in the basin throughout a (northern hemisphere) year, in which the river flows at a constant rate of 5 m^3 s^{-1} during both winter (December to February) and spring (March to May) and is zero in summer (June to August) and fall (September to November) and evaporation is zero in winter, 1 mm per day in spring, 4 mm per day in summer, and 5 mm per day in fall. The reader can adapt this spreadsheet to any more realistic data for a particular basin. We run the model starting in June with the salinity having an initial value of 35 and the ocean salinity constant at 35. This is a good guess for the initial condition, because the effects of winter rains will have subsided and remnant brackishness of the basin will have been offset by light evaporation in spring. Strictly, we should run the model for a few repeated years of identical data until the transients from this initial condition have been lost from the annual cycle. Table 3.4 shows the Excel code and Table 3.5 the Matlab code for this model. Figure 3.4 shows a plot of salinity S, against time t (in months).

As a practical application of the model, we might consider the controlled release of fresh water into the basin from a reservoir during summer and fall, designed to limit extreme hypersalinity. The model allows us to recommend a time series for a release rate, which is designed to stop the salinity exceeding a critical value of (say)

Table 3.4. *Excel code for box model of a coastal basin with summer evaporation and winter rainfall; the results are shown in Figure 3.4*

Column	Quantity	Units	Excel formula[a]
A	Time t	weeks	A2 = A1 + 1
B	Season		B2 = IF(A2 > 39, "spring", IF(A2 > 26, "winter", IF(A2 > 13, "autumn", "summer")))
C	R	$m^3 s^{-1}$	C2 = IF(A2 > 26,5,0)
D	Average R	$m^3 s^{-1}$	D3 = 0.5*(C3 + C2)
E	R/V	s^{-1}	E3 = D3*Tday /Volume
F	E	m per day	F2 = IF(A2 > 39,0.001, IF(A2 > 26,0, IF(A2 > 13,0.005,0.004)))
G	Average E/h	day^{-1}	G3 = 0.5*(F3 + F2)/h
H	Δt		H3 = 7*(A3 − A2)
I	P_e/h		I3 = E3 − F3
J	S_e	$kg\ m^{-3}$	J3 = IF(I3 > 0, L2,35)
K	ΔS	$kg\ m^{-3}$	K3 = − I3*J3*H3
L	S	$kg\ m^{-3}$	L3 = L2 + K3

[a] Based on the model $\Delta S = -(P_e/h)S_e\Delta t$. Tday = 86400, Volume = 60 10^6 m^3, $h = 1.5$.

$S = 50\,kg\,m^{-3}$ and yet keeps the total volume of reservoir water lost to the basin (and hence unavailable for irrigation purposes) to a reasonable minimum.

3.3.3 Ocean exchange

Our first two box models considered the effect of river flow and evaporation on the salinity of a basin maintained at constant volume. Salt entered, or left, the basin via water flow through the mouth in order that the volume of the basin remains constant. This flow is said to be *pressure-driven*, and is simply forced by the difference in surface level resulting from the inflow of river water or the outflow of water through the surface due to evaporation. In addition to these pressure-driven processes mentioned above, there is another class of processes by which the basin can lose salt to the ocean, or gain salt from the ocean, and this is called *ocean–basin exchange*. These processes tend to move the basin salinity towards that of the ocean. So, if the basin is fresh, salt flows in from the ocean and if the basin is hypersaline salt flows out from the basin to the ocean. If there is no river flow or evaporation, this exchange process results in the basin reaching the same salinity as the ocean in a time that we call the *ocean exchange time*, or *ocean flushing time* and denote by the symbol τ.

The processes responsible for this ocean exchange of salt are due to both advection and diffusion. The advective processes include tides, wind, and density forcing and

Table 3.5. *Matlab code for box model of basin with surface evaporation and winter river flow*

```
V = 60e6; t = 1:11; R = [0 1 5 10 15 20 15 10 5 0 0]; S(1) = 35;
for n = 2:11;
        R_average(n) = (R(n) + R(n − 1))/2;
        dt(n) = 3600*24*(t(n) − t(n − 1));
        dS = − S(n − 1)*R_average(n)*dt(n)/V;
        S(n) = S(n − 1) + dS;
end;
V = 60e6 ;h = 1.5; S(1) = 35;
for n = 1:60;
            if n > 26;R(n) = 5;else;R(n) = 0;end;
            if n > 39;
                E(n) = 0.001;
            else
                if n > 26;
                    E(n) = 0;
                else
                    if n > 13;
                        E(n) = 0.005;
                    else
                        E(n) = 0.004;
                    end;
                end;
            end;
            if n > 1;
                R_average(n) = (R(n) + R(n − 1))/2;
                R_overV(n) = 86400*R_average(n)/V;
                E_average_overh(n) = (E(n) + E(n − 1))/(2*h);
                dt(n) = 7;
                Pe_overh(n) = R_overV(n)−E_average_overh(n);
                if Pe_overh(n) > 0;Se(n) = S(n − 1);else;Se(n) = 35;end;
                dS = − Pe_overh(n)*Se(n)*dt(n);
                S(n) = S(n − 1) + dS;
            end;
    end;
```

must be either cyclic in time (such as the tide), or change direction spatially across the mouth (as for wind and density forcing) since the basin is of fixed mean volume. Wind forcing can drive currents that reverse in direction both horizontally and vertically. For both tidal and wind forcing, the incoming tidal water has ocean salinity and mixes with water in the basin before exiting with a changed salinity. Density forcing drives

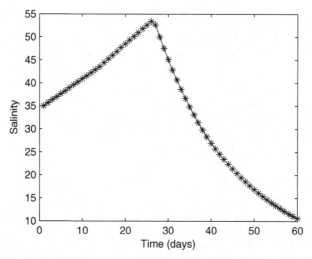

Figure 3.4 Salinity time series for box model driven by summer evaporation and winter rainfall. The time series starts at the beginning of summer (see Table 3.4).

surface and bottom currents which have slightly different salinities; if the basin is less dense than the ocean (a *classical* coastal basin), the surface current flows out of the basin and the bottom current flows into the basin, with the surface current carrying slightly less saline water than the bottom current. Such forcing means that the two counter-flowing currents have individual strengths which may greatly exceed the net, or total, flow through the mouth.

Consider the case of river flow R. We write the total outflow from the basin to ocean as $R+r$ where the inflow is r, so the salt flux inward to the basin is

$$F_S = -(R+r)S + rS_o = -r(S - S_o) \quad r>0 .$$ (3.15)

The salt continuity equation gives us

$$V\frac{\partial S}{\partial t} = F_S$$ (3.16)

where we have used partial derivatives to emphasize that we are now considering a number of processes that change salinity. Hence,

$$\frac{\partial S}{\partial t} = -\frac{R}{V}S - \frac{(S - S_o)}{\tau}$$ (3.17)

where τ is the time taken for the flow to fill the basin, i.e., $\tau = V/r$. The second term in (3.17) is the ocean–basin exchange. The basin–ocean exchange time is commonly calculated for tidal flow. Let us suppose that a volume of water V_{tidal} enters the basin on the flood tide (V_{tidal} is called the *tidal prism* because it is the volume of water contained in the prism formed by the perimeter of the basin, the low water level and the high water level).

This V_{tidal} comes from the ocean and so has salinity of S_o. We assume that the exiting water consists of a volume μV_{tidal} of basin water and the remainder $(1 - \mu)V_{tidal}$ is unmixed ocean water. When $\mu = 1$, we have total mixing of the tidal prism with the water in the basin at low tide. Hence, the amount of salt retained by the basin is

$$V\Delta S = \mu V_{tidal}(S_o - S). \tag{3.18}$$

This takes place in a time t_{tidal} which is the tidal period. Hence,

$$\frac{\Delta S}{\Delta t} = \frac{\mu V_{tidal}(S_o - S)}{V t_{tidal}} \tag{3.19}$$

or

$$\tau = \frac{V}{\mu R_{tidal}} \qquad R_{tidal} \equiv \frac{V_{tidal}}{t_{tidal}} \tag{3.20}$$

and in this case τ is known as the tidal flushing time. The time $V t_{tidal}/V_{tidal}$ is also known as the *volumetric tidal flushing time* although this is only a reasonable estimate of τ when there is strong mixing, and is usually very much shorter than the true tidal flushing time. We can devise a similar type of argument to derive the ocean–basin flushing time for wind or density-driven flow. The important quantities are μ and r, which is the volume flow between ocean and basin induced by wind or density differences, respectively.

The contribution of *diffusion*, or *dispersion*, to the ocean–basin exchange (the two terms are synonymous and mean a horizontal spreading or mixing of salt, or other material) will be treated in more detail later. The basic concept of dispersion of salt is that a flux of salt is established which is proportional to the spatial gradient of salinity and flows down the salinity gradient. For a box model that gradient must be represented by the difference in salinity between the basin and ocean. This clearly produces a form similar to (3.15), and so dispersion effects can be lumped in with other contributions to the ocean exchange time.

For ocean–basin exchange alone, (3.17) gives, in the absence of river flow,

$$S - S_o = (S_{t=0} - S_o) \exp(-t/\tau) \tag{3.21}$$

which shows that there is an exponential time decay of the difference between S and S_o. We can also solve the equation numerically. The ocean–basin exchange can be readily added to the numerical codes in Sections 3.3 and 3.4 if a value of τ is assumed. This is shown in Figure 3.5. Notice the effect of exchange on the model predictions made in Section 3.3 with the salinity returning to S_o at the end of the river flood.

The value of the exchange time τ is calculable by three-dimensional models, but for box models the value of τ is determined by comparing the predictions of the model with observed data. This process is referred to as *calibration* of the model. It is usual to calibrate the model by finding the value of exchange time that gives a best fit to the

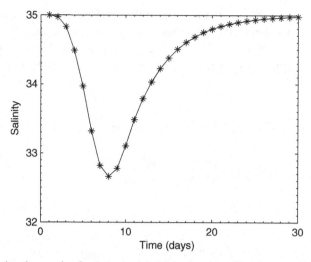

Figure 3.5 Salinity time series for box model driven by river flow from a storm, identical to Figure 3.3, but including an ocean exchange time of 5 days (see Tables 3.6 and 3.7).

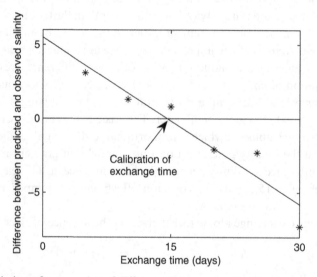

Figure 3.6 Variation of mean value of difference between predicted and observed salinity in a coastal basin during February (taken over the years 1984 to 2002) with the value of the ocean exchange time (in days) used in the model. Best fit is an ocean–basin exchange time of 15 days.

measured salinity of the basin over a specified period such as a month, a season or a year. A specimen calibration of this sort is shown in Figure 3.6. In this case the modeler is interested in trying to predict the value of S during the month of February, and so this is used as the criterion against which calibration is made; an alternative might be to take the root mean square of the difference between predicted and observed salinity.

Table 3.6. *Inclusion of ocean exchange in the Excel code for the box model of*
salinity of a coastal basin resulting from river flow (compare Tables 3.1 and 3.2)

Column	Quantity	Units	Excel formula[a]
A	Time t	days	A2 = A1 + 1
B	R	$m^3 s^{-1}$	Entered by user
C	Average R	$m^3 s^{-1}$	C3 = 0.5*(B3 + B2)
D	Time step Δt	days	1
E	R/V	s^{-1}	E3 = C3*Tday/Volume
F	τ	days	5
G	Salinity S	kg m^{-3}	G1 = 35; G2 = G1 + H2
H	ΔS	kg m^{-3}	H3 = (I2 − E3*G2)*D3
I	$(35 - S)/\tau$	kg m^{-3} per day	I2 = (35 − G1)/F2

[a] Based on the model $\Delta S = ((35 - S)/\tau - (R/V)*S))*\Delta t$. Tday = 86400, Volume = 60 10^6 m^3.

Table 3.7. *Inclusion of ocean exchange in the Matlab code for the box model*
of salinity of a coastal basin resulting from river flow (compare Table 3.3)

```
V = 60e6; tau = 5; dt = 1;
R_flood = [ 0 1 5 10 15 20 15 10 5 0 0];
S(1) = 35;
for n = 2:30;
        if n < 12; R(n) = R_flood(n); else; R(n) = 0; end;
        R_average(n) = 3600*24*(R(n) + R(n − 1))/2;
        dS = − S(n − 1) * R_average(n) * dt/V;
        dS = dS + dt*(35 − S(n − 1))/tau;
        S(n) = S(n − 1) + dS;
end;
```

Whether ocean exchange is important depends on the comparative values of $1/\tau$ and R/V and E/h. However, the latter two rates are often very seasonal, so there are usually seasons in which exchange is dominant. For density-driven exchange the exchange time τ is dependent on the density difference between the basin and ocean and therefore, on the salinity difference.

3.3.4 Surface heating

We use the same box model as employed for salt in Sections 3.3.1 to 3.3.3, but in this section consider heat. The box is again well mixed and has a temperature T. To convert the concentration of heat energy (J m^{-3}) to temperature we must divide by specific

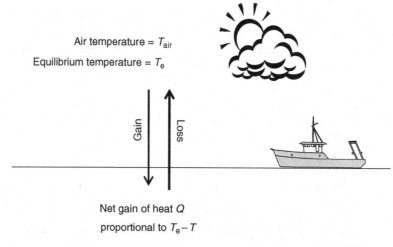

Air temperature = T_{air}

Equilibrium temperature = T_e

Gain

Loss

Net gain of heat Q

proportional to $T_e - T$

Water temperature = T

Figure 3.7 Schematic of processes involved in the net surface heat flux Q.

heat C_p (heat energy per degree per unit mass) and by density (mass per unit volume). The basin is assumed to be heated by solar radiation arriving at its surface at a rate Q (heat per unit time per unit area). This is converted into heat per unit volume by dividing by the depth h, so that

$$\frac{dT}{dt} = \frac{Q}{\rho C_p h}. \tag{3.22}$$

The ocean–basin exchange time in a box model is independent of the physical property being exchanged unless the property itself affects the rate of exchange; that usually only occurs as a consequence of the property affecting density. In the present chapter we are excluding such effects, and our present models are referred to as *barotropic* meaning (very loosely) that density does not affect motion. Strictly barotropic means that the isobaric surfaces (surfaces of constant pressure) are parallel with the surfaces of constant density which is a definition inherited from atmospheric physics. Hence, including ocean–basin exchange with characteristic exchange time τ,

$$\frac{dT}{dt} = \frac{Q}{\rho C_p h} - \frac{(T - T_o)}{\tau}. \tag{3.23}$$

Here, Q is the *net* rate of surface heating which is the difference between two terms. The first is the rate of solar heating and the second is a negative term due to heat loss from the basin back into the atmosphere. This is illustrated in Figure 3.7. This heat loss increases with T and so heating and cooling would eventually become equal (in a steady state) when T reaches a so-called *equilibrium temperature* T_e. This is the steady state

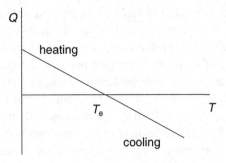

Figure 3.8 Schematic of the variation of the rate of heating/cooling Q of basin as function of temperature T of a coastal basin. The straight line is an approximation to the function valid near the point at which the line crosses the temperature axis; this point defines the *equilibrium temperature* T_e which is the demarcation between heat gain and heat loss by the basin.

temperature which would be found in a closed basin totally isolated from the ocean. The net rate of heating Q, when T is not too far removed from T_e, is illustrated in Figure 3.8.

If Q decreases linearly with T, it can be conveniently approximated as

$$Q = \rho C_p h \frac{T_e - T}{t_e} \tag{3.24}$$

where t_e is defined by

$$\frac{1}{t_e} \equiv -\frac{1}{\rho C_p h} \left(\frac{\partial Q}{\partial T} \right)_{T_e}. \tag{3.25}$$

The time t_e can be interpreted as the characteristic time for T to relax to T_e because (3.23) and (3.24) show that in the absence of exchange an (isolated) basin which has the ocean temperature T_o at time $t = 0$ relaxes to T_e as

$$T = T_e + (T_o - T_e) \exp(-t/t_e). \tag{3.26}$$

The steady state solution of (3.23) is obtained by making the time derivative zero,

$$T = T_o + \frac{T_e - T_o}{1 + t_e/\tau} \tag{3.27}$$

so that the steady state temperature T depends on the ratio t_e/τ with T varying between T_e and T_o, as t_e/τ increases from being much less than unity to much greater than unity. Essentially this determines the relative strengths of the two processes: heat exchange through the surface and heat exchange with the ocean. The value of T_e varies through the day–night cycle due to changes in solar radiation and air temperature. In a shallow basin of depth a few meters, typical values of t_e are of order a few days, so that typical shallow coastal lagoons, with ocean exchange times of order at least 10 days, have temperatures close to the daily average of the equilibrium temperature. However, according to (3.25), t_e is proportional to h, and so in deeper basins the temperature is determined by the strength of the ocean exchange. Furthermore such basins may be thermally stratified which invalidates our simple box models. The value of T_e depends on meteorological conditions. In very strong summer heating, it is elevated above air temperature, so shallow basins tend to be warmer than the ocean (*hyperthermal*). In winter when solar radiation is weaker, T_e may be close to air temperature, which is typically lower than the temperature of the open ocean in winter. So, coastal basins become *hypothermal*; strong winds increase evaporation. As a result, in winter T_e may be lower than air temperature.

3.4 One-dimensional models

The box models that we have studied so far in this chapter are often called *zero-dimensional* models, because they only tell us about changes with time of a mean quantity such as salinity or temperature; they do not give us information on the spatial variations through the basin. A one-dimensional model determines the variation of ocean properties along a line, and we note that a line has only one dimension. This is a long way from the real world (which has three dimensions), but in some coastal basins the major variation in properties occur along a line, so we can adequately model them one-dimensionally. That one dimension may be in the horizontal, and we shall consider such a model in this section. However, the one dimension could also be in the vertical and represent vertical mixing in the ocean.

3.4.1 River flow

We use x to denote the one-dimensional axis. So, the salinity S is a function of x and time t. Consider a plane of unit area which is perpendicular to the x axis, i.e., the axis is normal to the plane. The mass of salt which passes through a plane per unit time is called the salt flux and denoted by F_s. This flux of salt is created by diffusion and advection currents. The *salt continuity equation* in the absence of any sources or sinks follows from Chapter 2 as,

$$\frac{\partial S}{\partial t} = -\frac{\partial F_s}{\partial x} \tag{3.28}$$

so that for a steady state model (no time dependence) with no sources or sinks of salt, the flux of salt F_s is independent of position along the basin. To derive (3.28), we first need to consider the general case in which the flux does vary with position x. In that case, as the flux passes through a small distance dx, it must leave a quantity of salt in that element per unit cross-sectional area, equal to the difference in flux at x and $x + dx$, i.e., $F_s(x + dx) - F_s(x)$. This can be approximated by the negative of the derivative of F_s with respect to x (with increasing accuracy as dx decreases). Hence, if there are no *sources* or *sinks* for salt, i.e., there is no "external entry" of salt into the element dx (such as an external pipe), the rate of change of salt concentration with time at x, is given by (3.28), where partial derivatives are used here. Equation (3.28) is called the salt continuity equation, and if S is replaced by the concentration of any other *conservative* quantity (one that does not suffer changes other than those due to its physical movement within the basin) the continuity equation is again applicable, if the right-hand side of the equation is the gradient of the appropriate flux.

There are two components to the salt flux: advection and diffusion. The contribution of diffusion, to the salt flux density is $-\rho K_x dS/dx$ where K_x is the *diffusion coefficient* (which has units $m^2 s^{-1}$) or *coefficient of dispersion*. This produces a flux which flows from regions of high salinity to low salinity, and therefore tends to remove variations in salinity. If we have a current this carries salt, i.e., salt is advected along the basin. Such advection comes from river flow or evaporation. More generally, tides and winds can produce advection.

Let us take the case of a river, which puts a volume R of water per unit time into the basin at the head. The x axis runs from $x=0$ at the ocean end of the basin and increases with distance into the basin, and has value $x=L$ at the end of the basin, at which point the river flows into the basin. In Chapter 2 we established the continuity equation for volume in coastal basins and showed that in a steady state this implies that the volume flux density is constant in space. In the present model, the velocity (or current) u along the x axis is simply equal to the river flow R divided by the cross-sectional area of the basin. If we assume that the basin has constant depth and width, the current u_r is then constant. This current provides an advective term in the salt flux,

$$F_s = -Su_r - K\frac{dS}{dx}.$$

(3.29)

Notice that we are actually using salt concentration S rather than true salinity s. In (3.29), the advective flux has a negative sign to indicate that it runs in the negative x direction *outward* to the ocean. The form (3.29) is called the *advective–diffusion* equation.

Since the salt flux is constant in space, and zero at the head of the basin (the river inputs no salt, i.e., the river water has zero salinity),

$$0 = -Su_r - K\frac{dS}{dx}.$$

(3.30)

Equation (3.30) is easy to solve, and gives

$$S = S_o \exp\left(-\frac{u_r}{K}x\right)$$

(3.31)

where we have matched S at $x=0$ to the ocean salinity S_o. The units of u_r/K are inverse length. Hence, S decreases exponentially with distance x from the ocean. Equation (3.31) describes a penetration of salt into the basin forming a *salt-water wedge* at the mouth of the basin, which is characteristic thickness u_r/K. This thickness increases with increasing K and decreases with increasing u_r, showing that the wedge is due to diffusion counteracting the effect of the advection of river water. The salinity does not reach zero at $x=L$ (unless the basin is infinitely long); the reason is that the incoming river water must mix with salt diffusing from the mouth of the basin. The present model does not enter the river and is essentially just dripping water of zero salinity into the end of the basin and this does not of course produce zero salinity. The reader may wish to consider how the model might be extended into the river channel so that a point is reached at which the salinity does become zero. This requires inclusion of a cross-section for the flow which decreases with distance and this causes an increase in flow speed with x, that essentially increases the exponential decay of salinity with distance and may even produce a quasi-salinity front near the river mouth.

In this model, the effect of the river salinity is represented entirely through the condition of zero salt flux. If the river actually had a salinity of S_r, we would adjust (3.31) to

$$S = S_r + (S_o - S_r) \exp\left(-\frac{u_r}{K}x\right) \tag{3.32}$$

and if, for example, the river was replaced by a hypersaline discharge, the salinity would increase with distance x into the basin. The present model can be applied to any tracer material discharged into the basin. Let us call the concentration in this case C, and assume that the ocean value is zero so that (3.32) becomes

$$C = C_r\left[1 - \exp\left(-\frac{u_r}{K}x\right)\right]. \tag{3.33}$$

Typical values of K are 10 to 100 $\mathrm{m^2\,s^{-1}}$ while u_r (defined as discharge divided by cross-sectional area) is typically $100\,\mathrm{m^3\,s^{-1}}/10^4\,\mathrm{m^2} \sim 0.01\,\mathrm{m\,s^{-1}}$ so that u_r/K is typically 1 to 10 km. In much faster flows (such as fast streams), this distance may be reduced to just 10 or 100 m.

Comparing the present model with the box models of Sections 3.1 to 3.6, it is clear that the diffusion term in (3.29) is responsible for ocean–basin exchange (as well as mixing within the basin). Taking the average of (3.32) between $x=0$ and L, gives

$$\bar{S} - S_o = S_o\left[\frac{K}{Lu_r}\left\{1 - \exp\left(-\frac{u_r}{K}L\right)\right\} - 1\right] \tag{3.34}$$

and so assuming that K is sufficiently large to mix salt throughout the basin, (3.34) becomes

Table 3.8. *Microsoft Excel code for box model of basin with river flow and ocean exchange*

Column	Quantity	Units	Excel formula[a]
A	x	km	A2 = A1 + 1
B	K/u_r	km	Entered by user
C	Average K/u_r	km	C3 = 0.5*(B3 + B2)
D	x increment = Δx	km	1
E	ΔS	kg m^{-3}	E2 = $-$ F1*D2/B2
F	S	kg m^{-3}	F1 = 35, F2 = F1 + E2
G	S (analytical solution)	kg m^{-3}	G3 = 35*EXP($-$A3/B3)

[a] Model $\Delta S = - (u_r/K) S \Delta x$.

$$S_o - \bar{S} \to S_o \frac{u_r L}{2K} \tag{3.35}$$

and so matching the exchange term $(S_o - \bar{S})/\tau$ to the advective term $u_r S_o/L$ gives

$$\tau \to \frac{L^2}{2K} \tag{3.36}$$

which is proportional to the time taken for salt to diffuse through the basin. When $u_r = 0$ the rate of exponential decrease becomes zero, so the whole basin has a salinity equal to that of the ocean.

For a numerical model of the steady state salt distribution, the balance of advection and diffusion provides the finite difference equation

$$\Delta S = -\frac{u_r}{K} S \Delta x. \tag{3.37}$$

If we know S_o, the value of S at $x = 0$, (3.37) can be solved by integrating from $x = 0$ to L, taking small values of Δx (see Table 3.8 and Figure 3.9).

For $K/u_r = 2.3$ km the numerical solution with $\Delta x = 0.1$ km is very close to the analytical exponential solution. This agreement would become even better if we reduced the value of Δx. The requirement for numerical accuracy is clearly

$$\Delta x \ll K/u_r. \tag{3.38}$$

Figure 3.10 shows the case $K/u_r = 30$ km and $\Delta x = 1$ km so that Δx is $0.03 K/u_r$, which clearly satisfies the equation (3.38). Figure 3.9 shows the analytical solution and numerical errors (which are very small for this case). In practice, we could choose a larger value of Δx and still get reasonable results. Figure 3.10 also shows a repeat of the numerical model with the same value of Δx but a much smaller value of K/u_r, and we notice that there is a much greater error in the numerical values compared with the analytical form

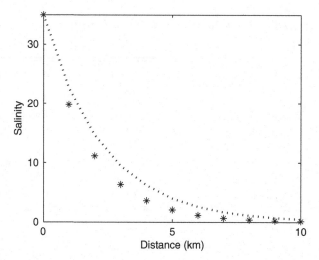

Figure 3.9 Steady state salinity variation along a one-dimensional basin due to river flow at the head and fixed salinity at the mouth. The characteristic length K/u_r (ratio of coefficient of dispersion to effective river flow velocity) is taken as 2.3 km. The broken line shows the analytical solution and the points are a coarse numerical model (see Table 3.8 for the Excel code).

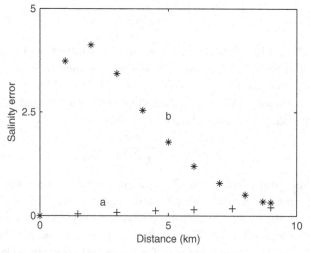

Figure 3.10 Numerical solutions of the one-dimensional model of variation of salinity in a coastal basin for two values of K/u_r when $\Delta x = 1$ km: (a) $K/u_r = 30$ km and (b) $K/u_r = 2$ km. For case (b) the numerical stability criterion is not satisfied and so substantial errors develop (compare Figure 3.9).

(which is an exact solution of the model). This is a consequence of Δx being 0.5 K/u_r which is rather high; Δx should be small compared to 0.5 K/u_r. Note that this produces a *numerical* error which is quite distinct from any limitations in the way that the model might represent a real basin.

3.4.2 River flow plus evaporation

In the presence of evaporation, the advective velocity in a one-dimensional basin varies with position x. The reason is that evaporation removes water from the surface of the basin and this must be replaced by an inflow from the ocean. Using our definition of the evaporative velocity E

$$\frac{\partial \eta}{\partial t} = -E \tag{3.39}$$

and the linearized continuity equation for a one-dimensional basin of constant depth h,

$$\frac{\partial \eta}{\partial t} = -h\frac{\partial u}{\partial x} \tag{3.40}$$

integration, and application of the boundary condition that u vanishes at $x=L$ (head of basin), the advective velocity now varies with position according to the law

$$u = -u_r + \frac{E}{h}(L - x). \tag{3.41}$$

The salt flux density now becomes

$$f_s = S\left\{-u_r + \frac{E}{h}L(1 - x/L)\right\} - K\frac{dS}{dx} \tag{3.42}$$

and since in the steady state this flux must be independent of x and vanish at the head

$$\frac{K\,dS}{u_r\,dx} - S\{-1 + \alpha_E(1 - x/L)\} = 0 \tag{3.43}$$

where the velocity u_E is defined by

$$u_E \equiv \frac{LE}{h}. \tag{3.44}$$

We derive the numerical form of the model as

$$\Delta S = -\frac{S\Delta x}{K}\left[u_r - u_E\left(1 - \frac{x}{L}\right)\right] \tag{3.45}$$

which reduces to (3.37) in the absence of evaporation (when u_E is zero) and in the absence of river flow becomes

$$\Delta S = \frac{u_E}{K}S\Delta x\left(1 - \frac{x}{L}\right). \tag{3.46}$$

Table 3.9. *Microsoft Excel numerical code for one-dimensional model of basin*

Column	Quantity	Units	Excel formula[a]
A	x	km	A2 = A1 + 1
B	K/u_r	km	Entered by user
C	$1 - x/L$		C1 = 1 − A1/L
D	$\Delta x =$ increment in x	km	1
E	ΔS	kg m^{-3}	E2 = u_E/K*F1*C2
F	S	kg m^{-3}	F1 = 35, F2 = F1 + E2

[a] Evaporation: $\Delta S = \frac{u_E}{K} S \Delta x (1 - x/L)$. $L = 20$ km, $u_E/K = 100$ km.

Table 3.10. *Matlab code for the one-dimensional model of equation (3.46) for the variation of salinity along a basin due to evaporation when $K/u_E = 100$ km and $L = 20$ km (figure 3.11).*

```
N = 25; L = 20e3; uEoverK = 1/100e3;
dx = L/N; S(1) = 35;DS(1) = 0;
for n = 2: N
    x(n) = (n − 1)*dx;
    DS(n) = uEoverK*S(n − 1)*dx*(1 − x(n)/L);
    S(n) = S(n − 1) + DS(n);
end;
```

This is coded in Excel in Table 3.9 for $K/u_E = 100$ km and $L = 20$ km and the results given in Figure 3.11 which shows hypersalinity of 38.5.

The analytic form is readily solved to give

$$S = S_o \exp\left[\frac{x u_E}{K}\left(1 - \frac{x}{2L}\right)\right]. \tag{3.47}$$

3.5 Simple models of chemical and biological processes

The hydrodynamic models discussed in this chapter establish a basis for a variety of water quality and chemical/biological processes in coastal basins. These are often called *ecological models* although they really model the perceived processes within the basin and not the structure of the ecology. The models of salt balance are seen as the foundation stone of these applied models because salt is a *totally conservative* quantity meaning that it has no internal interactions within the system and so its total mass is conserved. Salinity is one of the easiest quantities to measure in coastal basins and so it

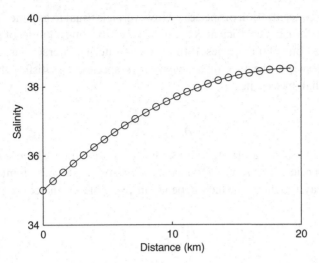

Figure 3.11 The prediction of the one-dimensional model (3.46) for the variation of salinity along a basin due to evaporation when $K/u_E = 100$ km and $L = 20$ km.

forms the natural basis for calibration of models, i.e., the determination of ocean exchange times and diffusion coefficients. These are the most difficult parameters to establish in box and one-dimensional models. Furthermore, models of salt balance contain none of the *internal* interactions found for non-conservative quantities, and this is the essential simplification that allows us to use *salt balance* for calibration of the hydrodynamic models.

The distinctive feature of *non-conservative* quantities in the water column is that they tend to either grow or decay, often as a result of interaction with one another and exposure to light. We will illustrate these processes by a study of flow through a channel, to which we apply the advection–diffusion equation (see Section 3.4.1)

$$F = Cu - K\frac{\partial C}{\partial x} \tag{3.48}$$

which states that the flux F of a quantity with concentration C comprises an advective term Cu (where u is the current in the x direction) and a diffusive term $(-K\partial C/\partial x)$. The usual practice is to use the term *constituent*, or *species*, for each of the quantities used in ecological or water quality models, and we will usually adopt the latter (but note the usage differs from that in *taxonomy*). Examples of species in a model are phytoplankton, zooplankton, nitrates, phosphates, oxygen, etc. Hence, we may have a range of species which we number 1, 2, ..., N and there is an advective–diffusion equation for each species, i.e.,

$$F_i = C_i u - K\frac{\partial C_i}{\partial x} \quad i = 1, 2, \ldots, N \tag{3.49}$$

where C_i is the concentration of species i. However, it is usual to assume that all species have the same diffusion coefficient K (derived by calibrating a model of salt balance). Salt is really just one of our species. If there are no internal interactions, or any source or sinks of species i, the flux of a *conservative* species, F_i satisfies the continuity equation for that species, i.e.,

$$\frac{\partial C_i}{\partial t} = -\frac{\partial F_i}{\partial x} \tag{3.50}$$

and this is also true for salt (note that [3.50] is only true in the absence of sources or sinks of that species). For other, *non-conservative*, species there is an interactive term or growth/decay function G which depends in principle on the concentration of all other species,

$$\frac{\partial C_i}{\partial t} = -\frac{\partial F_i}{\partial x} + G_i(C_1, C_2, C_3, \ldots). \tag{3.51}$$

Note that (3.51) replaces (3.50) with G_i representing source/sink terms in species i. A simple example of G is the self-decay process that afflicts all living creatures and has the form

$$G_i = -\frac{C_i}{\tau_i} \tag{3.52}$$

and this imposes a lifetime of τ_i on species i giving, with (3.51),

$$\frac{\partial C_i}{\partial t} = -\frac{\partial F_i}{\partial x} - \frac{C_i}{\tau_i}. \tag{3.53}$$

Equation (3.53) is a called a *first order* equation and the quantity $1/\tau_i$ is a *first order decay* constant (or growth constant if τ_i is negative). In the absence of any spatial variation in C_i, (3.53) has the solution for species 1,

$$C_1 = C_1(t = 0) \exp(-t/\tau_1). \tag{3.54}$$

Another example is the growth of phytoplankton given a supply of nutrients and availability of light. Phytoplankton (and most plant life in the ocean) tend to have their growth limited by both availability of light and their ability to take up nutrients from the water column. So, we need to introduce a *species* called *nutrients*, and we number that species as 2. A common form for the growth rate of species 1 due to uptake of nutrients and natural self-decay is

$$G_1 = -\frac{C_1}{\tau_1} + \frac{C_1}{\tau_1^{(2)}} \frac{C_2}{C_2^{(1)} + C_2} \tag{3.55}$$

where $\tau_1^{(2)}$ is a growth time of species 1 due to the nutrients (species 2) at very high concentrations of nutrients C_2, i.e., under conditions in which growth is not limited by that nutrient. The value of $\tau_1^{(2)}$ depends on light availability, and species 1 will either increase or decrease in concentration, according to whether $\tau_1^{(2)}$ exceeds or is less than τ_1. At lower values of C_2, the growth rate depends on C_2 and at very low concentrations, it is proportional to C_2. This situation is commonly, but loosely, referred to as *nutrient-limited growth*. The transition to nutrient-limited growth depends on the critical concentration $C_2^{(1)}$ in (3.55). The form (3.55) is essentially a *phenomenological* model, in that it is a convenient representation of the basic processes that control the dependence of growth on nutrient concentration but it is not necessarily based on any microscopic derivation of these processes. Assigning values for both $\tau_1^{(2)}$ and $C_2^{(1)}$ is a matter of *calibration*, although there is an increasing worldwide literature for suitable values of these parameters. If we adopt the form (3.55) the model requires some information on C_2. We can either assume a fixed value or model C_2 explicitly, i.e., develop a form (3.53) for species $i = 2$. This will involve C_1 since species 1 takes up nutrients. To develop this model we need to define units for C_1 and C_2, so that we can determine the change in C_1 due to changes in C_2. A convenient method is to use the mass of a particular element such as phosphorus. In that case, a model based on nutrient uptake would measure C_1 and C_2 in terms of mass of phosphorus. Hence in these units

$$\Delta C_1|_{2 \to 1} = -\Delta C_2|_{2 \to 1} \tag{3.56}$$

and so we can write down the growth/decay function for species 2 as equal and opposite to the nutrient uptake term in G_1, i.e., from (3.55)

$$G_2 = -\frac{C_1}{\tau_1^{(2)}} \frac{C_2}{C_2^{(1)} + C_2} \tag{3.57}$$

assuming that there are no other species in the basin, i.e., the nutrients are not being taken up by any other species. If we ignore any spatial dependence of species 1 and 2, i.e., we adopt a *box model*, and assume that $C_2 \ll C_2^{(1)}$ (so that we are in the nutrient-limited region) we have the coupled equations

$$\frac{dC_1}{dt} = -\frac{C_1}{\tau_1} + \frac{C_1}{\tau_1^{(2)}} \frac{C_2}{C_2^{(1)}}$$

$$\frac{dC_2}{dt} = -\frac{C_1}{\tau_1^{(2)}} \frac{C_2}{C_2^{(1)}}. \tag{3.58}$$

If we start with initial conditions in which C_1 and C_2 have finite values at time $t = 0$, the solution of the model (3.58) is simply that C_1 decays away to zero with increasing time as species 1 absorbs the nutrient pool described by species 2, and then experiences self-decay. Species 1 decays to *detritus* and this settles as sediment, which may then release nutrients back to the water column (we could introduce a third species to represent such detritus). The exponential decay of C_1 will be preceded by a period of rapid

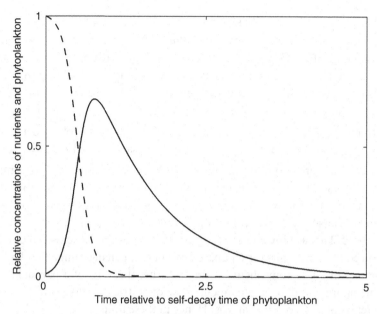

Figure 3.12 Modeled time variation of phytoplankton and nutrients (broken line) from an initial pulse of nutrients into the basin for $\tau_1 = 10\,\tau_1^{(2)}$.

growth if species 1 has a lifetime in excess of the time to consume species 2, i.e., $\tau_1 > \tau_1^{(2)}$. For phytoplankton, this is a simple model of an algal bloom which is terminated by C_2 becoming so depleted that self-decay exceeds growth. Blooms of this type are often found in coastal basins due to either an inflow of nutrients or changed physical parameters in the basin, such as the increase in solar radiation in spring (which creates the *spring bloom* of phytoplankton): nutrient pools in the ocean tend to build up over winter, and are then depleted by the spring bloom. In temperate oceans, this process is accentuated by basins being vertically mixed in winter and thermally stratified in spring/summer. Figure 3.12 shows a computed model prediction in which $\tau_1 = 10\tau_1^{(2)}$ and C_1 an initial value of $0.01C_2^{(1)}$ and C_2 starts at $C_2^{(1)}$.

This box model can be further developed by including exchange with the open ocean as in Section 3.3. It is common to use the calibration of the model against salt exchange (detailed in Section 3.3) to derive an exchange time τ, and then to add the terms $-(C_1 - C_o)/\tau$ and $-(C_2 - C_o)/\tau$, respectively, to the right-hand sides of the first and second equations in (3.58), where C_o is the concentration in the open ocean. As we mentioned before, this is a major simplification in the exchange term insofar as it is based on the mean concentration (rather than the concentration near the mouth of the basin). In Figure 3.12, the effect of such exchange would be to cause extra decay of both concentrations and its importance would be dependent on the ratio τ/τ_1. If $\tau/\tau_1 \gg 1$, exchange has little effect while exchange fast enough to produce $\tau/\tau_1 \gg 1$ would have the effect of bringing both concentrations toward C_o faster than the variations shown in the figure.

This type of box model of ecological processes, and water quality, has been widely used in coastal basins. Apart from the problems with the representation of exchange, box models of biological processes have other limitations that were not encountered for the simple modeling of salt and heat discussed in Section 3.3. These arise from the interactions between species. Such interactions are not well represented by the corresponding interactions of the mean values of the concentrations. It is also important to include the nutrient flux term in the continuity equation, otherwise the uptake of nutrients would totally deplete a region of the basin near (say) a seagrass meadow. This depletion is only partial (and usually difficult to measure), because nutrients advect and diffuse past the fixed uptake surface. To understand these effects it is necessary to include the first term on the right-hand side of (3.51), i.e., the gradient of the flux defined in (3.48). It is possible to continue with analytical solutions, or solutions coded in Microsoft Excel, provided that we limit the number of *independent* variables to one. An independent variable is a quantity that is varying (continuously) in the basin such as time t, or one of the spatial coordinates x, y, and z. We have seen examples of such models in this chapter.

For the case of nutrient uptake, we can find a solution with one independent variable if we ignore spatial variations (as in the box model development given above), or if ignore the time variation as in our earlier one-dimensional models. As an example of the solution of the nutrient uptake over a seagrass meadow or a coral reef, let us assume that the nutrient concentration varies in space but not time. This will provide a one-dimensional model that represents the change in nutrient concentration as water passes over the seagrass meadow or coral reef. Such changes in nutrient concentration are usually (but not always) small in coastal basins, but seagrasses and coral reefs can have very high nutrient uptake rates. The model essentially assumes that the nutrient flux into the *benthos* (the underlying biomass of coral or submerged aquatic vegetation) is provided by a change in the horizontal flux of nutrients. In order that the model be dependent on only one species, i.e., the only *dependent* variable is the nutrient concentration, we assume that the rate of uptake is independent of the chemical state of the benthos. This is known as the *mass transfer limited rate*, and represents a situation in which the coral or seagrass will maximize its nutrient uptake subject only to limitations imposed by the mass transfer of ions across its membrane surfaces, and the concentration of nutrients in the water column. It is usual to characterize the uptake rate by an *uptake rate constant* denoted here by S_r which has the dimensions of velocity and is defined as the rate of uptake per unit *horizontal* area (not the surface area of the membrane) for unit nutrient concentration. Hence, the steady state nutrient concentration (which we simply denote as C) becomes from (3.51)

$$\frac{\partial F}{\partial x} + G = 0. \tag{3.59}$$

We apply (3.51) with

$$G = -\frac{S_r C}{h} \tag{3.60}$$

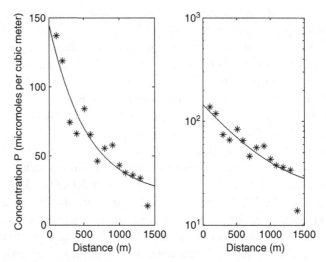

Figure 3.13 Variation (stars) of phosphorus concentration for water flowing over a coral reef (based on unpublished data for Kaneohe Bay, Hawaii, collected by M. J. Atkinson and the author in 2002). The data are plotted against both a linear and logarithmic scale. The full curves are obtained represent an exponential decrease with characteristic decay length of 550 m and a background phosphorus concentration of 20 μmoles m^{-3}.

where h is the water depth (which may vary with x) and since the nutrient flux F is given by (3.48)

$$K_x \frac{d^2C}{dx^2} - u\frac{dC}{dx} = \frac{S_r C}{h}$$ (3.61)

where we assumed that u and K are independent of x over the area of interest. The solution of (3.61) is

$$C = C_o \exp(-x/l) \qquad l \equiv \frac{K}{2u} \bigg/ \left\{ -1 + \sqrt{1 + \frac{S_r K}{hu^2}} \right\}$$ (3.62)

where C_o is the concentration at $x = 0$, i.e., at the open-ocean end of the basin. The simplest limit of (3.62) is for low dispersion, i.e.,

$$K \ll hu^2/S_r \qquad l \to uh/S_r$$ (3.63)

and so measurements of the decay of nutrient concentration over a reef provide, an apparently simple, way of finding a value for the *net* uptake constant S_r. An exponential decay of this type was first measured in Kaneohe Bay.[4] Figure 3.13 shows a latter data set collected on the same reef. When K is not negligible compared to hu^2/S, we must use the full expression for l as contained in (3.62) and for very large K,

[4] Atkinson and Bilger (1992).

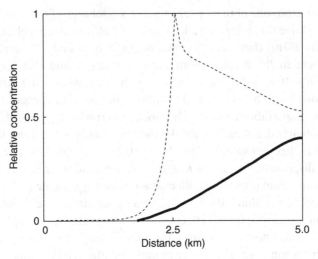

Figure 3.14 Modeled variation of phytoplankton and nutrients (broken line) for flow along a linear channel of length 5 km. A point source of nutrients is introduced at a distance of 2.5 km from the left-hand end of the channel ($x = 0$). The parameters of the model are given in the text.

$$K \gg hu^2/S \qquad l \rightarrow \sqrt{hK^3/4S}. \qquad (3.64)$$

Typical values of S_r, h, and u are $10^{-4}\,\mathrm{m\,s^{-1}}$, $1\,\mathrm{m}$, and $0.1\,\mathrm{m\,s^{-1}}$ respectively and so the *critical value of K* for transition from (3.63) to (3.64) is K of order $100\,\mathrm{m^2\,s^{-1}}$ which is larger than expected values of dispersion at the spatial scales of a reef (see Chapter 4) but we should note that this is a one-dimensional model in the horizontal and therefore averages over the both the vertical and lateral dynamics. Any averaging in a coastal basin has the effect of introducing effective dispersion into the dynamics of the basin. So, a depth-averaged one-dimensional model has what is called *Taylor shear dispersion* (named after the famous English fluid dynamicist G. I. Taylor) discussed in detail in Chapter 10. This increases K, and so dispersion is likely to affect the value of l. It is clear that increasing K has the effect of increasing the value of l, i.e., smoothing out the variation in nutrient concentration. So, the naïve use of the low dispersion limit (3.63) would have the tendency to underestimate the true value of S_r. Increasing dispersion clearly reduces the possibility of any variation in concentrations of nutrients throughout a coastal basin.

The models that we have discussed so far in this section are limited to single species, steady state solutions in one dimension, or to multiple species, time-dependent, box model (zero-dimension) solutions. It is evident that the spatial and temporal variations are essential to the dynamics of the ecology and water quality of a coastal basin. So, the step of including hydrodynamics into simple ecological models is an important one, and usually necessitates numerical solution. There are available a wide range of numerical packages that allow such models to be established. In Figure 3.14, we show a model solution from one of these packages[5] for a discharge of nutrients from a point source with a volume flux of magnitude $100 \, m^3 \, s^{-1}$ halfway along a channel (of length 5 km and width 300 m) that joins the open ocean at both ends. A current is flowing from left to right in the diagram with speed $0.05 \, m \, s^{-1}$, and there is a dispersion coefficient of $K = 10 \, m^2 \, s^{-1}$. The broken line shows a *plume* of nutrients flowing down the channel from the source, and the plume also has a tail penetrating upstream into the flow of length (about 500 m). The model was run for a period of 2 months and has essentially reached a steady state. In the upstream part of the plume, there is essentially no net flux and so there must be a balance between downstream advection and upstream dispersion, i.e., $uC \sim KdC/dx$. This provides an exponential solution which decays away from the source with characteristic length scale $K/u \sim 200 \, m$. In the downstream part of the plume the flux and hence concentration C would be approximately constant were it not for the effects of nutrient uptake by phytoplankton, which is shown by the full line in the figure. The model has been configured so that the nutrient concentration is $3 \, mg \, l^{-1}$ at the boundaries; this *ambient concentration* value has been subtracted from the nutrient concentrations shown in Figure 3.14.

The numerical scheme used in the model uses *up-winding* for the advection which means that the advection of nutrients is controlled by the concentration in the upstream edge of each finite difference cell. Hence, the concentration at the downstream boundary does not greatly affect the plume. If the nutrient concentrations for the plankton and water column are added, the result is almost spatially constant in the downstream plume as we would expect apart from a very obvious sharp decline immediately downstream of the source (which is due to diffusion into the tail of the plume) and a gradual, but small, decline due to natural morbidity of plankton; τ_1, the lifetime in (3.55), is taken as 10 days, and since it takes only $L/u = 5000/0.05 \, s \sim 28 \, h$ for particles to be advected out of the basin through the downstream boundary, the net morbidity is small. The nutrient uptake time $\tau_1^{(2)}$ is taken as 1 day, and the value of the concentration $C_2^{(1)}$ in (3.57) has the value of $2 \, mg \, l^{-1}$ compared with $10 \, mg \, l^{-1}$ for the water discharged by the source, and $3 \, mg \, l^{-1}$ at the upstream boundary. This means that the nutrient uptake rate in (3.57) becomes essentially independent of nutrient concentration (absorption of nutrients takes place at the maximum possible rate and is not restricted by their concentration) and, hence, plankton concentration should grow exponentially. The initial concentration of phytoplankton, and open ocean value, was set at $10 \, mg \, l^{-1}$ (this ambient value has been subtracted from the data displayed in the figure) and grows to 29.4 at the downstream boundary; this is an increase of a factor of 2.9 during the downstream transit time of 28 h. This compares favorably with the increase that would be predicted by simple exponential growth in that transit time t of $\exp(t/\tau_1^{(2)}) = 3.2$. Note that plankton growth is evidently occurring over most of the length of the channel.

---

[5] EcoLab from DHI Water and Environment.

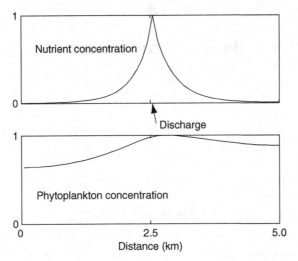

Figure 3.15 Results from model shown in Figure 3.14 when diffusion coefficient is increased by factor of 10.

This one-dimensional basin does have a well-defined exchange time given by the volume of the basin divided by the outward flux, or the length divided by the advective velocity (current) of $5000/0.05$ s $= 28$ h. However the outward flux of nutrients and plankton is not given by mean concentrations divided by this time; Figure 3.14 indicates that a box model based on this exchange time would underestimate exchange of plankton and nutrients by about 30% (although if the exchange time were calibrated against salinity that error would probably be reduced).

Figure 3.15 shows the model discussed above when the diffusion coefficient is increased by a factor of 10. This illustrates the importance of K to the model predictions.

3.6 Further reading

See further reading from earlier chapters. Additional texts for this chapter are:

Bloch, S. C. (2003). *Excel for Engineers and Scientists*, 2nd edn. New York: John Wiley.

Bourg, D. M. (2006). *Excel Scientific and Engineering Cookbook*. Sebastopol, CA: O'Reilly Media.

Day, J. W., C. A. S. Hall, and W. M. Kemp (2004). *Estuarine Ecology*. New York: Wiley-Interscience.

de Levie, R. (2004). *Advanced Excel for Scientific Data Analysis*. Oxford, UK: Oxford University Press.

Gottfried, B. S. (2002). *Spreadsheet Tools for Engineers Using Excel: Including Excel 2002*. New York: McGraw-Hill.

Smith, R. L. and T. M. Smith (1986). *Elements of Ecology*, 5th edn. New York: Harper & Row.

4

Basic hydrodynamics

4.1 Motion of a particle

As a particle moves, the values of its coordinates x, y, and z change with time t. So if we tag a particular particle, label it as 1 and call its position x_1, y_1, z_1, these quantities will be functions of time t. In the 1700s, Daniel Bernoulli investigated the forces present in a moving fluid. He considered the major problem of how the motion of a fluid differs from the motion of a set of discrete particles, or bodies (like the planets), as considered by Newton. In a sense, a fluid is like a collection of such particles. But in a fluid, one has to follow an infinite number of such particles and is often not interested in particular particles, but rather in the speed and direction of the fluid at a particular point. We explore this topic in more detail in Chapter 8. For the moment we only need the result that the acceleration of a particle in one direction while at the point x is

$$\text{acceleration} = \frac{\partial u}{\partial t} + u \frac{\partial u}{\partial x} \tag{4.1}$$

where $u(x, t)$ is the velocity of a particle whilst it is at the point x at time t. Note that in a steady state u is independent of time t, but the speed of a particle may be changing as its coordinates change, i.e., u is not the velocity of one particular particle but the velocity of a particle when it is at x. The first term in (4.1) is called the *local acceleration*, and the second term is referred to as the *advective* term.

The dynamical equation for a coastal basin involves equating the acceleration term to the sum of forces acting on that particle. In the dynamics of coastal basins, we tend to work with forces per unit area (which have the dimension of density times velocity squared). Often we divide through by density but still refer to the terms as *forces*. We consider the forces encountered in coastal basins in the following sections.

4.2 Basic dynamics in hydrodynamic models

If we denote the pressure gradient by F1, stress by F2, and Coriolis force by F3, Newton's law of motion gives using (4.1), in one dimension,

$$\frac{\partial u}{\partial t} + u\frac{\partial u}{\partial x} = F1 + F2 + F3. \tag{4.2}$$

This law is equivalent to the *conservation of momentum*. If u changes much more rapidly with time than it does with distance, the local acceleration is by far the more important term, and in the remainder of this chapter we shall write

$$\frac{\partial u}{\partial t} = F1 + F2 + F3. \tag{4.3}$$

As we shall see later there are many occasions when this approximation is not valid in coastal basins and in general, the validity of (4.3) is dependent on the spatial (and therefore temporal) scale at which we model the basin. Notice also that (4.3) is linear in u, while the full expression (4.1) has the non-linear advection term.

If the right-hand side of (4.3) is zero (there is a balance of all the forces in the coastal basin), the left-hand side becomes zero so that the currents are unchanging in time, and we say that we have a *steady state*. Many of our models only look at such steady states and are called *steady state models*. For steady state models in one dimension, it follows from the volume continuity equation that the volume flux is independent of the spatial coordinate x.

In two dimensions Newton's law and the continuity equation become

$$\frac{\partial u}{\partial t} + u\frac{\partial u}{\partial x} + \frac{\partial v}{\partial t} + u\frac{\partial v}{\partial y} = F1 + F2 + F3 \tag{4.4}$$

$$\frac{\partial \eta}{\partial t} = \frac{\partial hu}{\partial x} + \frac{\partial hv}{\partial y} \tag{4.5}$$

and again, a steady state model involves a balance of all the forces on the right-hand side of (4.4); the *divergence* of the flux $\partial hu/\partial x + \partial hv/\partial y$ vanishes.

4.3 Pressure

4.3.1 Pressure differences

The gradient of pressure in a fluid produces one of the most important forces controlling the movement of that fluid. Pressure is basically due to molecular motions which create a force in all three dimensions due to impact of the molecules. So, from pressure there is no *net* force on a fluid particle. However, the pressure does vary with position, and hence the pressure on the opposite sides of a plane may be unequal. This produces a net force called a pressure gradient. We experience the same effect if we dive in the ocean and find that the pressure due to the air in our lungs is not equal to the pressure of the surrounding water. So, it is the spatial gradient of pressure that produces a net force. We shall see that this usually arises from either the water surface of a coastal basin not being horizontal, or there being a spatial variation in the density of water. Newcomers to ocean dynamics are surprised by the very small gradients in surface elevation that can drive ocean currents, and at the same time the fact that the surface is not exactly horizontal. Some of these issues are connected with a common lack of realization of the density of water.[1]

Models which ignore density variations are very loosely called *barotropic*, while models that include density variation are very loosely called *baroclinic*.[2] It is usual to model a system initially by a barotropic model and later go on to a baroclinic model. We talk about *barotropic processes* (those that do not involve density variation) and also *baroclinic processes* (for which density variation plays a crucial part). Pressure gradients arising from density variations are often called *buoyancy effects*, or *buoyancy forcing*.[3]

4.3.2 Hydrostatic approximation

The *hydrostatic approximation* is a very basic assumption made in almost all models of coastal basins. It has a very high degree of validity (provided we do not work at small scales), and the approximation may only fail at very small spatial scales which are beyond the resolution of most present-day numerical models. Apart from pressure gradients, the most important force that acts on a particle in the ocean is its weight. However, this gravitational force is so large that it very quickly produces vertical movements that result in a pressure gradient that almost exactly matches these gravitational effects. This is called the *hydrostatic equilibrium*:

$$\frac{\partial p}{\partial z} = -g\rho \tag{4.6}$$

where $p(z)$ is pressure, g the acceleration due to gravity and $\rho(z)$ is the density of water, i.e., the gravitational force on an element of fluid from z to $z + \mathrm{d}z$ is $-g\rho\mathrm{d}z$ and this is balanced by the pressure difference $p(z + \mathrm{d}z) - p(z)$. The pressure profile that satisfies (4.6) is called the *hydrostatic pressure*. The weight of the water per unit area above a point at height z is $g \int_z^\eta \rho(z)\mathrm{d}z$ and produces a hydrostatic pressure,

$$p(z) = g \int_z^\eta \rho(z)\mathrm{d}z \tag{4.7}$$

relative to atmospheric pressure. For the case of constant density (4.7) reduces to

$$p = g(\eta - z). \tag{4.8}$$

---

[1] The author has a memory from childhood of first trying to lift a bucket of water.
[2] The strict definition is that a barotropic model has all its isobars parallel to isopycnals.
[3] We shall see later that even horizontal variations of density eventually lead to vertical variations.

Note that $(\eta - z)$ is the height of the water column (remember that z is negative below the height datum. Unless p satisfies (4.6) the water column will rapidly change until hydrostatic equilibrium *is* achieved. So, in most ocean dynamics we adopt the hydrostatic approximation. The equilibrium merely means that the pressure of the water exactly balances the weight of water above it. There is a distinction between the two types of forces involved here, although they are equal in magnitude. Pressure is due to internal molecular forces and acts equally in all directions, while weight is due to the Earth's gravity. In the above form we have ignored the pressure p_a due to the atmosphere; this can be easily included:

$$p = p_a + g(\eta - z). \tag{4.9}$$

Models that use the hydrostatic approximation are called *hydrostatic models*.

The hydrostatic approximation assumes that the force which causes vertical acceleration is very small compared with the almost exactly counterbalanced forces due the vertical pressure gradient and gravity. It is easy to prove that the comparative magnitude of the forces do satisfy this approximation provided that the spatial scale is not too small. For example, a current of $1\,\mathrm{m\,s^{-1}}$ driven along the surface of a closed basin by a wind stress will have an associated bottom current of the same magnitude in the reverse direction. If the basin is $10\,\mathrm{km}$ long the transit time of the current is 10^4 s. If the vertical current is spread over the basin and the basin depth is $10\,\mathrm{m}$, the vertical acceleration will be of order $10/(10^4)^2 = 10^{-7}$ $\mathrm{m\,s^{-2}}$. In fact the acceleration may only occur within 10% of the length of the basin and so the acceleration may be one or even two orders larger, i.e., $10^{-5}\,\mathrm{m\,s^{-2}}$. This is many orders smaller than the acceleration due to gravity. This example also serves to illustrate the dependence of the hydrostatic approximation on spatial scale. In most models of coastal basins we do not attempt to determine w but instead infer w from the volume continuity equation, i.e., from the divergence of the horizontal current.

Using the hydrostatic approximation (4.8) and (4.3) in one dimension,

$$\frac{\partial u}{\partial t} = -g\frac{\partial \eta}{\partial x} \tag{4.10}$$

i.e., by using the hydrostatic approximation and an assumption of constant density, the pressure gradient controlling the acceleration in the horizontal x direction is due to the surface slope $\partial \eta/\partial x$. An important aspect of (4.10) is that the pressure gradient is the same at all values of z. This is basically due to the assumption of constant density (and the hydrostatic approximation). If we use the linearized volume continuity equation in one dimension for constant depth, i.e.,

$$\frac{\partial \eta}{\partial t} = -h\frac{\partial u}{\partial x} \tag{4.11}$$

where u is the depth average current and differentiate (4.10) with respect to x,

$$\frac{\partial^2 u}{\partial t \partial x} = -g \frac{\partial^2 \eta}{\partial x^2} \tag{4.12}$$

and substituting (4.11)

$$\frac{\partial^2 \eta}{\partial t^2} = gh \frac{\partial^2 \eta}{\partial x^2} \tag{4.13}$$

which is a wave equation which can describe a wave of speed $c = \sqrt{gh}$.

4.3.3 Quasi-hydrostatic approximation

Let us write the pressure as $p(x, y, z, t)$ and consider the individual forces due to pressure on the six faces of a cube. Any surface surrounded by fluid with pressure p will experience a force equal to p per unit area normal to that surface. If the cube has sides dx, dy, dz, the total forces in the positive x direction are $p(x, y, z)dydz$ and $-p(x + dx, y, z)dydz$ since the force on the latter surface acts in the negative x direction. The total force per unit volume of the fluid is simply $-\partial p/\partial x$ in the x direction, $-\partial p/\partial y$ in the y direction, and $-\partial p/\partial z$ in the z direction. If we consider that pressure gradients are the only force acting on a particle in the water column and use (4.3) we obtain

$$\rho \frac{\partial u}{\partial t} = -\frac{\partial p}{\partial x}$$
$$\rho \frac{\partial v}{\partial t} = -\frac{\partial p}{\partial y} \tag{4.14}$$
$$\rho \frac{\partial w}{\partial t} = -\frac{\partial p}{\partial z} - \rho g$$

where we have included the *gravitational* force in the vertical which is *downwards*. In the hydrostatic approximation we assume that the right-hand side of the third of (4.14) is zero. We now wish to assume that the hydrostatic approximation is nearly true so that there are no *net* terms of magnitude g remaining on the right-hand side of (4.14). This is the *quasi-hydrostatic approximation*.

We find it convenient to define a reference pressure denoted by p_0:

$$p_0 \equiv p_a - g\rho z \tag{4.15}$$

which is (for constant density ρ) the *hydrostatic pressure*, i.e., the pressure which would be given by the hydrostatic approximation at height z when the surface is at $z = 0$ (p_a denotes atmospheric pressure). We are going consider a small perturbation in the level of the ocean to $z = \eta$, which is responsible for causing the pressure at z to change to $p(z)$ where

$$p(z) = p_0 + p'(z, \eta). \tag{4.16}$$

We shall attempt to find the form of p' to order η. We could do this immediately to the accuracy of the hydrostatic approximation which would give $p'(z, \eta) = g\rho\eta$, but there are clearly other terms in p' of order η which are *non-hydrostatic*. We would expect these to be controlled by the timescale of variations in η (becoming increasingly small as these timescales increase). Apart from the hydrostatic component of $p'(0, \eta)$ (which is $g\rho\eta$), there may be non-hydrostatic corrections in $p'(0, \eta)$ but these will clearly be of at least second order in η, and so to the order to which we are working, $p'(0, \eta) = 0$.

To find the form of p' we need to consider the dynamic equations obtained by substituting (4.16) in (4.14), i.e.,

$$\rho \frac{\partial u}{\partial t} = -\frac{\partial p'}{\partial x}$$
$$\rho \frac{\partial v}{\partial t} = -\frac{\partial p'}{\partial y} \tag{4.17}$$
$$\rho \frac{\partial w}{\partial t} = -\frac{\partial p'}{\partial z}$$

or if we adopt the vector notation for velocity $\boldsymbol{u}$, (4.17) is

$$\rho \frac{\partial \boldsymbol{u}}{\partial t} = -\nabla p' \tag{4.18}$$

where the right-hand side of (4.18) is the negative *gradient* p'. The continuity equation expressed in vector notation, is that the divergence of $\boldsymbol{u}$ vanishes, i.e.,

$$\nabla.\boldsymbol{u} = 0 \tag{4.19}$$

As we saw in Chapter 2 provided that the velocity field is *irrotational*, i.e., the curl of $\boldsymbol{u}$ vanishes, it is possible to write $\boldsymbol{u}$ as the gradient of a scalar potential W,

$$\boldsymbol{u} = -\nabla W. \tag{4.20}$$

Taking the divergence of (4.18)

$$\rho \frac{\partial \nabla.\boldsymbol{u}}{\partial t} = -\nabla.\nabla p' \tag{4.21}$$

and using (4.19) we see that p' satisfies the Laplace equation,

$$\nabla^2 p' = 0, \tag{4.22}$$

where the left-hand side of (4.22) is the Laplacian operator,

$$\nabla^2 \equiv \frac{\partial^2}{\partial x^2} + \frac{\partial^2}{\partial y^2} + \frac{\partial^2}{\partial z^2}. \tag{4.23}$$

Let us look for a solution in three dimensions of (4.22) and (4.18) that produces a surface gravity wave propagating in the x direction with wave number k and angular frequency ω. Since the quasi-hydrostatic model that we are studying is entirely linear we can assume that both $\boldsymbol{u}$ and p have a dependence on t, x, and z of the form

$$\boldsymbol{u}, p \sim e^{i(k_x x + k_z z + \omega t)} \tag{4.24}$$

The surface elevation η is a function of x and t only and so can be written as

$$\eta \sim e^{i(k_x x + wt)}. \tag{4.25}$$

Equations (4.22) and (4.24) give

$$k_x^2 + k_z^2 = 0 \tag{4.26}$$

of which there are two solutions:

$$k_x \rightarrow k \quad k_z = \pm ik. \tag{4.27}$$

Hence we can rewrite (4.24) as

$$w = e^{i(kx + \omega t)} \left(a e^{kz} + b e^{-kz} \right) \tag{4.28}$$

where a and b are arbitrary constants. The vertical velocity w must vanish at the bottom $z = -h$ and this fixes the ratio of a to b so that (4.28) must have the form

$$w = w_0 e^{i(kx + \omega t)} K \quad K(kh, z/h) \equiv \frac{\sinh(kh\{1 + z/h\})}{\sinh(kh)} \tag{4.29}$$

where we have introduced the function K in (4.29) as unity at $z = 0$ and so w_0 is the vertical speed at $z = 0$. The quantity $(z + h)$ should be read as *distance from the bottom* and $k(z + h)$ as *distance from the bottom divided by wavelength times 2π*. As we move up the water column, the argument $k(z + h)$ increases from zero to kh. It is important to know whether kh is smaller or much greater than unity. In either case, the function K in (4.29) becomes unity at $z = 0$ but for small kh, K decreases linearly with z, while for large kh, K decreases exponentially; this is shown in Figure 4.1.

Note that we are only working to first order in η so that the surface is essentially at $z = 0$. We can either use the real or imaginary term from the complex exponential in (4.29), and so let us write the vertical component of velocity w as

$$w = w_0 \sin(kx + \omega t) K(kh, z/h). \tag{4.30}$$

At the surface w must be equal the rate of increase of the elevation η with time, so (4.30) implies that the elevation must have a cosine variation with (x, t),

$$\eta = \eta_0 \cos(kx + \omega t) \tag{4.31}$$

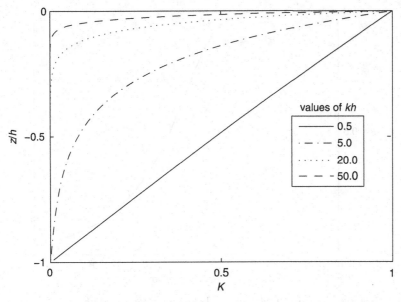

Figure 4.1 The function $K(kh, z/h)$ for values of $kh = 0.5$, 5, 20, and 50, illustrating the linear decrease from unity at the surface when $kh < 1$ and exponential decrease when $kh \gg 1$. Those two limits correspond to shallow water and deep water waves, respectively.

and so fixes that value of w_0 allowing us to rewrite (4.30) as

$$w = -\omega\eta_0 \sin(k_x x + \omega t) K(kh, z/h), \tag{4.32}$$

hence

$$\frac{\partial w}{\partial t} = -\omega^2\eta_0 \cos(k_x x + \omega t) K(kh, z/h). \tag{4.33}$$

Notice that w is first order in η and also proportional to ω, so that it tends to zero as we move to longer timescales, i.e., ω tends to zero. The implication from (4.33) and the third of (4.17) is that p' is proportional to ω^2, i.e.,

$$\frac{\partial p'}{\partial z} = \omega^2\rho\eta_0 \cos(k_x x + \omega t) K(kh, z/h) \tag{4.34}$$

and integrating with respect to z,

$$p' = \frac{\omega^2\rho\eta_0}{k_x} \cos(kx + \omega t) \frac{\cosh(kh\{1 + z/h\})}{\sinh(kh)} \tag{4.35}$$

and so equating p' at $z = 0$ from (4.35) to $g\rho\eta$ from (4.31) we obtain

$$\omega^2 = gk_x \tanh(k_x h), \tag{4.36}$$

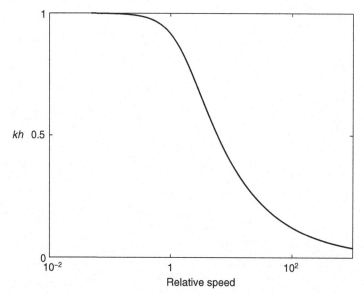

Figure 4.2 Relationship between relative speed c/c_0 and wave period kh.

which is the necessary relation between frequency ω and wave number k called the *dispersion relation* because it gives us the (phase) velocity c:

$$c \equiv \frac{\omega}{k} = c_0 \sqrt{\tfrac{\tanh(kh)}{kh}} \quad c_0 \equiv \sqrt{gh}. \tag{4.37}$$

Figure 4.2 shows c/c_0 and the wave period as a function of kh, and Figure 4.3 shows the relative against the relative wavelength λ/h and relative period defined as $c_0 T/h$.

When $kh \ll 1$, i.e., the wave length $2\pi/k$ is much greater than the water depth h, $c \to \sqrt{gh}$ which is the *usual* case in coastal basins (although short waves can exist for which this is not true). In this limit c is non-dispersive, i.e., it is independent of k. In the opposite limit of $kh \gg 1$, $c \to \sqrt{g/k}$ which is the more usual case in the deep ocean for surface gravity waves (but not, for example, for tidal waves which have very small k) and means that waves with small k (long wavelength and long period) travel more quickly than short period waves, i.e., c is dispersive. It is worthwhile emphasizing that the appearance of the depth h in the dispersion relation is entirely due to the vanishing of w at the seabed; there is of course no consideration of friction or viscosity in this treatment.

Equations (4.37) and (4.34) give

$$\frac{\partial p'}{\partial z} = g\rho\eta_0 k \cos(kx + \omega t)\frac{\sinh(k\{z+h\})}{\cosh(kh)}. \tag{4.38}$$

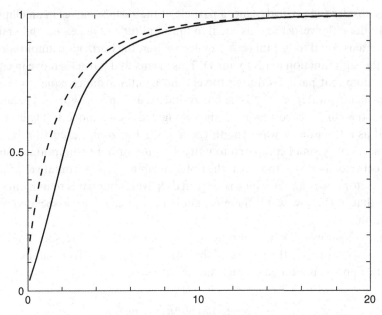

Figure 4.3 Plot of relative speed c/c_0 against relative wavelength λ/h (dashed line) and relative period $c_0 T/h$ (solid line).

Using the continuity equation and (4.32) we obtain

$$\frac{\partial u}{\partial x} = \omega \eta_0 k \sin(kx + \omega t) \frac{\cosh(k\{z+h\})}{\sinh(kh)}, \qquad (4.39)$$

and integrating with respect to x,

$$u = -\omega \eta_0 \cos(kx + \omega t) \frac{\cosh(k\{z+h\})}{\sinh(kh)}. \qquad (4.40)$$

Integrating (4.32) and (4.40) we can obtain the position of a particle as a function of time:

$$z = \eta_0 \cos(kx + \omega t) \frac{\sinh(k\{z+h\})}{\sinh(kh)} \qquad (4.41)$$

$$x = -\eta_0 \sin(kx + \omega t) \frac{\cosh(k\{z+h\})}{\sinh(kh)} \qquad (4.42)$$

so at the surface

$$z = \eta_0 \cos(kx + \omega t) \qquad (4.43)$$

$$x = -\eta_0 \sin(kx + \omega t) / \tanh(kh). \qquad (4.44)$$

So, if we consider the vertical plane, a particle at the surface moves in an ellipse with amplitude along the vertical z axis equal to the amplitude of η, i.e., η_0, and amplitude along the x axis equal to $\eta_0/\tanh(k_x h)$. In deep water, the horizontal amplitude is also η_0 (since the tanh function tends to unity). This means that particles move in circles of radius η_0; note that particles do *not* travel a horizontal distance equal to the wavelength. In shallow water, $k_x h \ll 1$ the horizontal excursion is $\eta_0/k_x h$, i.e., greater than η_0, and so the ellipse becomes increasing elongated as we move to shallow water. Measured as a fraction of wave length $(2\pi/k_x)$, the horizontal excursion is $\eta_0/(2\pi h)$ which is necessarily small compared to unity, because of our assumption of small η_0. The vertical velocity drops to zero at the bottom, whilst the horizontal velocity drops by only a factor of $\cosh(k_x h)$, which is large in deep water but tends to unity in shallow water, so that in shallow water there is almost no vertical shear associated with the wave motion.

The analysis shows that the non-hydrostatic terms in the pressure gradient are inversely proportional to the square of the timescale, i.e., ω^2. If we had assumed the hydrostatic approximation based on a surface at $z = \eta$,

$$\frac{\partial p'}{\partial x} = -g\rho\eta_0 k_x \sin(k_x x + \omega t), \tag{4.45}$$

compared with

$$\frac{\partial p'}{\partial x} = -g\rho\eta_0 k_x \sin(k_x x + \omega t)\frac{\cosh(k_x\{z+h\})}{\cosh(k_x h)}, \tag{4.46}$$

so the *non-hydrostatic difference* to order η_0 is

$$g\rho\eta_0 k_x \sin(k_x x + \omega t)\left[1 - \frac{\cosh(k_x\{z+h\})}{\cosh(k_x h)}\right] \tag{4.47}$$

which vanishes at $z = 0$. At the bottom, the non-hydrostatic difference is in general non-zero but for $k_x h \ll 1$ the difference tends to zero as $(k_x h)^2/2$. So in shallow basins we can assume that the hydrostatic approximation is valid. In deep basins, the actual pressure gradient tends to zero below the surface as compared with the hydrostatic approximation, in which pressure is constant with depth (for constant density).

Waves with $kh < 1$ are commonly called shallow water, or *long* waves, whereas waves with $kh > 1$ are termed deep water, or *short* waves. Storms over the deep ocean produce a spectrum of waves, and simple models indicate that the maximum speed of the waves may approach the maximum wind speed. If we assume that $kh \gg 1$, the appropriate limit of (4.37) is $c \to \sqrt{g/k}$ and $c \to gT/2\pi \sim 1.56T$. So, a wave of period 15 s has speed of $23.4\,\mathrm{m\,s^{-1}}$. Our model would predict a maximum wavelength and period, and these waves will propagate away from the storm most rapidly and at some distance away from the storm we will only see these fast waves. It is of importance to distinguish between the phase speed c and the so-called group velocity, which is the speed at which the center of a disturbance moves through the ocean. Standard wave theory shows that this is given by $\partial\omega/\partial k$ and from (4.36) this is $c/2$. Hence, the time taken for a front of swell waves from a storm to propagate over a distance of 5000 km is about 6 days. The wave *period*, or frequency, tends to be used by modelers to characterize waves in preference to the wave *length*. The reason is that frequency and period are preserved as the wave propagates over the ocean, whereas the wavelength is dependent on the speed, which changes as the wave enters shallow water. If we analyze the frequency of surface waves near the mouth of a coastal basin, we usually find peaks in the wave spectrum between periods of 10 and 20 s, corresponding to swell waves from distant storms and much shorter period waves from local winds. The two types of waves are easily recognized visually: swell waves typically have heights of at least 1 m and are regular in shape and period whereas wind waves (or *sea*) have a broad distribution of periods (from 10 s downwards) and are choppy and irregular.

Swell waves are able to travel very long distances across the deep ocean because they are very weakly damped. This weakness of damping is surprising at first sight, and is a result of the waves having orbital velocities that do not extend to the bottom of the ocean. When swell enters the shallow coastal regions, it makes a transformation from a short wave ($kh > 1$) to a long wave ($kh < 1$) and the orbital velocities are then damped by their interaction with the bottom. This interaction takes place only in a relatively thin boundary layer called the *wave boundary layer* controlled by the distance that the stress generated by the bottom orbital velocity can penetrate into the water column.

Coastal basins and the deep ocean experience both long and short waves. For example, tidal motion in the deep ocean involves long waves ($kh < 1$), and boat wakes in coastal basins create short waves ($kh > 1$). Bow waves from a boat are very dispersive, i.e., the speed depends on wavelength (or k), so that we see the characteristic longer waves arriving at the coast as the boat passes.

4.4 Shear stress

Horizontal *shear stress* is a force in the horizontal plane (per unit area of the water column). Figure 4.4 illustrates *internal, surface,* and *bottom* stress. The figure shows infinitesimally thin layers at the surface ($z = 1$) and at $z = 0, -1$, and the bottom of the water column ($z = -3$). The vector A is the surface wind shear stress in the x direction. The vector E is the bottom shear stress (due to bottom friction). Vectors $B, C,$ and D are internal shear stresses at $z = 0, -1, -2$ respectively. The box with dotted sides represents a particle of water centered at $z = -0.5$. The net force on this box from shear stress is $B - C$ shown by the vector on the right-hand side of the box. There is also a net force on the box on its left-hand side due to the horizontal pressure gradient and this is labeled F. In hydrodynamic models we shrink the box with dotted sides to a thickness in the z direction of dz so that $B - C$ involves the vertical gradient of shear stress times dz. The total force on the water column is $A - E - (\text{sum } F \text{ over } z)$.

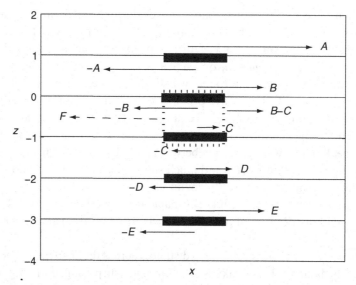

Figure 4.4 Schematic of shear stress and pressure gradient in the water column of a coastal basin with surface elevation $\eta = 1$ and water depth $h = 3$ (see text).

All stresses in coastal basins are basically due to some of some form of friction or drag. Surface stress is usually due to winds acting on the surface, i.e., there is a drag between the air moving over the ocean surface and surface water; we call this *wind stress*. Bottom stress is due to drag between the bottom of water column and the bed of the basin. We call this *bottom friction* or *bottom drag*. Wind stress and bottom friction have their most important effects near the surface and bottom of the water column, and we talk about the *surface* and *bottom boundary layers* in which these stresses are dominant. For example, in deep ocean basins wind stress creates *surface currents*. In shallow water these boundary layers occupy all, or most, of the water column. For example, in an estuary of depth 10 m, surface shear stress and bottom friction may be important throughout the water column. Internal stresses are created by friction, or viscosity within the water column and control the variation of current with height, i.e., the forms of *u(z)* and *v(z)*. Stresses applied at the surface (wind stress) propagate down through the water column as internal stress, and accelerate the water below the surface. Bottom stress similarly propagates upward. If the current varies with height, we say that we have *velocity shear*. Shear stress is continuous throughout the water column. There is no difference between the wind stress and the *internal shear stress* just inside the water column at the surface, or between the internal stress at the bottom and bottom stress. There is sometimes some confusion about the direction of the force, and it is necessary to consider the element of the water column on which the force is acting.

We use the symbol τ for the horizontal shear stress to conform to usual convention. Stress is a measure of the internal distribution of force unit area within any body that balances and reacts to the loads applied to it. To be more precise in our definition of shear stress, we should define an elemental volume of side dx, dy, dz and say that the shear stress τ_{ij} is defined as the shear stress on face i (where $i = 1, 2, 3$ corresponding to the x, y, and z components respectively) in the j direction. So our τ is properly called τ_{zx} or τ_{zy}, and we shall drop the subscript in one dimension.

The net force from shear on an element from z to $z + \mathrm{d}z$ is $\tau(z + \mathrm{d}z) - \tau(z)$, so that for stress alone the dynamical equation is

$$\frac{\partial u}{\partial t} = \frac{\partial \tau}{\partial z}. \tag{4.48}$$

If we integrate (4.48) with respect to z over the water column, the left-hand side is

$$\int_{-h}^{\eta} \frac{\partial u}{\partial t} \mathrm{d}z = \frac{\partial}{\partial t} \int_{-h}^{\eta} u \mathrm{d}z, \tag{4.49}$$

and recalling the definition of depth average velocity

$$\bar{u} \equiv \frac{1}{h + \eta} \int_{-h}^{\eta} u \mathrm{d}z \tag{4.50}$$

we obtain

$$(h + \eta) \frac{\partial \bar{u}}{\partial t} = \int_{-h}^{\eta} \frac{\partial \tau}{\partial z} \mathrm{d}z \tag{4.51}$$

and so finally

$$(h + \eta) \frac{\partial \bar{u}}{\partial t} = \tau(\eta) - \tau(-h), \tag{4.52}$$

i.e., the acceleration of the water column is controlled by the difference between the surface and bottom stresses. Note that the stress is defined as positive in the x direction on the upper surface of an element of fluid.

We associate the surface stress with wind blowing over the basin. In the absence of wind forcing the surface stress is often modeled as zero. The bottom stress is due to the roughness of the bottom, and we shall see that both surface and bottom stress are usually related to the square of the flow velocity: at the surface this is the wind speed, and at the bottom it is the bottom current. The reason for this quadratic form of stress at the surfaces is that the surface currents are responsible for both velocity shear and turbulence, and produce boundary layers described by a logarithmic dependence of current on distance from the surface. So, it is conventional to write surface stress as a product of density, drag coefficient, and square of speed where the density refers to the moving medium (air for the surface stress and water for bottom stress), hence

$$\tau(-h) \equiv \tau_b = \rho C_d u^2(-h) \tag{4.53}$$

$$\tau(\eta) \equiv \tau_w = \rho_a C_w W^2 \tag{4.54}$$

where ρ is the density of water, C_d the bottom drag coefficient, ρ_a the density of air, C_w the surface wind drag coefficient, and W the wind speed (properly measure at $10\,\text{m}$ above water level supposed to be beyond the boundary layer formed in the moving air by the surface of the basin).

4.5 Oscillators

4.5.1 Simple harmonic oscillator

In order to understand many of the models of coastal basins and some aspects of the ideas of this chapter it is important to revise some aspects of the theory of the *simple harmonic oscillator*. Consider a particle moving in one dimension subject to a restoring force directed towards the origin $x=0$, with components in the x direction proportional to x. The dynamical equation is

$$\frac{d^2 x}{dt^2} = -\omega_0^2 x \tag{4.55}$$

where ω_0^2 is the positive coefficient of the linear dependence of the restoring force on distance. For example, (4.55) exactly describes the small-amplitude motion of a pendulum, in one dimension, on a non-rotating Earth with the origin $x=0$ immediately below the point of rest of the pendulum. The restoring force comes from the weight of the pendulum bob, which can be resolved into a force along the pendulum cord (which is balanced by the tension of the cord), and a force at right angles to the cord proportional to the sine of the angle between the cord and the vertical. For small displacements this reduces to a horizontal force proportional to the horizontal displacement of the bob from the origin.

Equation (4.55) ignores friction, and we know that the solution consists of a periodic oscillation exactly determined by the initial conditions, usually involving the pendulum being pulled to one side and released from rest. It is referred to as the undamped simple harmonic oscillator. The solution obtained by releasing the oscillator from rest at non-zero x is really a transient solution, which has infinite lifetime in the absence of friction. Equation (4.55) is readily solved. The simplest mathematical formulation is

$$x = x_0 e^{i\omega_0 t} \tag{4.56}$$

where x_0 is the displacement at time $t=0$ and our solution involves only the real part of the complex numbers on the right-hand side of (4.56). Differentiating (4.56) with respect to time multiplies the right-hand side by $i\omega$:

$$\frac{dx}{dt} = x_0 i \omega_0 e^{i\omega_0 t} \tag{4.57}$$

and repeating this process reproduces (4.55). Note that the real part of the right-hand side is zero at time $t = 0$, consistent with the oscillator being released from rest. The general solution involves

$$x = x_0 e^{\pm i\omega_0 t} \tag{4.58}$$

which means that we can choose a solution consisting of a sum of $\sin(\omega_0 t)$ and $\cos(\omega_0 t)$ to fit any initial conditions.

4.5.2 Damped harmonic oscillator

If we include linear friction proportional to speed, (4.55) becomes

$$\frac{d^2 x}{dt^2} = -\omega_0^2 x - 2\omega_f \frac{dx}{dt} \tag{4.59}$$

where ω_f is a *frictional damping* frequency. The solution to (4.59) is

$$x = x_0 e^{mt} \tag{4.60}$$

and substituting (4.60) in (4.59) gives

$$m^2 = -\omega_0^2 - 2\omega_f m \tag{4.61}$$

$$m = -\omega_f \pm \sqrt{(\omega_f)^2 - \omega_0^2}. \tag{4.62}$$

If $\omega_f < \omega_0$ the solution has the form

$$x = x_0 \exp\left(-\omega_f t + i\omega_0 t \sqrt{1 - \omega_f^2/\omega_0^2}\right) \tag{4.63}$$

involving a damped oscillation of reduced frequency. Notice that as damping increases the frequency of the oscillation decreases, which is a commonly observed phenomenon in nature. If $\omega_f = \omega_0$, the oscillator is said to be *critically damped* and the frequency has effectively become zero. For $\omega_f > \omega_0$,

$$x = x_0 \exp\left[-\omega_f \left(1 \pm \sqrt{1 - \omega_0^2/\omega_f^2}\right) t\right], \tag{4.64}$$

there being two damped solutions and the solution with the smaller negative exponent, i.e., $\exp\left[-\omega_f \left(1 - \sqrt{1 - \omega_0^2/\omega_f^2}\right) t\right]$ is the longest lived.

4.5.3 Forced damped oscillator

If we apply a forcing function $F(\omega)$ of angular frequency ω to the damped harmonic oscillator,

$$\frac{d^2x}{dt^2} = -\omega_0^2 x - 2\omega_f \frac{dx}{dt} + F(\omega), \tag{4.65}$$

the unforced solution discussed previously will be complemented by a so-called *particular integral* corresponding to a forced response. We can assume that this response will occur at frequency ω since the system is linear. Hence, writing

$$F(\omega) = Ae^{i\omega t}$$
$$x = Be^{i(\omega t - \phi)} \tag{4.66}$$

where ϕ is a phase lag between the forcing function and response, and so substituting in (4.65),

$$Be^{i\phi} = \frac{A}{-\omega^2 + \omega_0^2 + 2i\omega_f\omega}, \tag{4.67}$$

i.e.,

$$B = \frac{A/\omega_0^2}{\sqrt{\left(1 - \omega'^2\right)^2 + 4\left(\omega_f'\omega'\right)^2}}$$
$$\phi = \tan^{-1} \frac{2\omega_f'\omega'}{1 - \omega'^2} \tag{4.68}$$
$$\omega' \equiv \omega/\omega_0$$
$$\omega_f' \equiv \omega_f/\omega_0.$$

When ω_f/ω_0 is small, the amplitude of the response is maximal near $\omega = \omega_0$ (as we illustrate in Figure 4.5). For this reason ω_0 is called the *resonance frequency* of the oscillator.

To keep a real harmonic oscillator (or pendulum) in motion it must be given a periodic "push" (a good example is the push that we give to a child's swing). The timing of the push is not critical, because the response of the pendulum is so peaked at the resonance frequency that its response at that frequency dominates all other motion. This is a very important scientific principle that is responsible for many processes in the physical world around us.

A coastal basin has many degrees of freedom and is subject to many types of internal forces, and it behaves as a set of harmonic oscillators in terms of its response to external forcing. So when we analyze its frequency response, we see sets of peaks each of which resembles the peak in Figure 4.5.

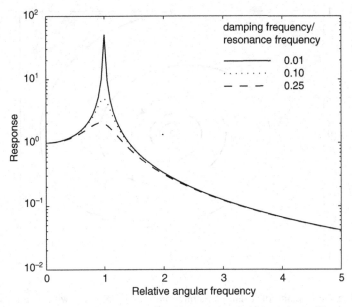

Figure 4.5 The relative (forced) response of a damped simple harmonic oscillator to forcing at various frequencies (relative to the resonant frequency of the oscillator) for various values of the ratio of the damping frequency to the resonance frequency. As the damping frequency decreases, so the response becomes increasingly sharply peaked at the resonant frequency.

4.5.4 Phase planes

Equation (4.59) is also susceptible to a type of modeling used elsewhere in this text known as *phase plane modeling*. Let us non-dimensionalize t by $1/\omega_0$ and x to some characteristic amplitude x_0,

$$\frac{d^2x}{dt^2} = -x - 2\omega_f \frac{dx}{dt}, \tag{4.69}$$

and ω_f is now measured relative to ω_0. Let us write the time derivative of x, i.e., the velocity, as y:

$$y = \frac{dx}{dt}, \tag{4.70}$$

so that (4.69) becomes

$$\frac{dy}{dt} = -\omega_0^2 x - 2\omega_f y$$
$$\frac{dx}{dt} = y. \tag{4.71}$$

Basic hydrodynamics

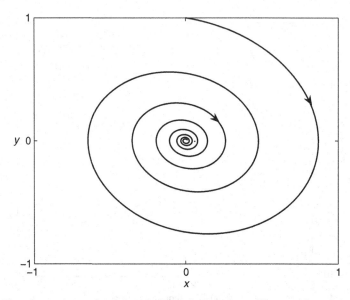

Figure 4.6 The phase plane of a damped simple harmonic oscillator with a ratio of damping frequency to resonance frequency of 0.1. The variables x and y are defined in the text and refer to the displacement and speed of the particle represented by the oscillator. The figures shows only one trajectory for the particle. All trajectories have the same shape and spiral into the point $x = 0$, $y = 0$.

We can solve (4.71) in a *phase plane* (Figure 4.6) that consists of a two-dimensional space with two axes called x and y. Each *state* of the system is described by the value of x and y, i.e., (x, y) which is a point on the phase plane, and tells us the position of the particle x and its velocity $y = dx/dt$. Equation (4.71) tells us the rate of change of x and y with time t. So if we take a small change in time dt, we move to a new point $(x + dx, y + dy)$ given by

$$y \rightarrow y + dt(-\omega_0^2 x - 2\omega_f y)$$
$$x \rightarrow x + y dt$$

$$(4.72)$$

and since we know the value of (x, y), we can find the new (x, y). Table 4.1 shows a simple code for the phase plane of a simple harmonic oscillator. Equation (4.71) is said to be autonomous in time, which means that time t does not enter explicitly, but is just a parameter that varies along the path of the particle. If we introduced a forcing function on the right-hand side of the first of (4.71) which is a function of time t, the system would be non-autonomous, and the change in x and y in time dt would depend not only on x and y but also on the value of t. For an autonomous system we can eliminate t and for example, (4.71) is equivalent to

$$\frac{dy}{dx} = -\frac{\omega_0^2 x + 2\omega_f y}{y}.$$

$$(4.73)$$

Table 4.1. *Matlab code for phase plane of a simple harmonic oscillator*

```
xinitial = 0; yinitial = 1; t = 0;
dt = 0.01; wf = 0.1; xseries = xinitial; yseries = yinitial; x = xinitial; y = yinitial;
while t < 45;
        ynew = y − dt *(x + 2 * wf * y);
        xnew = x + dt * y;
        t = t + dt;
        xseries = [xseries; x];
        yseries = [yseries; y];
        x = xnew; y = ynew;
end; figure; plot(xseries, yseries)
```

4.5.5 Non-linear oscillators

The code in Table 4.1 can easily be modified to change the nature of the restoring force. For example, changing the value of y_{new} to $y_{new} = y − dt *(x * abs(x) + 2 * w_f * y)$ corresponds to a restoring force that increases as the square of displacement, i.e., (4.69) is replaced by

$$\frac{d^2x}{dt^2} = -x|x| - 2\omega_f \frac{dx}{dt}. \tag{4.74}$$

The phase plane for (4.74) is shown in Figure 4.7 for the special case of an undamped oscillator.[4] A very noticeable difference between this figure and Figure 4.5 is the flattening of the *trajectory* (x, y), so that y has comparatively less variation through much of the cycle than for a simple harmonic oscillator and seems to switch comparatively suddenly from a positive to a negative value. A powerful tool in modeling is that we are able to construct extreme situations as an artifact to aid our understanding of processes. So, in Figure 4.8 we show a model of an oscillator in which the restoring force is increased by the sixth power of displacement. This greatly accentuates the flattening of the trajectory (top right panel), and in the lower right panel we see the time series of x and y. These panels can be compared with those on the left-hand side of the figure, which show the model results for a simple harmonic oscillator. The bottom right panel demonstrates that y switches suddenly from 1 to –1, causing x to vary almost linearly with time either positively or negatively. The reason for this behavior is that the restoring force in (4.72) can be viewed as the gradient of a potential $V(x)$. A damped particle within this potential comes to rest at $x = 0$. For a simple harmonic oscillator this potential is quadratic, and most real potential wells in nature are quadratic for small x as we would expect for a potential which is continuous and

---

[4] A good introduction is Minorsky (1974).

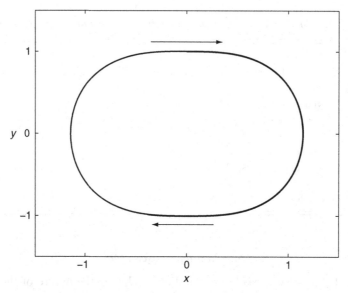

Figure 4.7 The phase plane for a undamped harmonic oscillator with the usual linear restoring replaced by a quadratic restoring force. The variables x and y are defined in the text and refer to the displacement and speed of the particle represented by the oscillator. All trajectories have the same shape and are closed orbits (due to the lack of friction). Notice that the speed remains relatively constant for a large part of the cycle and then changes direction very quickly.

differentiable at $x=0$. However, many potentials are singular at the equilibrium point. The examples provided here show potentials that are illustrated in Figure 4.8. Equation can be solved for $y(x)$ using V, i.e.,

$$y^2 = 2 - V(x) \tag{4.75}$$

where we have chosen $y=1$ at $x=1$ and $V(1)=1$. This gives $y = \pm\sqrt{2}$ at $x=0$. Let us take $V = x^{n+1}$. For large n, y decreases slowly as x increases from 0 with an increasingly rapid change in sign of y as $x =$ passes through unity. A more accurate representation of a non-linear oscillator is to assume that the potential is quadratic at small displacements and becomes increasing non-linear at large displacements (Figure 4.9), as for example the restoring force

$$\frac{d^2 x}{dt^2} = -x\left(1 + |x|^6\right) - 2\omega_f \frac{dx}{dt}. \tag{4.76}$$

As a further example consider the case of a simple harmonic oscillator with simple sinusoidal forcing. This must be damped, and we consider two types of damping: one is simple linear damping as in (4.69) and the other is so-called quadratic damping in which the damping force is proportional to the velocity squared, so that (4.69) becomes with sinusoidal forcing at frequency ω_a

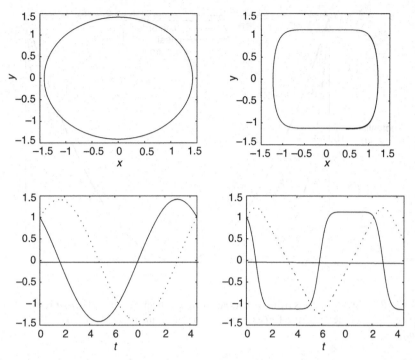

Figure 4.8 Comparison of the phase planes for undamped harmonic oscillator with a simple linear restoring force and a highly non-linear restoring force. The two panels on the left-hand side show a simple harmonic oscillator: the upper panel is the phase plane as in Figure 4.6 but without damping; the lower panel gives the time series of x (full line) and y (broken line). The two panels on the right-hand side show corresponding features for an undamped oscillator with restoring force increased by the sixth power of displacement. Notice that the very non-linear oscillator *flips* the direction of its motion at almost constant speed.

$$\frac{d^2x}{dt^2} = -x - 2\omega_f \frac{dx}{dt}\left|\frac{dx}{dt}\right| + \sin(\omega_a t). \tag{4.77}$$

Note that we must *not* write the damping force literally as $(dx/dt)^2$ because the force must be directed in the opposite direction to the velocity. Equation (4.77) provides a phase plane:

$$\frac{dy}{dt} = -\omega_0^2 x - 2\omega_f y|y| + \sin(\omega_a t)$$
$$\frac{dx}{dt} = y. \tag{4.78}$$

Equation (4.78) is not strictly autonomous and the trajectories do depend on the value of t at the initial point of a trajectory, but this is not important once the transients have decayed away. Figure 4.10 shows the phase planes and times series

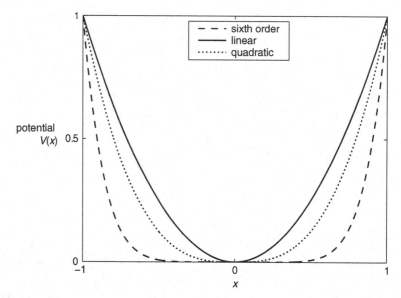

Figure 4.9 Potential functions from which the linear, quadratic, and sixth order restoring forces can be derived. As the restoring force becomes more non-linear so the potential becomes more *flat-bottomed*.

for linear and quadratic damping. In both cases the forcing occurs at the resonance frequency.

4.5.6 Square root singularity

Quadratic damping has a similar effect to non-linearity in the restoring force: y remains almost constant through a large part of the cycle. Notice in particular that the curve for y crosses the time axis *almost* at right angles (dy/dt is close to divergence). The reason for that divergence is that there is close to balance between the forcing function and damping force, so that the velocity y is *slaved* to the forcing function. Hence, y becomes almost proportional to the square root of the forcing function, so its derivative is about equal to the inverse root of the forcing function; this is called a *square root singularity* and is important to our studies of tidal jets in Chapter 8.

If we think of y as the current in a basin and x as the horizontal displacement of a particle, we can model the tidal displacement by an equation of type (4.69). In such a system it is possible that the restoring force is created by a balance of pressure gradient and bottom friction which may be quadratic. Consequently the tidal current can have an interesting time dependence: the tide flows with almost constant strength in one direction, and then suddenly switches to the opposite direction with the same strength. A boat moored in such a flow pulls against the mooring in one direction, and then suddenly becomes slack and the boat changes direction to set in a current of opposite

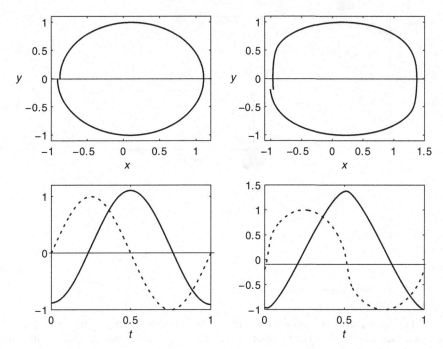

Figure 4.10 Phase planes and time series for a damped harmonic oscillator with sinusoidal forcing and linear restoring force. The two panels on the left-hand side apply to linear damping and those on the right to quadratic damping: the upper panels are the phase planes as in Figure 4.8 and lower panel are the time series of x (full line) and y (broken line). Only one cycle is shown in the phase plane; the ends of the trajectory do not meet and instead the trajectory would spiral inward to the point $x = 0$, $y = 0$.

direction. This is in contrast to a boat moored in a sinusoidally varying current which gradually changes direction.

4.6 Effects of a rotating Earth

4.6.1 Coriolis force

The Coriolis force is an apparent force that we see in coordinate systems that are attached to the Earth and are due to the rotation of the Earth around its axis and to the shape of the Earth. The Coriolis force acts to the right of the motion of a particle in the northern hemisphere and to its left in the southern hemisphere. It is proportional to the speed of the particle. Models that ignore the Coriolis force are called *irrotational models* (since they ignore all effects of the rotation of the Earth); they are often used in shallow water.

Gaspard-Gustave de Coriolis (1792–1843) showed that the effective force in the horizontal, now called the *Coriolis force*, can be written as the vector cross-product

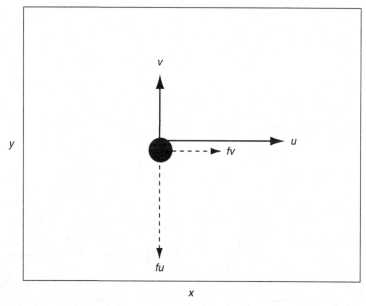

Figure 4.11 Components of Coriolis force for a particle with velocity components (u, v) in the (x, y) plane are (fv, $- fu$). The value of f is positive in the northern hemisphere (illustrated here).

$$F = -2\boldsymbol{\omega}_\phi \times \mathbf{v} \tag{4.79}$$

$$\boldsymbol{\omega}_\phi = \boldsymbol{\omega} \sin \phi \tag{4.80}$$

where v is the velocity in the horizontal plane, $\boldsymbol{\omega}$ is the angular rotation of the earth about its axis, ϕ the latitude of the particle, and the direction of $\boldsymbol{\omega}_\phi$ is the local vertical. A cross-product acts at right angles to both of the constituent vectors and is normal to the plane containing those vectors, and so the Coriolis force is horizontal at right angles to the vector v, while $\boldsymbol{\omega}_\phi$ is the component of the Earth's rotation in the vertical at latitude ϕ. Notice there is no Coriolis effect at the equator ($\phi = 0$), in some sense equivalent to the cylindrical Earth.

Let us include the Coriolis force in the two-dimensional dynamic equations assuming no pressure gradient or other forces. Using Figure 4.11,

$$f \equiv 2\boldsymbol{\omega}_\phi \tag{4.81}$$

$$\frac{\partial u}{\partial t} = fv \tag{4.82}$$

$$\frac{\partial v}{\partial t} = -fu. \tag{4.83}$$

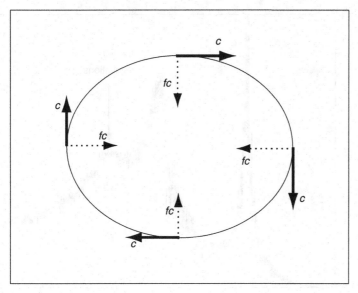

Figure 4.12 Illustration of the Coriolis force fc acting towards the center of an inertial circle around which travels a particle of speed c in the northern hemisphere (in the southern hemisphere motion is anticlockwise).

4.6.2 Inertial oscillations

Differentiate (4.82) with respect to t:

$$\frac{\partial^2 u}{\partial t^2} = f\frac{\partial v}{\partial t}, \tag{4.84}$$

and use (4.84) for $\partial v/\partial t$:

$$\frac{\partial^2 u}{\partial t^2} = -f^2 u, \tag{4.85}$$

a solution of which is

$$u = c\cos(ft) \tag{4.86}$$

$$v = -c\sin(ft) \tag{4.87}$$

which describes a particle traveling in a circle, and note that the direction of travel depends on the sign of f. The phase in (4.86) and (4.87) has been chosen so that so that at time $t = 0$, the particle has a positive component of velocity u along the x axis and zero component v along the y axis. With increasing time, u decreases and v becomes increasingly negative (for positive f). This means that the particle is moving counterclockwise starting at the top of the orbit, shown in Figure 4.12. We could suppose that

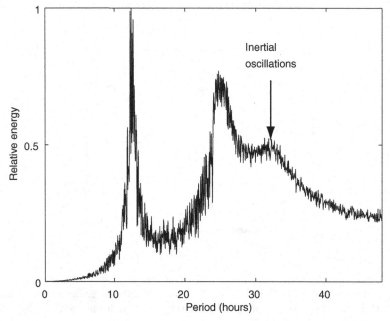

Figure 4.13 Frequency analysis of the record from a current meter deployed at a latitude of $22°$. Notice the inertial response at a period of 32 h indicated by arrow.

u and v are independent of position in the coastal basin, and so from the viewpoint of particle trajectories, all particles are traveling in circles of identical radius (but each particle orbit has a different center). The basin is clearly not in a steady state. Because u and v are independent of position in the basin, it follows that there are no non-linear advection terms and hence, this non-steady state is an exact solution of the dynamical equations. It is to be emphasized that the Euler and Lagrangian approaches to the dynamics of basins will give identical solution, but only under conditions in which the non-linear advections in the Euler equations vanish. That is true in the present case (albeit for a solution which is almost trivial). When we look at solutions of the dynamic equations in the absence of friction we can always add an inertial oscillation. The magnitude of that oscillation depends on the initial conditions.

Inertial oscillations take place at the *inertial frequency* that is entirely characteristic of the latitude at which the particle is located, and in a rotational sense determined by the hemisphere. The record from any instrument deployed in the ocean, when subjected to a frequency analysis, may show a peak at the inertial frequency. In many coastal basins friction is too strong for such a peak to exist (friction exceeds that required for critical damping), and otherwise the magnitude of the peak is controlled by the amount of frictional damping in the basin (Figure 4.13). The inertial peak is really a consequence of the response of the basin to a forcing, such as that due to the wind or tides. Interestingly, numerical models show initial oscillations to a degree that reflects the algorithms used for friction, and this is often not a fruitful area for comparison with field data. The period of inertial oscillations is 12 h at the poles (not 24 h as for the Foucault pendulum), and increases with decreasing latitude to become 24 h at a latitude of 30° and infinitely long at the equator. The paths of particles performing inertial oscillations which extend over a range of latitudes are not circular, because of the variation of the Coriolis force with latitude. Apart from the inertial period we also refer to a quantity called the *inertial time*, which is the inertial period divided by 2π, or the time for an inertial oscillation to move through 1 radian, i.e., the inverse of the angular frequency.

4.6.3 Inertial circle

Suppose a water particle is given a sudden velocity and acted upon only by the Coriolis force, and assume that it is in the northern hemisphere (Figure 4.12). If the initial speed of the particle is c, there will be a force of magnitude fc, per unit mass of the particle that acts at right angles and to the right of the motion. The parameter f is called the Coriolis parameter and has the dimensions of inverse time. Notice that the Coriolis force always acts at right angles to velocity, and therefore the speed of the present particle is constant in time. Furthermore, since the acceleration is at right angles to the velocity and of constant magnitude, the trajectory of the particle is a circle. This is the natural tendency of particles in the ocean on a rotating Earth. The radius of the circle only depends on the speed of the particle. The same trajectory would occur for a particle on land, except that it is difficult to arrange matters so that only the Coriolis force acts on such a particle. Usually on land, friction is more important. The ratio of friction to Coriolis force is called the *Ekman number*, and in order to see these circular paths we need to have a low Ekman number.

A particle of velocity c in a circle of radius r has an acceleration towards the center of c^2/r. This acceleration equates to the force per unit mass,

$$fc = \frac{c^2}{r},$$ (4.88)

and so the radius of the circle is proportional to the initial speed c,

$$r = \frac{c}{f},$$ (4.89)

giving the time for the particle to travel around the circle T_{inertial} as

$$T_{\text{inertial}} = \frac{2\pi r}{c}$$ (4.90)

and using (4.89) and (4.90) this becomes

$$T_{\text{inertial}} = \frac{2\pi}{f}.$$ (4.91)

The value of the Coriolis parameter f is a function of latitude ϕ from (4.91) and (4.93):

$$f = 1.45 * 10^{-4} \sin \phi. \qquad (4.92)$$

The above analysis assumed that the particle was in the northern hemisphere. In the southern hemisphere f is negative so that the Coriolis force acts to the left of the velocity, and a particle moves around an inertial circle counterclockwise (as opposed to clockwise in the northern hemisphere).

The Earth spins around its axis with a period of one sidereal day (23.93 h), which we denote by T_{day} (it is not *exactly* 24 h since our conventional day of 24 h is the time for the Sun to reach its zenith each day, and this is affected by the orbit of the Earth around the Sun). Because of the factor of 2 in (4.79), the inertial period at the pole is $T_{\text{day}}/2$, or close to 12 h and at other latitudes it is

$$T_{\text{inertial}} = \frac{T_{\text{day}}/2}{\sin\phi} \qquad (4.93)$$

which is about T_{day} at a latitude of 30° and tends to infinity at the equator; for example, at a latitude of 10° it is 2.9 T_{day} and at 1° it is 29 T_{day} (so the divergence of T_{inertial} with ϕ is fairly weak). The inertial period is exactly 12 h at a latitude of 85.78°, which is about 500 km from the Pole.

4.6.4 Foucault pendulum

We turn now to a pendulum whose bob is free to move in two dimensions, x and y. We analyze the motion in the x and y directions in a similar manner to our previous analysis for the one-dimensional motion in Section 2.14. On a *non-rotating* Earth, the motions in the x and y directions would be independent and (4.55) simply replaced by

$$\frac{d^2 x}{dt^2} = -\omega_0^2 x$$
$$\frac{d^2 y}{dt^2} = -\omega_0^2 y \qquad (4.94)$$

with solutions

$$x = x_0 e^{i\omega_0 t}$$
$$y = y_0 e^{i\omega_0 t} \qquad (4.95)$$

with the analysis in the x and y directions being identical to that which we presented for the x axis alone in (2.14); that case really just reduces to a one-dimensional problem, since we could choose the plane of motion of the pendulum to be the x axis.

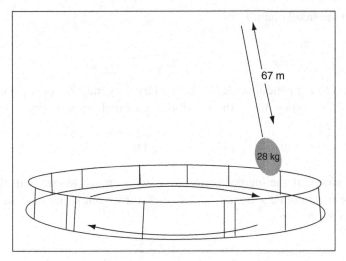

67 m

28 kg

Figure 4.14 Drawing of a Foucault pendulum. This classical experiment demonstrates that the Earth rotates so that the floor and observers are rotating, while the plane of the pendulum remains fixed in space. The arrows show the apparent motion of the pendulum for the northern hemisphere.

In the middle of the nineteenth century Foucault[5] experimented with a means of observing the rotation of the Earth through the motion of a pendulum which could be sustained in motion over many hours. This culminated, in 1851, with the 28-kg *Foucault pendulum* suspended experimentally from the 67-m-high dome of the very prestigious Pantheon in Paris, France (Figure 4.14). The pendulum had a period of 16.4 s and the plane of its motion, with respect to the Earth, rotated slowly clockwise due to the rotation of the Earth. We shall see that the Foucault pendulum remains in a fixed plane relative a non-rotating local reference plane, and therefore appears to move through $360°$ in a period of $2\pi/\omega_\phi$, which is twice the inertial period, i.e., 24 h at the poles. The rotation is in the same sense as an inertial rotation, but at only half the angular speed.

On a rotating Earth, the Coriolis force must be included and so (4.94) becomes

$$\frac{d^2x}{dt^2} = -\omega_0^2 x + f\frac{dy}{dt}$$
$$\frac{d^2y}{dt^2} = -\omega_0^2 y - f\frac{dx}{dt}$$

(4.96)

where f is the Coriolis parameter. We look for solutions of form e^{mt} and substituting in (4.96)

$$\left(m^2 + \omega_0^2\right)^2 = -f^2 m^2$$
$$m^4 + \left(2\omega_0^2 + f^2\right)m^2 + \omega_0^4 = 0$$

(4.97)

---

[5] Jean-Bernard-Léon Foucault (1819–68), a French physicist best known for his pendulum experiment, but who also made an early measurement of the speed of light, and invented the gyroscope. The Foucault Crater on the Moon is named after him.

and defining the small quantity ε

$$\varepsilon \equiv \frac{f}{2\omega_0} \qquad (4.98)$$

which is equal to the pendulum period divided by the rotational period (of order the ratio 20 s to 24 * 3600 s $\sim 10^{-4}$), the solution of the quadratic equation in m^2 is

$$m^2 = \omega_0^2 \left[-(1 + 2\varepsilon^2) \pm \sqrt{(1 + 2\varepsilon^2)^2 - 1} \right]. \qquad (4.99)$$

We assume now that the inertial period $2\pi/f$ is very much longer than the natural period of the pendulum $2\pi/\omega_0$, i.e., $\varepsilon \ll 1$ so that if we retain only the highest order terms in ε, (4.99) becomes

$$m^2 = -\omega_0^2 [1 \mp 2\varepsilon] \qquad (4.100)$$

and substituting back (4.98)

$$m = i\omega_0 [1 \pm \varepsilon] \qquad (4.101)$$

where we have approximated the square root in (4.101) by the first order expansion in ε. Hence the solution for a pendulum on a rotating Earth is

$$x \sim \exp^{[i(\omega_0 \pm f/2)t]}. \qquad (4.102)$$

Using (4.102) we start the pendulum at $t = 0$ swinging parallel to the x axis:

$$x \sim e^{i\omega_0 t} \cos\left(\frac{1}{2}ft\right) \qquad (4.103)$$

and substituting (4.103) in the first of (4.96) we have

$$f\frac{\partial y}{\partial t} = \omega^2 x + \frac{\partial^2 x}{\partial t^2} = -\left(\frac{f}{2}\right)^2 e^{i\omega_0 t} \cos\left(\frac{f}{2}t\right) \qquad (4.104)$$

and integrating with respect to t,

$$y = e^{i\omega_0 t} \sin\left(\frac{f}{2}t\right), \qquad (4.105)$$

showing that the pendulum continues to swing with an angular frequency of ω_0 plus a rotation in the (x, y) plane. The time for the plane of the pendulum to turn through 360° is $2\pi/(f/2) = 4\pi/f$, which is twice the inertial period and equal to $2\pi/(\omega_\phi)$. So as we move toward the equator, the plane rotates more slowly and at the latitude of Paris rotates 270° in 24 h. The Foucault pendulum is one of the most obvious demonstrations of the rotation of the Earth, and is installed as a scientific demonstration in many major buildings around the world (but not *too* near the equator). Often, magnets concealed beneath the floor provide a stimulus to keep the pendulum swinging. This is equivalent to the occasional *push* that we give a child on a swing, and is effective because the response of the basic oscillator constituted by the pendulum is sharply peaked at w_0 as we saw in section 2.14. The Foucault pendulum is recognized as a truly important scientific experiment, because it demonstrated in a very direct way that the Earth is rotating and not stationary.

Figure 4.5 shows the response curve for a damped harmonic oscillator. The Foucault pendulum reduces to that case in the absence of rotation and damping. If we spectrally analyze (4.105) we obtain a response curve that is the *analogue* of Figure 4.5, except that there is a peak at angular frequency of $f/2$ (not f) in addition to the peak at ω_0.

Inertial oscillations are a natural consequence of the rotation of the Earth around its axis, and result from the apparent motion of the Earth around a vertical axis drawn at any point on the Earth. The name *inertial oscillations* is perhaps a little obscure, and describes the need for a particle to obey Newton's laws of motion in what we call *an inertial frame*, given that it is acted on by Earth's gravity and the reaction of the Earth's surface to the associated weight. An inertial frame is simply a reference system in which Newton's laws of motion are obeyed. To a good first approximation, a reference frame fixed relative to the Sun is an inertial frame. A local reference frame fixed relative to the spinning Earth is *not* an inertial frame, and the reader is acted on by several forces in order to remain locally stationary, to continue reading this chapter. Once any of the particles around us are given a local velocity, the illusion that we are living in an inertial frame vanishes. The particle will (in the total absence of friction) appear to move in a circular orbit called an inertial cycle and perform inertial oscillations. We do not usually see these oscillations in our daily lives, because we live in an environment dominated by friction, and (as we saw for the simple harmonic oscillator) friction damps away the motion before a single oscillation is performed. By contrast, the deep ocean is a low friction environment and inertial oscillations are observed. This difference in friction is only true for large bodies of water and, of course, a person swimming in the ocean experiences far more friction than when walking to the ocean for that swim. A particle executing inertial oscillations is *not* (as sometimes stated) the dynamic counterpart of a particle in *free space*, tending to move in a straight line in the inertial reference frame external to the Earth. That *is* a motion associated with Foucault pendulum, but is not true for inertial oscillations. The physical reason for particles tending to perform inertial oscillations is also treated in a later section.

4.6.5 Inertial circle from space

It is worth considering in detail the magnitude of the Coriolis force, and in particular the reason why the inertial period is not 24 h at the poles (since a particle fixed in space would appear to rotate around the poles with that period). The details of the dynamics are given below. To understand the force, we need firstly to understand some aspects of the rotation of the Earth.

Consider yourself standing vertically on the Earth in the northern hemisphere with your arms stretched out horizontally with your right arm towards the North Pole and the left directly away from it. Although you may not realize it, the rotational speeds of points on the Earth below your right and left hands are different. If you think about the Earth rotating on its axis, you can see that the rotation speed of points at the poles is zero, and rotational speed at the equator is maximal. The reason is that the rotational speed increases with distance from the rotational axis of the Earth. So, the ground below your left hand is rotating more slowly than below your feet, and that below your right hand more rapidly than that below your feet. When you are in the northern hemisphere, it appears to you the ground below your right hand is moving away from you and that below your left hand is moving toward you. So the Earth appears to be rotating counterclockwise around you. When in the southern hemisphere, the rotation is clockwise. The effect is due to the change of tangential speed with position, and so this rotational effect disappears at the equator where both hands sense the same tangential speed on either side of the equator. It is maximal at the poles: stand at either of the poles and the Earth rotates around you in 24 h. The period of rotation increases as you move towards the equator and becomes infinite at the equator.

Let us start by considering Newton's equation of motion on a rotating plane. The acceleration seen by an observer in a rotating polar coordinate system (with angular speed $\boldsymbol{\omega}$ about a normal axis) for a particle moving at velocity $\boldsymbol{v}$ in that rotating system provides an apparent force given by the vector cross-product

$$F = -\boldsymbol{\omega} \times \boldsymbol{v}. \tag{4.106}$$

A particle at rest in a stationary coordinate system appears to an observer in the rotating coordinate system to rotate around the axis with angular speed $\boldsymbol{\omega}$. So, if the coordinate system is rotating counterclockwise, ω is positive and the observer sees the particle as rotating with angular speed $-\boldsymbol{\omega}$, i.e., clockwise. If the particle is given a constant velocity in the stationary frame, it appears in the rotating system to move in a series of circles with an additional constant velocity. This motion is also easily determined by solving (4.106) which is Newton's law of motion in the rotating system. This system is the basis of laboratory experiments called *rotating tables*. Particles in the rotating coordinate system tend to execute circular motions with angular speed $-\omega$. This is the basic difference from a stationary coordinate system, in which particles given an initial acceleration move in a straight line. If we consider the dynamics of a particle, it clearly makes no difference to its actual trajectory whether we use a stationary or rotating coordinate system. The choice of coordinate system is, in a sense, arbitrary (although in practice one system may be far easier to use than the other).

Suppose we look at the dynamics of a particle on Earth. It is now necessary to take into account the force of gravity, so that a set of coordinates fixed on Earth have very great advantages. Consider, firstly, a hypothetical Earth that consists of a cylinder rotating around its axis. What are the equations of motion for people living on the surface of this cylinder? In a plane tangential to the surface of the cylinder, which is their local *horizontal* plane, there is no additional force equivalent to (4.106). However, there is a force in their local *vertical* direction, i.e., normal to the tangential plane, but this only changes the effective weight of particles. We shall see the same result is true for people who live at the equator on our own (nearly) spherical Earth.

If our Earth were truly spherical, we would find ourselves tending to slide towards the equator. In the molten Earth this force (which is the *horizontal* component of the centrifugal force) produced an Earth slightly flattened at the poles and bulged at the equator. Consider the local horizontal plane at any point (a plane tangential to the surface), and define the local *vertical* as normal to this plane. The gravitational force acts through the surface directed towards the center of the Earth, but this force is not parallel to the *local vertical*. Instead, we can resolve this force into local *vertical* and *horizontal* components. The centrifugal force acts outward and also has components in the vertical and horizontal planes, and finally there is a reaction on the surface of the Earth to the weight of the particle. The *vertical* component of true gravity and the vertical component of the centrifugal force are combined to form an effective weight that is exactly matched by the reaction; the slope of the *horizontal* plane is such that the component of gravity and centrifugal force exactly cancel.

So, we do not find ourselves sliding equatorward and provided we keep still, we see no effect of the Earth's rotation. For this reason, we humans had no idea that the Earth was rotating until we tried to understand the dynamics of our Solar System and the stars. For a particle that is stationary on our Earth, Newton's law appears to be exactly satisfied in the sense that there is a balance of forces. When a particle moves relative to the rotating *horizontal* plane, it upsets this balance and this produces a pseudo force. Basically there are two components to the Coriolis force on our Earth. The first is the effect of the rotating coordinate system as we found for a rotating plane (the *flat Earth* case), and the second is the change in the effective rotational speed with latitude. It is easy to visualize how a particle moving towards either pole will change its rotation speed, and hence accelerate at right angles to its original velocity in an easterly sense (because the Earth is rotating west to east). Similarly, if a particle moves parallel to the equator at any latitude (and say in an easterly direction) it changes its centrifugal force, and therefore creates a force towards the equator (and a change in its weight).

The factor of 2 in (4.79) is due to there being two components to the Coriolis force. A consequence of this factor is that a particle on Earth with no other forcing tends to perform circular motions at angular speed $2\omega_\phi$ (*not* ω_ϕ), which we call inertial oscillations. In the previous section we showed that these oscillations are obtained by solving (4.79). It is easy to understand the dynamics of these oscillations. Consider a particle close to the North Pole. Newton's law is exactly satisfied if the particle is stationary. Viewed from a stationary coordinate system in space, this particle is rotating from west to east at the angular rotational speed of the Earth, i.e., looking down from above

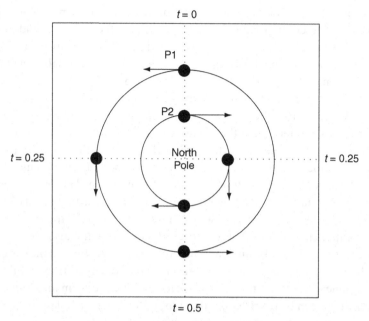

Figure 4.15 The path of two particles at the North Pole as viewed from above in a reference frame fixed in space. Particle P1 is at rest on the Earth, and so moves *anticlockwise* in the fixed frame with a period of one sidereal day. Particle P2 is moving *clockwise* in the fixed frame at the same speed as P1. Dynamically both particles are identical, and on a frictionless Earth would experience the same set of forces. An observer on Earth describes particle P2 as executing an inertial oscillation. The figure shows the position of the two particles at time $t=0$, ¼ day, ½ day. P2 meets P1 every half sidereal day so that is its period of oscillation relative to an observer on Earth who ascribes this behavior to the Coriolis force. Particles performing inertial oscillations are not traveling in straight lines in a fixed frame.

at the pole the particle is rotating counterclockwise (Figure 4.15). Suppose that we alter the particle's sense of rotation to clockwise but maintain the same angular speed. Newton's law is still exactly satisfied because the centrifugal force does not change direction with the sense of rotation. In the coordinate system of the Earth, the particle is now performing a circular path at angular speed $-2\omega_\phi$. We see that one of the factors of ω_ϕ comes from the centrifugal effect, and the ω_ϕ factor from the rotation of the coordinate system. Hence, there are two equally satisfactory states for the particle: either stationary or rotating *clockwise* with angular speed $2\omega_\phi$. Of course the latter state (the inertial oscillation) is damped by friction with the Earth.

4.6.6 Ekman number

The *Ekman number* is named after the famous Swedish theoretical and experimental oceanographer V. Walfrid Ekman (1874–1954). It measures the importance of the Coriolis force relative to friction (or viscosity) in the same environment. The magnitude of the inertial oscillation signal is reduced by frictional damping, and in extreme damping the effects of the Coriolis force are virtually zero. A coastal basin can be viewed as containing a simple damped harmonic oscillator at the inertial frequency. In situations where there exists continuous forcing, such as those due to the tides and winds, the inertial oscillator is constantly stimulated but its response depends on frictional damping (as well as the frequencies of the forcing functions). To evaluate the comparative effects of Coriolis and frictional forces, we use a measure called the *Ekman number*, E, which is defined in a variety of ways in the literature, according to the formulation of friction, it basically measures the strength of friction (or more generally frictional or viscous *stress*) relative to the Coriolis force and for a frictional damping time t_d, has the form

$$E = \frac{1}{t_d f} \qquad (4.107)$$

so that rotational effects are only felt if $t_d \gg 1/f$. We can reformulate (4.107) for the case of quadratic friction (4.53) by making

$$t_d \sim \frac{hu}{\tau_b} \qquad (4.108)$$

being the time for mean momentum hu to be overcome by bottom stress. Using (4.53) $\tau_d = h/(C_d u)$ where u is an average current speed, (4.108) gives

$$E = \frac{C_d \bar{u}}{hf}. \qquad (4.109)$$

An Ekman number of unity provides a fuzzy demarcation criterion between systems dominated by friction, i.e., $E > 1$ and systems dominated by rotation. Our everyday world of the terrestrial environment is one with a high Ekman number, although the effects of the Coriolis force must be considered in the trajectory of rockets. The deep ocean is a low Ekman number environment, as is the atmosphere away from the immediate vicinity of the Earth's surface, and there usually develops a balance between Coriolis forces and pressure known as *geostrophic balance*. Coastal basins are midway between the extremes of the deep ocean and terrestrial environment in having Ekman numbers close to unity, and indeed one of the first steps in modeling a coastal basin should be to estimate the Ekman number. One of the important factors that distinguish models that work well in the deep ocean and in the coastal environment is the Ekman number. For example, consider a bay at latitude 30° with mean tidal current speed of 1 m s^{-1}, a mean depth of 5 m, and a bottom drag coefficient C_d of 0.0025 (sand). The inertial period is 24 h and so $f = 2\pi/(24 * 3600) = 7 * 10^{-5}$. Hence $E = 6.8$ which means that friction is dominant, with very reduced rotational effects. The bottom drag coefficient C_d in most coastal basins is higher than that for pure sand, and we would anticipate that our estimate of E is higher. However, surface movement of buoyant plumes typically experiences much reduced friction, and the Coriolis effect is often seen in the tendency of river plumes to swing to the right in the northern hemisphere or left in the southern hemisphere. However, these effects may often be caused by the flow of local alongshore currents.

4.7 Further reading

See further reading from earlier chapters. Additional texts for this chapter are:

Infeld, E. and G. Rowlands (2000). *Nonlinear Waves, Solitons and Chaos*, 2nd edn. Cambridge, UK: Cambridge University Press.

Lamb, H. (1993). *Hydrodynamics*. New York: Dover Publications.

Mellor, G. L. (1996). *Introduction to Physical Oceanography*. New York: Springer.

Schetz, J. A. and A. E. Fuhs (1999). *Fundamentals of Fluid Mechanics*. New York: Wiley-Interscience.

Stommel, H. M. (1972). *The Gulf Stream; A Physical and Dynamical Description*. Berkeley, CA: University of California Press.

Stommel, H. M. and D. W. Moore (1989). *An Introduction to the Coriolis Force*. New York: Columbia University Press.

5

Simple hydrodynamic models

5.1 Wind blowing over irrotational basin

When a wind blows over the surface of the ocean it produces a *stress* τ_w on the surface (Figure 5.1); τ_w is a force per unit area of surface and has units of $N\,m^{-2}$. Wind stress is due to the friction between the moving air and water surface and has the form

$$\tau_w = \rho_a C_w W^2 \tag{5.1}$$

where W is the wind speed $10\,m$ above the water surface, C_w is the wind drag coefficient (usually about 0.001), and ρ_a is the density of air ($\sim 1.2\,kg\,m^{-3}$). For example, if $W = 10\,m\,s^{-1}$ $\tau_w = 1.2 * 0.001 * 100 = 0.12\,N\,m^{-2}$. The use of a surface drag coefficient C_w is a simplification (since the actual roughness of the surface depends on the wave state), but is nevertheless a good starting approximation for many models of coastal basins.

5.1.1 Wind blowing over a closed basin of uniform depth

Table 5.1 shows an example of the current ($m\,s^{-1}$) in the water column for a *closed* basin that is 10 m deep. Current is shown at various times after the wind starts to blow (at $t = 0$). Notice that the surface elevation is about $\eta = 1\,m$. The table refers to currents at a particular value of x which are also shown in Figure 5.2.

Initially the water column is entirely at rest, so $u = 0$ everywhere. After a short time the surface water moves in the direction of the wind. A return current also develops in the deeper parts of the water column (this is a consequence of the basin being closed). Note that a negative sign associated with u means that the current is moving in the negative x direction. The positive x axis is chosen in the direction of the wind. After a time of 7200 s (2 h) a steady state is reached.

We can calculate the total volume flux by just adding the contributions of the currents in all the layers and multiplying by their thicknesses (which are all 1 m). Notice that the volume flux *density* is positive at the surface and negative in the lower water column, i.e., downwind at the surface and upwind at the bottom. When the total flux is non-zero and positive, i.e., the water is moving downwind, the surface must be

139

Table 5.1. *The currents (m s⁻¹) in the closed basin shown in Figure 5.1 at ten levels down through the water column. The wind stress is turned on at time* t = 0. *The table shows the way that the currents increase with time becoming positive (downwind) at the surface and negative (upwind) at the bottom with the total volume flux being initially positive (causing the surface elevation to increase down wind and decrease upwind) but becoming almost zero after a time of 5400 seconds. The data ignore seiching motions.*

	Time t (s)				
	0	1800	3600	5400	7200
Height z (m)					
1	0	0.10	0.13	0.14	0.14
0	0	0.09	0.12	0.13	0.13
−1	0	0	0.10	0.11	0.12
−2	0	0	0.06	0.08	0.08
−3	0	0	0.04	0.06	0.06
−4	0	0	0.01	0.03	0.03
−5	0	0	−0.03	−0.05	−0.05
−6	0	0	−0.06	−0.09	−0.09
−7	0	−0.02	−0.10	−0.15	−0.15
−8	0	−0.01	−0.12	−0.17	−0.20
−9	0	0	−0.03	−0.05	−0.07
−10	0	0	0	0	0
Q_x volume flux per unit width (m² s⁻¹)	0	0.16	0.12	0.02	0.00

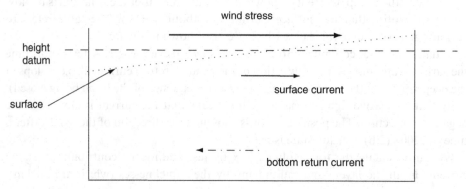

Figure 5.1 A wind blows over a coastal basin. Currents flow downwind at the surface and the water surface sets up in the downwind direction due to the surface shear stress exerted by the wind. Current returns upwind at the bottom.

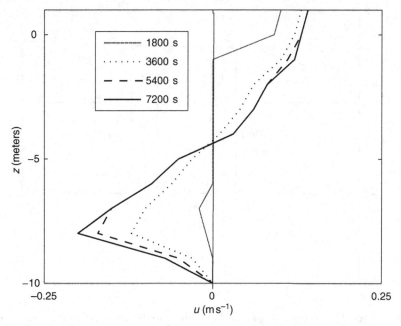

Figure 5.2 Currents in a water column of elevation approximately 1 m, at various times t, due to a wind stress turned on at time $t=0$ over a closed basin of depth 10 m as given in Table 5.1.

changing with time. This change consists of a titling of the surface in which the surface rises downwind and drops upwind. Before the steady state is achieved the currents are said to be *transient*. In the steady state the total volume flux Q_x is zero; the surface flux downwind being matched by a flux upwind in lower part of water column. The *transient* Q_x must be a function of the distance downwind, because the surface elevation is increasing in some places and decreasing elsewhere. The bottom of the water column moves horizontally in the opposite direction to that at the surface, because a pressure gradient pushes water in the opposite direction to the wind; it is a result of the increase in surface elevation with distance downwind. If we use the hydrostatic approximation (Chapter 4), pressure p at height z,

$$p(z) = g\rho(\eta - z) \tag{5.2}$$

where $\eta - z$ is the vertical length of the water column above a point at height z, ρ is the density of seawater (taken as 1026 kg m^{-3}), and g is the acceleration due to gravity (9.81 m s^{-2}). The variation of pressure with x (at any height) produces a force (per unit volume) parallel to the x axis equal to the pressure gradient $-\partial p/\partial x$ as can be seen in Figure 5.3

$$-\frac{\partial p}{\partial x} = -g\rho\frac{\partial \eta}{\partial x} \tag{5.3}$$

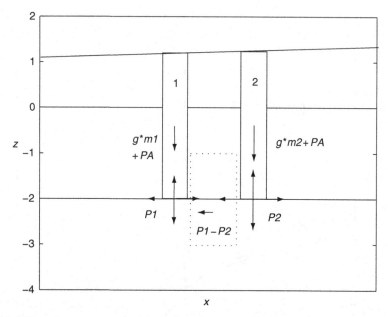

Figure 5.3 Effect of pressure varying along the x axis in producing a net force on a water particle represented by the box with dotted sides. The pressure on either side of the box is represented by the vectors of magnitude $P1$ and $P2$. In the hydrostatic approximation $P1$ and $P2$ are such as to balance the weight of the water column above $gm1$ and $gm2$ (per unit cross-sectional area) respectively where $m1$ and $m2$ are their masses and g is the acceleration due to gravity (plus the effects of atmospheric pressure represented by the symbol PA). The net force on the particle is $P1 - P2$ in the x direction (which is here negative). Analytically, we take the limit of a box of thickness dx in the x direction so that the net force involves the derivative of pressure with respect to x.

The *constancy of pressure force with z* in (5.3) is due to density being constant (and our use of the hydrostatic approximation). Notice that the force due to the pressure gradient is negative and forces water back upwind. The total force due to pressure, i.e., summed over the entire water column is

$$-\int_{-h}^{\eta} \frac{\partial p}{\partial x} dz = -g\rho(h+\eta)\frac{\partial \eta}{\partial x}. \tag{5.4}$$

This must balance the wind stress τ_w provided that the bottom stress is small, i.e., the sum of (5.4) and the wind stress is zero, i.e.,

$$\tau_w = g\rho(h+\eta)\frac{\partial \eta}{\partial x}. \tag{5.5}$$

For example, taking $g = 9.81\,\mathrm{m\,s^{-2}}$, $\rho = 1026\,\mathrm{kg\,m^{-3}}$, $\rho_a = 1.2\,\mathrm{kg\,m^{-3}}$, $C_w = 0.001$, $W = 5\,\mathrm{m\,s^{-1}}$, and $h = 100\,\mathrm{m}$, we find $\partial \eta / \partial x = 3\ 10^{-8}$, and so over a distance of 30 km the elevation of the water surface changes by 0.90 mm. If we reduce h to 1 m the change would be 90 mm. The change in elevation due to the forcing is called the *surface set-up*. We shall find that surface set-up is a very common immediate response of coastal basins to various forms of forcing.

5.1.2 Storm surge

The increase in surface elevation (the surface set-up) at the coast due to onshore storm winds over a basin is called the *storm surge* given by (5.5) as

$$\Delta\eta = \frac{\tau_w L}{g\rho h} \quad \eta \ll h \tag{5.6}$$

and using (5.1)

$$\Delta\eta = \frac{\rho_a C_w W^2 L}{g\rho h} \quad \eta \ll h \tag{5.7}$$

where L is the fetch of the wind (the distance over which the wind blows). If we take depth $h = 10$ m near the coast and consider a wind blowing onshore with speed $W = 20 \, \text{m s}^{-1}$, the slope from (5.5) is $5 \, 10^{-6}$, so taking a distance of $L = 100$ km we get an elevation change of 0.5 m. This elevation change is called the *storm surge*. Equation (5.6) does not include any effects of the Coriolis force and ignores friction (both of these forces are important in determining actual storm surges). Equation (5.7) only refers to the steady state surge. In practice, a coastal basin tends to oscillate, or *seiche*, under the influence of an imposed wind stress, and so there can be an overshoot effect. Equation (5.7) is only valid if the surface elevation change is small, and so this is often invalid for extreme surges in coastal basins.

In cases of extreme storm surge it is important to consider the magnitude of the drag coefficient C_w. This characterizes the roughness of the ocean surface and is dependent on wave conditions which may be generated by the storm or have entered the coastal basin as swell. The use of numerical models to determine storm surges considers both onshore winds and alongshore currents which produce a surge due to the Coriolis effect. An important part of that modeling is the determination of local wind fields. An example of a numerical model is *SLOSH* (Sea, Lake and Overland Surges from Hurricanes) which is a computerized model run by the US National Hurricane Center to estimate storm surge heights and winds resulting from historical, hypothetical, or predicted hurricanes. Figures 5.4 to 5.6 illustrate the storm surge from a tropical cyclone (Hurricane Frances hitting Florida in 2004).

5.1.3 Friction in wind-driven motion

If we include bottom friction, the simple balance of pressure with surface stress is no longer valid. Friction will only be present if the bottom of the water column is moving.

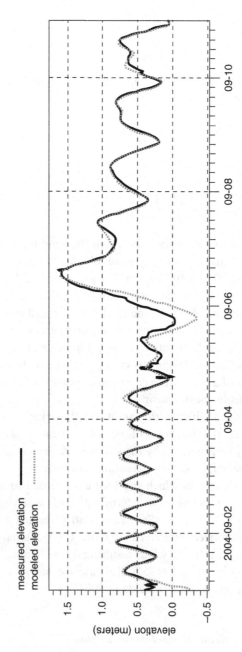

Figure 5.4 Time series of surface elevation in meters in Tampa Bay, Florida during early September 2004 (time axis shows month–day), due to the passage of tropical cyclone Hurricane Frances (see path in Figure 5.5). The full line is taken from a water level sensor run under the National Oceanic and Atmospheric Administration (NOAA) Physical Oceanography Real Time System (PORTS) and the lighter broken line is a numerical model of Tampa Bay (US Geological Survey).

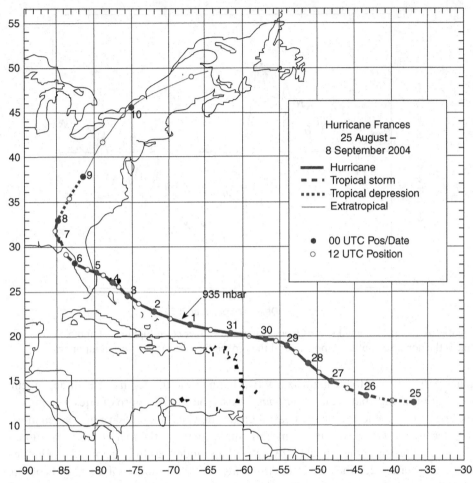

Figure 5.5 Path of tropical cyclone Hurricane Frances (see surface elevation in Tampa Bay, Florida in Figure 5.4). The various line styles used for the path indicate the category assigned to the storm and the numbers show the day of month for each position of the center of the storm. The minimum pressure was 935 mbar when the storm was at the location shown by the arrow.

The usual situation is that the bottom water moves upwind and the surface water downwind, so that the *total* volume flux is zero. This would produce a bottom frictional stress that acts against the bottom flow, i.e., in the same direction as the wind and so increases the surface set up.

5.1.4 *Wind blowing over a closed basin of varying lateral depth*

This model considers a wind stress directed along the axis of a long irrotational channel (closed at both ends) in which the depth varies in the cross-sectional (lateral)

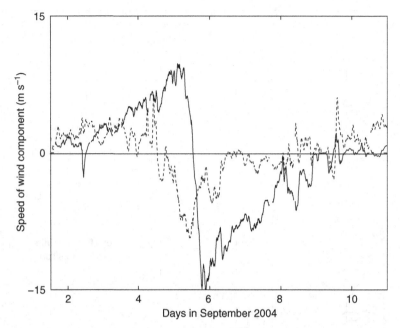

Figure 5.6 Illustrative wind speeds from Hurricane Frances (see Figures 5.4 and 5.5). The full line is the north-to-south component and the broken line the east-to-west component.

plane perpendicular to the main axis of the channel, and the depth is itself independent of position along that main *longitudinal* axis. The model approximates a common bathymetry for regions of many coastal basins, in which the circulation is much more affected by the *lateral* variations in bathymetry than by the *longitudinal* variations. Figure 5.7 illustrates the model bathymetry. This model was first proposed by Gabriel Csanady (born 1925, and one of the founding fathers of models of lakes and coastal basins) for circulation in long lakes but is equally applicable to many coastal basins.[1]

The model in Section 5.1.1 represents the special case of the present model in which the lateral cross-section has constant depth. The analysis of the flow is very similar, and we treat the present model as a sum of vertical slices taken parallel to the longitudinal axis. In each of these slices there is likely to be downwind motion at the surface and upwind near the bottom. Again, we look for a steady state solution. The steady state circulation in each of these slices is solved independently, and we assume that there is no lateral current in the bulk of the channel, although there are evidently lateral currents at the ends of the basin which ensure that the total volume flux down the channel is zero in the steady state (since the flux cannot pass out through the ends). These end regions are not included explicitly in the model, but provide a bulk condition that the total volume flux is zero across any cross-section in the steady state, with

---

[1] See for example Csanady (1982).

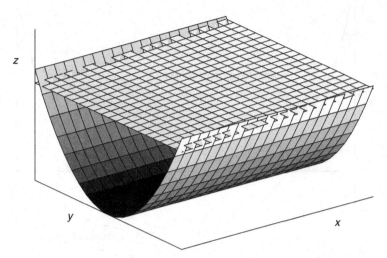

Figure 5.7 Bathymetry of a long narrow basin in which depth is only a function of the lateral variable y and independent of the longitudinal variable x. The figure illustrates the position of the water surface in the presence of forcing by a wind blowing along the x axis (from left to right).

some slices carrying current downwind and others carrying current upwind. Again, in the steady state, there is a balance between wind stress and pressure force. As a consequence of the symmetry of the channel and lack of lateral current, the lateral surface slope is zero. This means that the longitudinal pressure gradient is the same along any of the slices. However, the slices do vary in depth, and the pressure gradient has more influence in the slices that are deeper. The depth-averaged longitudinal current in each slice will be non-zero and the general rule is that the depth average current in the shallow parts of the basin are downwind, but upwind in the deeper portions. This is illustrated in Figure 5.8 and the depth average circulation is shown in Figure 5.9.

The circulatory motion seen in Figure 5.9 is called a *topographic gyre* because the water has a gyratory motion induced by the topography (the depth variation of the bottom of the basin). Figure 5.9 contains two gyres. This type of gyre is one of many types of gyre that occur in coastal basins; all gyres involve a closed eddy motion in the circulation. Shown in Figure 5.10 is a possible lateral cross-section of the currents in which the dots are the points of arrows coming towards the observer and crosses are the rear ends[2] of the arrows moving away from the observer. The surface current in all slices has a tendency to be downwind with a reversal of current with depth in the deeper slices. So the topographic gyre is a gyre in the total volume transport, and may not be evident at a single depth. Such gyres are a common feature of wind-forced motion in coastal basins and often have a significant effect on circulation.

---

[2] Sometimes likened to the tail feathers attached to arrows in archery.

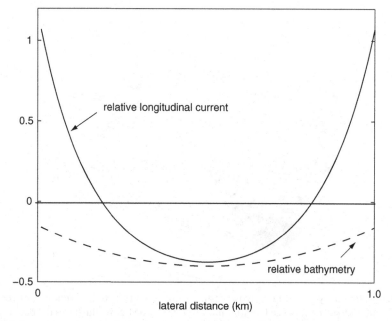

Figure 5.8 Lateral variation of the depth-averaged longitudinal current for the long basin in Figure 5.7 for which the laterally varying bathymetry is the same for all sections across the basin. A wind is being directed along the axis of the basin (away from the reader) and produces positive (downwind) current (away from the reader) in the shallow edges of the basin and a negative return current in the center of the basin.

Our present description of the processes that lead to topographic gyres is a good example of the power of simple models. Much more detailed models lead to greater accuracy in terms of details of the circulation. Although such models may be just two-dimensional, there are some advantages in using three-dimensional models, since they can describe bottom friction more effectively. Bottom friction is an important ingredient of models of wind-driven flow in irrotational coastal basins of variable depth, and is necessary in order to obtain steady states. Let us place the x axis along the longitudinal direction with the y axis across the lateral section of the basin. The surface is assumed to lie close to the height datum, and so the depth averaged dynamical equation in the x direction for each slice is

$$h(y)\frac{\partial u}{\partial t} = -gh(y)\frac{\partial \eta}{\partial x} + \frac{\tau_w - \tau_b}{\rho} \tag{5.8}$$

where u is the depth average current, and τ_b is the bottom frictional stress, so that in the steady state

$$\tau_b = \tau_w - g\rho h(y)\frac{\partial \eta}{\partial x}. \tag{5.9}$$

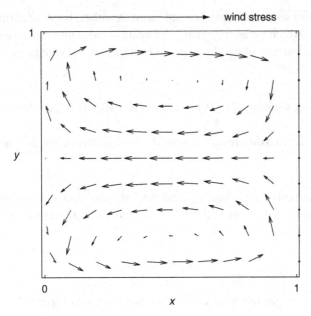

Figure 5.9 A plan view of the depth-averaged circulation in a long basin corresponding to the lateral view in Figure 5.8 with the wind directed along the x axis. This plan view also shows the circulation at the ends of the basin and the scale of the x axis has been changed for ease of presentation.

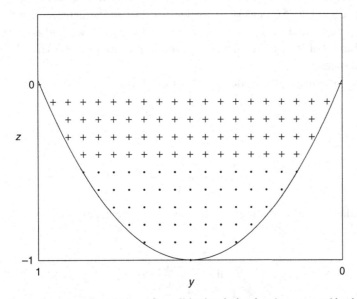

Figure 5.10 Lateral cross-sectional view of possible circulation in a long coastal basin to which a wind stress is applied in the longitudinal direction parallel to the long axis of the basin and away from the reader. The small points represent arrowheads which are traveling towards the reader and the crosses are the tails of arrows pointing away from the reader. The exact form of the current direction across the plane depends on the vertical eddy viscosity.

The models of this section assume a closed basin. Another simple situation is that of an open channel in which we assume there can be no steady state pressure gradient. This means there is a balance of wind stress and friction, i.e., (5.9) reduces to

$$\tau_b = \tau_w \tag{5.10}$$

and we refer to the solution of (5.10) as u_0 which is called the open channel wind-driven current.

We will consider two different forms for the bottom stress. The first is linear friction

$$\tau_b \rightarrow \rho C_d \, u_f \, u \tag{5.11}$$

where u_f is a constant positive background current speed due to waves or tides that are averaged on much smaller timescales than those involved in the dynamics of u in (5.8). Hence, for this case

$$u_0 \equiv \frac{\tau_w}{\rho u_f \, C_d} \tag{5.12}$$

and if we apply (5.1) for the wind stress as a function of wind speed W, (5.12) becomes

$$u_0 \equiv \frac{\rho_a C_w}{\rho u_f C_d} \, W^2 \tag{5.13}$$

and taking some typical values for surface drag coefficient ($C_w \sim 0.001$), bottom drag coefficient ($C_d \sim 0.0025$), density of air ($\rho_a \sim 1.2 \, \text{kg m}^{-3}$), density of marine water, ($\rho \sim 1026 \, \text{kg m}^{-3}$), and background speed ($u_f \sim 0.15 \, \text{m s}^{-1}$) we find $u_0 \sim 0.003 \, W^2$ so that a wind speed of $10 \, \text{m s}^{-1}$ (fresh breeze) produces a current $u_0 \sim 0.3 \, \text{m s}^{-1}$ which is 3% of the wind speed.

The quadratic form for the frictional bottom stress is

$$\tau_b \rightarrow \rho C_d |u| u \tag{5.14}$$

and this leads to

$$\tilde{u}_0 \equiv \sqrt{\frac{\tau_w}{\rho C_d}} \tag{5.15}$$

where the tilde is used to distinguish this case from the linear case in (5.13) and again using (5.1)

$$\tilde{u}_0 \equiv \sqrt{\frac{\rho_a C_w}{\rho C_d}} W \tag{5.16}$$

the typical values for surface drag coefficient ($C_w \sim 0.001$), bottom drag coefficient ($C_d \sim 0.0025$), density of air ($\rho_a \sim 1.2 \, \text{kg m}^{-3}$), and density of marine water ($\rho \sim 1026 \, \text{kg m}^{-3}$) lead to $u_0 \sim 0.02 \, W$ which is 2% of the wind speed.

Returning to the closed basin, linearization of the friction makes the mathematics of the model a little more tractable. Solving (5.9) with (5.11) gives

$$u = u_0(1 - sf(y))$$

(5.17)

where u_0 is the open channel speed (5.12) and $f(y)$ is the lateral variation of depth, i.e.,

$$h_0 f(y) \equiv h(y)$$

(5.18)

in which h_0 is the depth at the deepest point of the basin and

$$s \equiv \frac{\rho g h_0}{\tau_w} \frac{\partial \eta}{\partial x}.$$

(5.19)

We note that s is the ratio of the total pressure gradient summed over the deepest water column to the wind stress. If we take a simple linear depth variation and a basin width of b, i.e.,

$$f(y) = \frac{y}{b}$$

(5.20)

and make a change of variable to

$$y' = \frac{sy}{b},$$

(5.21)

(5.17) becomes

$$u = u_0(1 - y')$$

(5.22)

and so u changes sign at $y' = 1$. The total volume flux is positive (downwind) for $y' < 1$ and negative (upwind) for $y' > 1$ (so that we have current flowing into the wind), and must be equal in *magnitude* (opposite in sign) since the basin is closed, i.e., s must be such that

$$\int_0^1 y'(1 - y')dy' = \int_1^s y'(y' - 1)dy' \quad s \neq 0$$

(5.23)

where we have transposed the upper limit of the right-hand integral from $y = b$ to $y' = s$ using (5.21). Evaluating the integrals, we find

$$\frac{s^3}{3} - \frac{s^2}{2} = 0,$$

(5.24)

i.e., $s = 1.5$ and so (5.21) shows that the boundary between downwind and upwind current occurs at $y = \frac{2}{3}b$ and generally

$$u = \frac{\tau_w}{\rho u_f C_d}\left(1 - \frac{3y}{2b}\right).$$

(5.25)

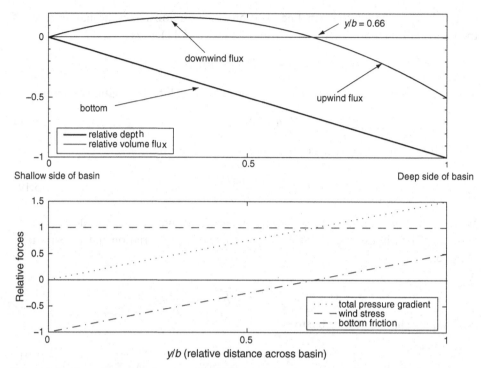

Figure 5.11 Upper panel shows the variation of longitudinal volume flux with lateral position across a basin whose depth varies laterally. The basin is forced by a longitudinal wind stress and the model employs linear friction (text gives results for quadratic friction). Lower panel shows the balance of relative forces.

Since $s = 1.5$ the total pressure gradient summed over the water column is $1.5\tau_w$ in the deepest water and since the depth h varies linearly with y, the average of the longitudinal pressure gradient term taken across any lateral section of the basin in (5.8) is 75% of τ_w. The average of (5.25) shows that bottom friction supplies the other 25% of τ_w, opposing the wind stress for $y < \frac{2}{3}b$ and opposing the pressure gradient for $y > \frac{2}{3}b$. The strength of currents in the basin is entirely controlled by the friction coefficient and at $y = 0$, the wind stress is totally balanced by friction. Notice that there is a total net frictional force that opposes the wind. Figure 5.11 shows the current across the basin for linear friction, and the comparative values of the force on the vertical longitudinal slices due to wind, stress pressure gradient and bottom friction.

For the case of quadratic friction, (5.22) is replaced by

$$u = \tilde{u}_0 \sqrt{1 - sf(y)}. \tag{5.26}$$

The volume fluxes in this case are

$$\int_0^1 y'\sqrt{1-y'}dy' = \int_1^s y'\sqrt{1-y'}dy' \quad s \neq 0 \tag{5.27}$$

and evaluating the integrals,

$$-(2+3y')(1-y')^{3/2}\Big|_0^1 = (2+3y')(y'-1)^{3/2}\Big|_1^s \tag{5.28}$$

which leads to

$$(2+3s)(s-1)^{3/2} = 2. \tag{5.29}$$

Taking the square of (5.29)

$$(2+3s)^2(s-1)^3 = 4 \tag{5.30}$$

and expanding

$$9s^5 - 15s^4 - 5s^3 + 15s^2 - 8 = 0 \tag{5.31}$$

the only real root of (5.31) is $s = 1.46$. So, the case of quadratic friction differs only slightly from that of linear friction with the average pressure gradient being 73% of the wind stress, so that friction is marginally greater.

The assumption of a pressure gradient that is independent of lateral position does not apply at the end of the basin, since the current most either turn from downwind to upwind or vice versa. This requires a component of velocity v in the y direction which must be balanced by friction. The problem becomes soluble with a two-dimensional numerical model.

The model that we have outlined might be termed the model of vertical longitudinal slices, in which each slice is effectively independent, but subject to the constraint of zero total longitudinal volume flux. The direction of the current in each slice depends only on its depth, and so we often find several topographic gyres in different parts of the basin. Each gyre is associated with a deeper channel running past shallows. These gyres are always evident in two-dimensional wind-driven models. They are a feature of the depth-averaged flow and very important to transport properties. If the model is run with tidal motion present, the topographic gyres may not be evident until the tidal motion is filtered out or simply removed. Whether the gyres could be evident in the surface flow of a basin is discussed in a later section. For a rotational basin, the Coriolis force affects the topographic gyres (as will be discussed later). The strength of the Coriolis force depends on the magnitude of the current, and as we have seen, the current in the present models increases as the friction coefficients decrease. This shows that the influence of Coriolis forces is controlled by the Ekman number. In the next sections we discuss two further balances of forces. The first involves wind stress (and or any form of shear stress) and the Coriolis force (called the Ekman balance). The second is a balance between the Coriolis force and pressure gradient (called *geostrophic balance*). These balances of pairs of forces are susceptible to analytical, or simple numerical models, whereas situations with more than two types of force can often only be treated by more detailed numerical models.

5.2 Ekman balance

The *Ekman balance* is a balance between wind stress and Coriolis force, and one of the most fundamental force balances in ocean models. It always plays a critical role in the surface currents in the deep oceans and continental shelf. It can also be important in coastal basins but, as we shall see, is often obscured by the dominant effects of friction, i.e., the Ekman balance requires low Ekman number.

5.2.1 Ekman transport

Suppose a two-dimensional water column in the northern hemisphere has a wind stress applied to the surface, and is affected by the Coriolis force, and there is *no pressure gradient*. The steady state solution of this problem is that the current flows at right angles to the wind, with the wind stress on its left (in the northern hemisphere).

This steady state is called the Ekman balance after V. W. Ekman's pioneering work on the effect of the Coriolis force on wind-driven currents. This result is quite different from our previous model, in which the steady state involves a balance between wind stress, pressure gradient, and possibly friction. However, there are many situations in which the pressure gradient is not the dominant force that opposes wind stress, and the Ekman balance is very common in deeper basins. It is an example par excellence of the influence of the Earth's rotation.

The linear depth-averaged dynamical equations in two dimensions for a rotating Earth, and ignoring bottom friction, are:

$$\frac{\partial u}{\partial t} = -\frac{\partial p}{\partial x} + fv + \frac{\tau_w \cos \theta}{\rho h} \tag{5.32}$$

$$\frac{\partial v}{\partial t} = -\frac{\partial p}{\partial y} - fu + \frac{\tau_w \sin \theta}{\rho h} \tag{5.33}$$

where θ is the angle between the direction of the wind stress and the x axis, and we have dropped the over-bar on the components of depth average current (u, v). Let the x axis be aligned along the direction of the wind and assume that there is no pressure gradient, and the system reaches a steady state: (5.32) and (5.33) become

$$v = -\frac{\tau_w}{\rho f h} \tag{5.34}$$

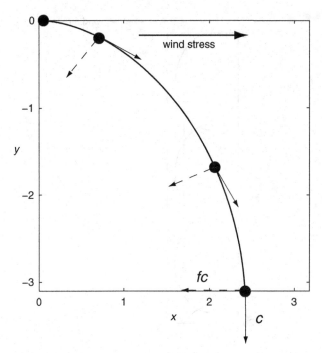

Figure 5.12 Illustration of the evolution of the steady state depth-averaged currents in Ekman balance showing the direction of the wind stress and steady state current in the northern hemisphere. The wind stress arrow is directed along the *x* axis. The diagram shows the trajectory of a particle released from rest at the origin; its speed increases with time until the steady state is reached. The dashed arrows represent the Coriolis force per unit mass. This trajectory ignores inertial oscillations, i.e., they have been filtered out (compare Figure 5.13). Only wind stress and the Coriolis force are present, so that in the steady state there is a balance between ρhfc and the wind stress.

$$u = 0 \qquad (5.35)$$

as shown in Figures 5.12 to 5.14. The component of current along the *y* axis, *v*, is negative in the northern hemisphere ($f > 0$) and becomes positive in the southern hemisphere, because f is then negative, i.e., the current flows to the left of the wind in the southern hemisphere. Friction does affect Ekman balance.

The quantity $-\tau_w/\rho f$ is known as the Ekman transport and if we use (5.1) and $f = 2\pi/T_{inertial}$

$$v = -\frac{\rho_a C_w T_{inertial} W^2}{2\pi \rho h} \qquad (5.36)$$

the typical values for surface drag coefficient ($C_w \sim 0.001$), density of air ($\rho_a \sim 1.2\,\mathrm{kg\,m^{-3}}$), density of marine water ($\rho \sim 1026\,\mathrm{kg\,m^{-3}}$), depth $h \sim 5\,\mathrm{m}$, inertial period ~ 24 h, $W = 10\,\mathrm{m\,s^{-1}}$, we find v $\sim 0.3\,\mathrm{m\,s^{-1}}$, which is 3% of wind speed. So, we notice that

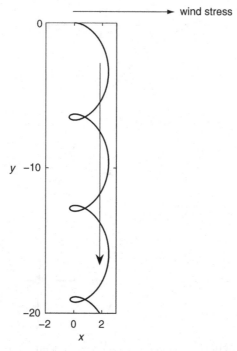

Figure 5.13 A repeat of Figure 5.12 but including inertial oscillations without frictional damping. The arrow indicates the mean current.

all of our estimates of wind-driven currents in this chapter are typically a few percent of wind speed although this depends on depth and other parameters. Nevertheless modelers typically estimate wind currents by this *3% rule*.

5.2.2 Vertical profile of Ekman current

The model of the last section tells us only that the vertical profile of current must be such that its *depth average is a current at right angles to the wind*. This is not, of course, the direction of the surface current which can only be obtained from a model of the profile of the current through the water column. Ekman's interest in this current stems from an observation made in 1893 by the Norwegian oceanographer Fridtjof Nansen (1861–1930). Nansen was a professor of zoology and later oceanography at the Royal Frederick University in Oslo, and contributed with groundbreaking works in the fields of neurology and fluid dynamics. Nansen loaded 6 years' supply of food and 8 years' supply of fuel on the specially made ship *Fram*, and left Oslo port, heading for the North Pole. He let his ship be frozen into the Artic ice and drift to the North Pole. Nansen noticed that *Fram* drifted not downwind but at an angle of 20° to 40° to the right of the wind direction. Nansen's observation led to Ekman's pioneering work on the effect of the Coriolis force on wind-driven currents.

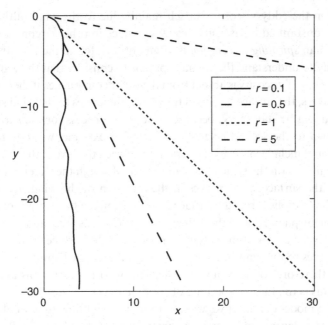

Figure 5.14 A repeat of Figure 5.13 including linear friction. The parameter r is the Ekman number defined as $r = C_d u_f / fh$, where u_f is the background friction velocity and h the height of the water column. The simple Ekman balance applies approximately when $r < 0.1$ but for $r > 5$ there is an approximate balance between wind stress and friction and the current flows almost parallel to the wind. Inertial oscillations are only evident for r less than about 0.1.

The exact form of vertical profile of current, and hence the angle of the surface current, depends on the relation between internal stress and velocity shear, i.e., the way that the horizontal momentum input at the surface by the wind is mixed down through the water column. Observation of the angle between the wind and surface current is, in a sense, a way of investigating the downward mixing of surface momentum. This is usually modeled by a *coefficient of eddy viscosity*, which arises from the turbulence of the water in the coastal basin, but is otherwise akin to the ordinary viscosity of a fluid although of much greater magnitude at the spatial scales at which we usually work in coastal basins. Whatever the form of the viscosity, it is clear that the direction of the current vector must change with distance below the surface. The reason is that on average the direction of the current vector is normal to the wind, while at the surface the *stress* is parallel to the wind (although the surface current will not be parallel to the wind because of the effects of viscosity). This rotation of the direction of the current with height was originally discovered by Ekman and is commonly called the *Ekman spiraling effect or Ekman rotation*. There is a special case that Ekman solved analytically, as an illustration of the spiral effect, and this involves a viscosity that is independent of height and gives a regular spiral known as the *Ekman spiral*. In reality, that is a very unlikely form for the eddy viscosity, especially when the eddy viscosity is itself mainly due to the wind (although it may approximate constant eddy viscosity near the surface in other circumstances). In the literature, Ekman *spiraling* or *rotation* is often abbreviated to the Ekman *spiral*.

A useful way to understand the Ekman spiral is to consider that the water column is divided into layers. If viscosity is absent from the water column, i.e., if there is no friction between the layers, the surface layer will travel at right angles to the wind stress and the layers beneath will be stationary, because there is no process for the surface stress to propagate down to the second layer. If, at the other extreme, we were to ignore the Coriolis force and include viscosity, i.e., friction between the layers, the wind stress does penetrate downward (in the steady state all the way through the water column). In this extreme case, the surface water moves in the direction of the wind. If we have both friction and the Coriolis force, the surface layer must move to the right of the wind (in the northern hemisphere) due to the influence of the Coriolis force and produces a shear stress on the layer beneath it in the same direction as the surface current. The second layer down moves to the right of the layer above it due to the Coriolis force and so on. The currents therefore rotate around to the right with increasing distance below the surface, and this is the Ekman spiraling or rotation effect.

To make the model quantitative we need to understand how to write the dynamical equation for each layer. Let us assume that each layer is infinitesimally thin and of thickness dz. The stress (force per unit area) on the upper face of the layer at z is $\tau_x(z)$ in the x direction and $\tau_y(z)$ in the y direction. On the lower face the corresponding shear stresses are $\tau_x(z-dz)$ and $\tau_y(z-dz)$ so that the net forces per unit volume are $\partial \tau_x/\partial z$ and $\partial \tau_y/\partial z$ and converting to force per unit mass we get (ignoring advection of momentum)

$$\frac{\partial u}{\partial t} = fv + \frac{1}{\rho}\frac{\partial \tau_x}{\partial z} \qquad \frac{\partial v}{\partial t} = -fu + \frac{1}{\rho}\frac{\partial \tau_y}{\partial z} \tag{5.37}$$

where the stresses are derived from the velocity shear as

$$\tau_x = \rho K_z \frac{\partial u}{\partial z} \qquad \tau_y = \rho K_z \frac{\partial v}{\partial z} \tag{5.38}$$

and K_z is the eddy viscosity. In the steady state

$$\frac{1}{\rho}\frac{\partial \tau_x}{\partial z} + fv = 0 \qquad \frac{1}{\rho}\frac{\partial \tau_y}{\partial z} - fu = 0 \tag{5.39}$$

substituting (5.38) into (5.39),

$$K_z \frac{\partial^2 u}{\partial z^2} + fv = 0 \qquad K_z \frac{\partial^2 v}{\partial z^2} - fu = 0 \tag{5.40}$$

where we have assumed that K_z is independent of z (an assumption which is not a good one for surface currents but is simplest to deal with analytically). The units of K_z are $m^2 s^{-1}$ and so $\sqrt{K_z/f}$ is a length. A dimensionless height z' defined as

$$z' \equiv \frac{z}{L_E} \tag{5.41}$$

where

$$L_E \equiv \sqrt{\frac{2K_z}{f}} \tag{5.42}$$

and L_E is called the *Ekman length* which can also be expressed in terms of the inertial period as

$$L_E \equiv \sqrt{\frac{T_{inertial} K_z}{\pi}}. \tag{5.43}$$

In dimensionless form (5.40) becomes

$$\frac{\partial^2 u}{\partial z'^2} + 2v = 0 \quad \frac{\partial^2 v}{\partial z'^2} - 2u = 0. \tag{5.44}$$

The simplest way of solving (5.44) is to use a complex variable to represent the current in the (x, y) plane,

$$V \equiv u + iv \tag{5.45}$$

and this is conveniently written as

$$V = se^{i\theta} \tag{5.46}$$

where s is the speed of the current and θ the angle between the x axis (along which the wind blows) and the current vector. Hence (5.44) and (5.45) give

$$\frac{\partial^2 V}{\partial z'^2} - 2iV = 0 \tag{5.47}$$

and looking for a solution of form

$$V = e^{\alpha z'} \tag{5.48}$$

(5.47) and (5.48) give

$$\alpha^2 = 2i \tag{5.49}$$

and so

$$\alpha = 1 + i \tag{5.50}$$

hence (5.50), (5.48), and (5.46) give

$$V = V_0 e^{z'} e^{i\theta} \quad \theta = z' \tag{5.51}$$

where V_0 is complex, or in terms of u and v

$$u = s_0 e^{z'} \cos(z' + \gamma) \quad v = s_0 e^{z'} \sin(z' + \gamma) \tag{5.52}$$

where

$$V_0 = s_0 e^{i\gamma}. \tag{5.53}$$

We apply the boundary condition that at the surface, which we take as $z' = 0, \tau_y = 0$, and by (5.38)

$$\left.\frac{\partial v}{\partial z'}\right|_{z'=0} = 0 \tag{5.54}$$

and using (5.52)

$$\frac{\partial v}{\partial z'} = s_0 e^{z'} [\sin(z' + \gamma) + \cos(z' + \gamma)] \tag{5.55}$$

and so

$$\left.\frac{\partial v}{\partial z'}\right|_{z'=0} = s_0(\sin \gamma + \cos \gamma). \tag{5.56}$$

Hence, (5.54) and (5.56) give $\gamma = -\pi/4$ so that the current vector is 45° to the right of the x axis along which the wind blows (angles in complex analysis being measured positive counterclockwise), i.e.,

$$V = s_0 e^{z/L_E} e^{i\theta} \quad \theta = z/L_E - \pi/4 \tag{5.57}$$

where we have returned to a dimensional height and s_0 is the (dimensional) current vector on the surface. Note that z decreases as we move down from the surface, and so the current speed decreases as we move away from the surface. The value of θ also decreases, meaning that the current vector turns counterclockwise as we move downward.

The boundary condition at the surface along the x axis is $\tau_x = \tau_w$, and by (5.38)

$$\left.\frac{\partial u}{\partial z'}\right|_{z'=0} = s_0 \left[\sin\left(\frac{\pi}{4}\right) + \cos\left(\frac{\pi}{4}\right)\right]. \tag{5.58}$$

Note that in the above equation z' is dimensionless, so that both sides of the equation have the dimension of velocity. Hence we have

$$s_0 = \frac{\tau_w L_E}{\rho K_z \sqrt{2}} \tag{5.59}$$

and using the definition of the Ekman length,

$$s_0 = \frac{\sqrt{2}\tau_w}{\rho f L_E} \tag{5.60}$$

$$u_0 = \frac{\tau_w}{\rho f L_E} \qquad v_0 = -\frac{\tau_w}{\rho f L_E}. \tag{5.61}$$

It is interesting to check that the depth average of (5.57) does give a current at right angles to the wind stress, i.e., along the negative y axis. This means that if we integrate u down through the water column, the integral should be zero, and using (5.52)

$$I \equiv \int_{-\infty}^{0} e^{z'} \cos\left(z' - \frac{\pi}{4}\right) dz'. \tag{5.62}$$

If we change the variable of integration to p, where

$$p \equiv z' - \frac{\pi}{4}, \tag{5.63}$$

then (5.62) becomes

$$I = e^{\pi/4} \int_{-\infty}^{-\pi/4} e^{p} \cos p \, dp = e^{\pi/4} \frac{e^{p}}{2} [\cos(p) + \sin(p)] \Big|_{-\infty}^{-\pi/4} = 0 \tag{5.64}$$

as required.

We can calculate the Ekman transport by a similar integral of $v(z)$ from (5.62):

$$\phi_w \equiv s_0 L_E \int_{-\infty}^{0} e^{z'} \sin\left(z' - \frac{\pi}{4}\right) dz' \tag{5.65}$$

and again using (5.63)

$$\phi_w = s_0 L_E e^{\pi/4} \int_{-\infty}^{-\pi/4} e^{p} \sin p \, dp = s_0 L_E e^{\pi/4} \frac{e^{p}}{2} [\sin(p) - \cos(p)] \Big|_{-\infty}^{-\pi/4} \tag{5.66}$$

$$= -\frac{s_0 L_E}{\sqrt{2}}$$

which with (5.60) recaptures the Ekman transport, i.e.,

$$\phi_w = -\frac{\tau_w}{\rho f}. \tag{5.67}$$

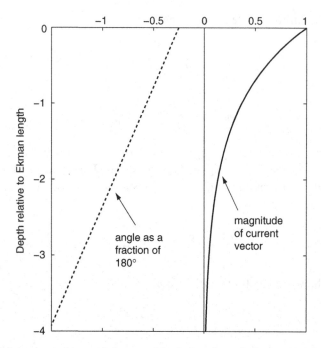

Figure 5.15 Height dependence of the speed and direction of the current vector in Ekman spiraling with constant eddy viscosity.

If the wind stress has a general direction in the (x, y) coordinate plane the resultant transport is

$$\phi_y = -\frac{\tau_x}{\rho f} \tag{5.68}$$

$$\phi_x = \frac{\tau_y}{\rho f}. \tag{5.69}$$

The solution (5.57) is usually called the *Ekman spiral* and is illustrated in Figures 5.15 to 5.17. Shortly after Ekman's analysis of the problem, C. A. Bjerknes did some simple experiments to illustrate Ekman rotation, and this is easy to repeat as a laboratory demonstration. The apparatus consisted of a low glass cylinder (12 to 17 cm high and 36 to 44 cm wide) resting on a horizontal table that could be rotated at a rate of about seven turns per minute. A stream of air was blown diametrically across the surface of the water and the direction of the current measured at different depths using a small vane (4 cm long and 1 mm high). It was found that the angle of deflection from the wind increased with depth, so illustrating Ekman spiraling. The physical basis of the spiraling is simply that each layer of water tends to be deflected by the Coriolis force relative to the net shear stress applied to it form above and below. Ekman spiraling is

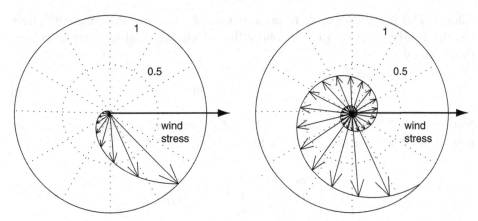

Figure 5.16 The left-hand panel shows the magnitude and direction of the current in Ekman spiraling relative to the direction of the wind stress. This is a polar plot and the dotted circles and lines show angles. The right-hand panel shows a similar plot but the length of the vector is the fourth root of the current, and this enables the reader to see much more of the spiral. So, for example, a vector of length 0.5 in the right-hand panel represents a current of $(1/2)^4 = 1/16$ of the surface current and occurs near an angle of $210°$ to the wind.

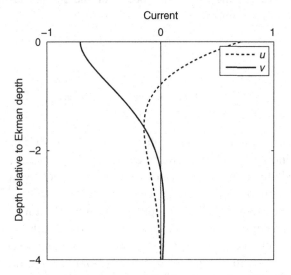

Figure 5.17 The components of current (u, v) in the (x, y) plane for Ekman spiraling (or rotation). The wind stress is in the positive x direction. Notice that the depth average of u (dotted line) is zero.

not limited to the effect of wind stress and can occur whenever the Coriolis force acts under situations of vertical shear in horizontal currents.

The Ekman balance has associated with it the possibility of upwelling or downwelling, i.e., the vertical advection of water upward or downward in the water column. This is a consequence of the variation of u and v in the horizontal plane, i.e., the divergence of the surface volume flux which must lead to a vertical volume flux given by

$$\frac{\partial u}{\partial x} + \frac{\partial v}{\partial x} + \frac{\partial w}{\partial x} = 0. \tag{5.70}$$

Notice that (5.70) refers to u and v at the surface, and *not* their depth average. If we now vertically average (5.70) over the Ekman layer

$$\int_{-L_E}^{0} \frac{\partial w}{\partial x} dz = - \int_{-L_E}^{0} \left(\frac{\partial u}{\partial x} + \frac{\partial v}{\partial x} \right) dz. \tag{5.71}$$

Since we are in a steady state, w must essentially vanish at the surface so that its value at the bottom of the Ekman layer is

$$w_E = \frac{\partial \phi_x}{\partial x} + \frac{\partial \phi_y}{\partial y} \tag{5.72}$$

where ϕ_x and ϕ_y are volume fluxes summed over the Ekman layer. Using vector notation with U as the volume flux

$$\phi_x = \frac{\tau_y}{\rho f}, \quad \phi_y = -\frac{\tau_x}{\rho f} \tag{5.73}$$

$$w_E = \frac{1}{\rho} \left(\frac{\partial \tau_y}{\partial x} - \frac{\partial \tau_x}{\partial y} \right) = \frac{\hat{z}}{\rho} . \nabla x \frac{\tau}{f} \tag{5.74}$$

where $\hat{z}$ is the unit vertical vector. The right-hand side of (5.74) is the curl of the wind stress. Positive wind stress curl causes divergence in the Ekman layer, known as *Ekman pumping* or upwelling, and negative wind stress curl is associated with downwelling. Maps of wind curl are produced for the deep ocean.

It is clear that the speed in the Ekman spiral reduces by a factor of $1/e$ at a depth L_E equal to the Ekman depth below the surface, and over this depth the angle changes by 1 radian (57°). The characteristic region into the water column over which the stress essentially decreases to almost zero is called the *Ekman layer*, and has a thickness proportional to the Ekman length L_E provided that this is less than the depth of the water; we have assumed infinite water depth at the moment, i.e., no effect from the bottom. The *Ekman layer* is the layer of influence of the surface stress. Below the Ekman layer no shear stress or friction results from the surface stress. It is important to understand the physical reason that the surface stress only penetrates a finite distance into the water column on a rotating Earth. The mixing of momentum into the water column is due to eddy viscosity, and L_E increases with K_z. The Coriolis force inhibits that mixing by reducing the component of the current in the direction of the wind, and as f increases L_E is reduced. As an example, at latitude 30° if $K_z = 0.001 \text{ m}^2 \text{s}^{-1}$, $T_{\text{inertial}} = 24 * 3600$ (24 h) so by (5.43) $L_E = 5.2 \text{ m}$. However, coastal basins can experience very low wind conditions, and in the absence of tidal forcing may experience very low values of vertical mixing coefficient K_z. An example of a situation of this type involves very shallow drogues deployed just below the water surface during an episode of a slight breeze over an extended period with maximum speed of 3 m s^{-1}. In Chapter 7 we explore some basic ideas of vertical mixing in basins, which indicates that there is a variation of the mixing coefficient beneath the surface of the form

$$K_z(z) = \gamma u_*(\eta - z) \qquad (5.75)$$

where

$$u_* \equiv \sqrt{\frac{\tau_w}{\rho}} \qquad (5.76)$$

in which τ_w is the surface wind stress and ρ the density of water in the basin. Using the form (5.1) we find

$$K_z(\eta - z) = AW(\eta - z) \quad A = \sqrt{\frac{\rho_a}{\rho}} C_w \approx 8 * 10^{-5} \qquad (5.77)$$

so that for a wind speed of $W = 3 \text{ m s}^{-1}$, K_z is of order 10^{-5} at a depth of 0.1 m, and our example has an Ekman length based on a mean K_z of $L_E = 0.3 \text{ m}$. Care should be taken with the application of (5.77) near surfaces in strong wave conditions because of uncertainty in the thickness of the roughness layer. The form (5.77) naturally raises the issue of solving the Ekman problem (5.40) with viscosity that increases linearly with distance from the surface. There are a variety of possible forms for K_z, and its functional dependence on z and derivatives of u and v called *turbulence closure systems*. These obviously change the form of the spiral but do not fundamentally alter the basic phenomenon. However, the process itself is not easy to observe in coastal basins because of the presence of pressure gradients and bottom stress. A second bottom Ekman layer can form at the bottom due to bottom frictional stress. The thickness of the bottom layer has the same function form (5.43) as the surface boundary layer. It spirals clockwise *into the water column* (in the northern hemisphere) going upward, i.e., counterclockwise as viewed descending down through the water column. In shallow water the two layers may overlap and the rotations cancel one another. The prediction of the Ekman solution that the wind-driven surface current runs at 45° to the wind direction is not generally valid because of the complex nature of the surface interaction.

Ekman also published solutions for the wind-driven motion on a rotating Earth in the presence of a coastal boundary condition which effectively adds a pressure gradient. The problems that we are considering are linear in the forcing functions, which means that if we consider a wind stress as one forcing function and pressure gradient as a second forcing function, the solution for the two forcing functions applied together is simply the sum of these two solutions. Both solutions must be obtained for a rotating Earth. Introducing a coast in the Ekman problem involves a boundary condition of zero current across the coastline. As we have seen previously, this is achieved by a set-up of water surface against the coast. The form of the set-up is entirely dictated by the currents generated by the wind stress and the set-up affects only the *pressure-driven currents*. Furthermore, those pressure-driven currents are not affected by the shear stress due to the wind. In reality, of course, there is no distinction within the water column between the two types of currents, and our description is simply a modeler's view of the processes in a coastal basin. This way of treating individual processes within linear systems is the quintessential pillar of all our understanding of physical systems. In most systems it is not totally valid, and in some systems it is totally invalid. In simple systems, it works quite well and trades only a little accuracy for a great deal of understanding. That is true in the present case, because the two types of current do interact through their influence on the coefficient of vertical mixing, and perhaps through the non-linear advection terms. But, nevertheless, the linear model provides a good basis for our understanding. Problems that are totally non-linear are usually not helped by linear approximations, and such problems play a major part in science. Non-linearity is the basis of *interaction between processes* as opposed to one process merely forcing another process. For example, incoming waves may create sand ripples which increase bottom roughness, and hence reduce wave activity so that the ripples may be ameliorated leading to increased wave activity.

5.3 Geostrophic balance

This section presents a model known as *geostrophic balance* which involves a balance between a pressure gradient and the Coriolis force. This is one of the most important balances of forces in environmental dynamics and applies to both the ocean and atmosphere. In coastal basins, it is a common balance in the cross-shore direction. The balance assumes that pressure gradient and Coriolis force are much greater than other forces such as friction, wind stress etc. It is a model of the steady state.

Consider a water column acted on by only the Coriolis force and pressure gradients as in Figure 5.18, where the pressure gradient is acting as a force in the positive x direction and a particle at the origin is released from rest. The pressure gradient experienced by the particle is constant here, but the Coriolis force increases and changes direction as the particle accelerates and moves until it exactly balances the pressure gradient. The particle is then traveling in the negative y direction, i.e., at right angles to the pressure gradient, keeping the region to high pressure to its right (in the northern hemisphere).

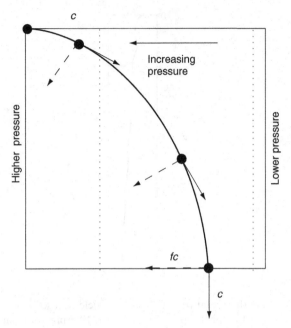

Figure 5.18 Illustration of the evolution of the steady state depth-averaged currents in geostrophic balance showing the direction of the pressure gradient and steady state current in the northern hemisphere. The contours of pressure (dotted straight lines) are parallel to the *y* axis. The diagram shows the trajectory of a particle released from rest at the origin; its speed increases until a steady state is reached. The dashed arrows represent the Coriolis force. This trajectory ignores inertial oscillations, i.e., they have been filtered out. Only a pressure gradient and the Coriolis force are present, so that in the steady state there is a balance between *ρhfc* and the pressure gradient.

Writing the pressure gradient as $-\partial p/\partial x$ the geostrophic balance with the Coriolis force has the form

$$-\rho f v = -\partial p/\partial x. \tag{5.78}$$

In (5.78) we can see that the steady state current is parallel to the contours of constant pressure with the high pressure on the right of the velocity (northern hemisphere). Figure 5.19 is a repeat of Figure 5.18, but with the inertial oscillations included and some friction. The friction damps out the oscillations in a few cycles and also causes the trajectory to move slightly to its right (a different scaling is used for the two axes in the figure which exaggerates the deflection).

Consider a low pressure region in the oceans as shown in Figure 5.20 (which may, for example, be due to a dip in the ocean level). This dip in the potential is also termed a *potential well*. The analytic form for the pressure is chosen as

$$P = -P_0 \exp\left(-\frac{(x - x_0)^2 + (y - y_0)^2}{L^2}\right) \tag{5.79}$$

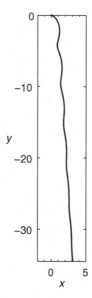

Figure 5.19 Figure shows the actual trajectory of a particle released from rest in the system shown by Figure 5.18 (northern hemisphere) without the filtering out of inertial oscillations, and including some friction. The frictional forces damp out the inertial oscillations after about three cycles and cause the trajectory to settle into a steady state path that is rotated (by friction) about 5° towards the lower pressure region, i.e., to its left.

where L is the characteristic size of the potential well and P_0 is its strength. In Figure 5.20, the center of the well is taken as $x_0 = y_0 = 200$ km and $L = 100$ km. In a moment, we shall see that the size L of the potential determines whether geostrophic balance is actually achieved. The figure shows the circles of constant potential, i.e., the contours of the potential field and arrows (directed towards the center of the potential field) that represent the force due to the potential which accelerates particles *downhill* towards the center of the potential field. The choice of $L = 100$ km made here means that the potential has a spread from about 100 to 200 km in both the x and y directions. The figure shows a model experiment in which a particle is released from rest at $x_0 = 300$ km, $y = 200$ km at time $t = 0$ in a rotating basin in the northern hemisphere. The particle travels around the trajectory counterclockwise, keeping the low pressure center to its left. Figure 5.21 shows a close-up of the start position of the particle and its position at the end of the model run (the Coriolis force accelerating the particle towards the center of the small inertial circles). The undulations in the orbit are inertial oscillations, and notice that these are clockwise. Figure 5.22 shows the time series of the x coordinate of the particle conveniently plotted as a function of tf. The period of the orbit is seen to be $425/f$ compared with the inertial oscillations (the small undulations figure 5.22) which have period $tf = 2\pi$.

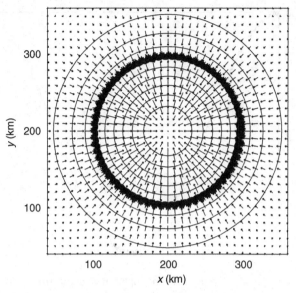

Figure 5.20 The trajectory of a particle (thick line) released from rest in a potential well in a coastal basin on a rotating Earth (northern hemisphere). The potential has circular symmetry: with circular contours of potential and the arrows represent the direction and magnitude of the inward potential gradient. The particle travels around the trajectory counterclockwise, keeping the low pressure center to its left. Hidden within the orbit in this figure are undulations due to inertial oscillations; Figure 5.21 shows a close up of the start position of the particle and its position at the end of the model run.

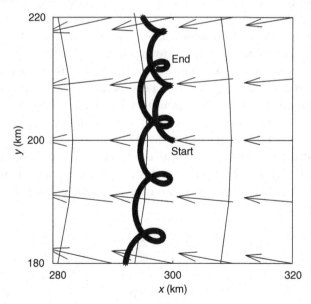

Figure 5.21 A close-up from Figure 5.20 showing the start position of the particle and its position at the end of the model run. The undulations in the orbit are inertial oscillations.

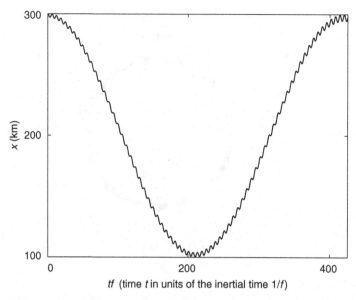

Figure 5.22 A time series of the x coordinate of the particle in Figure 5.20. The period of the circular orbit around the potential is close to $425/f$. The small undulations are due to inertial oscillation with period $2\pi/f$. The period of the circular orbit relative to the inertial time $(1/f)$ depends on the strength of the potential gradient and increases as that gradient decreases.

The balance shown by Figure 5.20 is one of the most fundamental in all of ocean dynamics. Currents circulate counterclockwise (clockwise in southern hemisphere) around a low pressure region and this is called *cyclonic* motion. Around a high pressure system the circulation is *anti-cyclonic* (clockwise in the northern hemisphere). This produces the familiar wind flows around highs and lows on weather maps. In oceanography we call such a circulation a *geostrophic gyre*. If such a gyre is created it can persist for an extended time only being dissipated by friction. Large-scale ocean currents can spin off mesoscale eddies which have either low or high pressure regions at their centers. These usually have a core of water with a different temperature. A warm core eddy has higher pressure at its center, and a cold core eddy has lower pressure so that the circulation is respectively anti-cyclonic and cyclonic. A current obtained from (5.78), i.e., by dividing the local pressure gradient by the Coriolis parameter, is called the *geostrophic current* since it is the current that would result from geostrophic balance, i.e.,

$$-v = -\frac{1}{\rho f}\frac{\partial p}{\partial x} \tag{5.80}$$

$$u = -\frac{1}{\rho f}\frac{\partial p}{\partial y}. \tag{5.81}$$

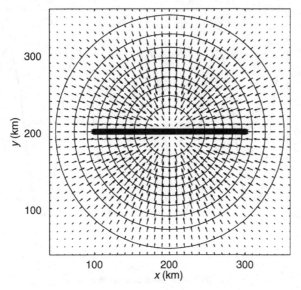

Figure 5.23 Repeat of Figure 5.20 for a non-rotational basin, i.e., with the Coriolis force turned off.

This was a classical method of determining currents in the deep ocean prior to modern current meters and is a still a fundamental tool of deep ocean dynamicists; its value lies in the applicability of geostrophic balance in the deep ocean and also on the continental shelf.

As an example of geostrophic balance, consider a coastal ocean of constant depth and constant density $\rho = 1026\,\mathrm{kg\,m^{-3}}$, with a perfectly straight coast. Suppose that there is a ribbon, of width 10 km, of steady current flowing parallel to this coast with a speed $0.1\,\mathrm{m\,s^{-1}}$. Assuming that the current is in geostrophic balance and that the elevation is everywhere small, we can find the surface slope at the coast using $g = 9.81\,\mathrm{m\,s^{-2}}$ and $f = 10^{-4}\,\mathrm{s^{-1}}$ with (5.80), by adopting the hydrostatic approximation for p,

$$\frac{\partial \eta}{\partial x} = \frac{fv}{g}. \tag{5.82}$$

So in an offshore distance of 10 km the elevation changes by 0.01 m.

In a non-rotational system, the Coriolis force is absent and therefore a system would not move towards geostrophic balance. For a non-rotational system, particles released from rest in the potential well illustrated in Figure 5.23 would simply oscillate back and forth along a straight line as a (non-linear) harmonic oscillator (in the absence of friction).

If the particle were given an initial radial velocity, there would be a corresponding oscillation with different phase and amplitude and the oscillation would occur in two dimensions. Figure 5.24 shows the time series of the x coordinate of the particle in Figure 5.23, and for convenience we have also shown this as a function of tf (although

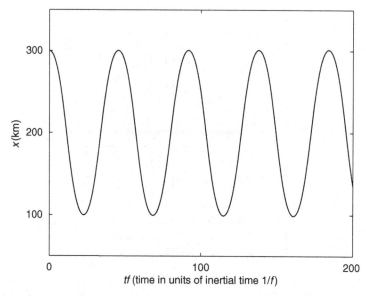

Figure 5.24 A time series of the x coordinate of the particle in Figure 5.23.

the Coriolis force is actually turned off). The period of this oscillation is $tf = 45.6$, and so the angular frequency corresponds to $\omega = 2\pi f/45.6$, i.e., $\omega = f/7.3$. This is purely a function of the strength of the potential and its spatial scale L. The magnitude of the acceleration of particles in this potential well is of order ωu.

The model considered above is not a realistic evaluation of the effect of decreasing potential on geostrophic balance, because it employs a particle that is not affecting the potential. If instead we use a water particle, we are effectively creating a disturbance in the surface elevation. This will spread over the potential in a time of order L/c where c is the speed of a surface wave $c = \sqrt{gh}$, where h is the depth of the basin. The Rossby radius of deformation is the distance moved by a gravitational wave in the *inertial time* $1/f$,

$$R_{rossby} = \frac{c}{f} \tag{5.83}$$

and this often called the *barotropic Rossby radius*. In a typical coastal basin of depth $h = 1$ m, $R_{rossby} \sim 30$ km and for a depth $h = 10$ m, $R_{rossby} \sim 100$ km. Consider a coastal basin without friction. If one creates a disturbance in the water surface of the basin, there results a system of waves and eventually a steady state may be reached. Suppose the initial disturbance is a surface dip or potential well. In the steady state there will be a remnant of the original dip which is in geostrophic balance with currents circulating around the potential well. The magnitude of the potential decreases from a maximum to zero over a distance that is essentially the Rossby radius. So if L is originally much larger than the Rossby radius, the velocity field around the potential adjusts to accommodate the potential. In other words, the equalization process provided by dispersal of the water particles as waves is much slower than the influence of the Coriolis force. If L is much smaller than the Rossby radius, then the potential well essentially fills up, leaving only a broad dip in surface elevation and currents.

The *baroclinic–Rossby radius* is used in stratified systems and replaces the speed of surface gravity waves by the speed of internal gravity waves, which is typically 10 to 100 times slower, making the appropriate Rossby wave smaller by a corresponding factor. In either case, the Rossby radius is the length scale at which current starts to follow pressure contours rather than moving at right angles to them. So, as f decreases (moving towards the equator) the Rossby radius becomes larger. We can think of it in terms of the direction of the flow which would, in the absence of the Earth's rotation, be straight down the pressure gradient. Instead, in geostrophic flow, the streamlines of the flow are *deformed* by the Coriolis force, becoming parallel to the isobars (lines of constant pressure) instead of across them. When an ocean basin is much larger than the radius of deformation, the circulation in the interior of that ocean will be in geostrophic balance. Alternatively, the radius of deformation gives us the scale of features such as fronts (distance across the front) and the width of ocean currents. We expect these to be *of the same order as* the radius of deformation.

5.4 Isostatic equilibrium

Isostatic equilibrium is a balance between pressure gradients in the atmosphere and ocean. It involves the ocean surface adopting an elevation that conforms to the pressure exerted by the atmosphere, so that the pressure within the ocean is spatially constant. This type of balance is usually associated with the large spatial scales over which atmospheric pressure varies significantly, but is a much weaker balance than the geostrophic balance discussed in Section 5.3.

In the steady state the model assumes that the ocean adjusts its surface until the net *pressure gradient* is zero. If the atmospheric pressure is written as $p_a(x)$ in a one-dimensional model, and the elevation as $\eta(x)$, the hydrostatic pressure gradient at any depth is

$$-\frac{\partial}{\partial x}(p_a + g\eta) \tag{5.84}$$

and if this is zero, so that

$$p_a + g\rho\eta = \text{constant} \tag{5.85}$$

which means that a change in atmospheric pressure of Δp_a produces a change in surface elevation of $\Delta\eta$,

$$\Delta\eta = -\frac{\Delta p_a}{g\rho} \tag{5.86}$$

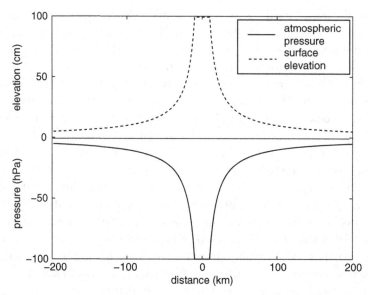

Figure 5.25 The pressure depression due to a tropical cyclone (hurricane) with a core of radius 10 km (full line) and the associated isostatic elevation change of the ocean (broken line).

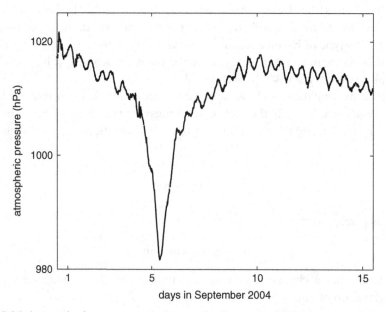

Figure 5.26 Atmospheric pressure during early September 2004, measured in Tampa Bay, Florida, due to the passage of tropical cyclone Hurricane Frances (see path in Figure 5.5).

and this is called the *isostatic* change in elevation. In the SI system the unit of pressure is the pascal (Pa), equivalent to $1\,N\,m^{-2}$, named after Blaise Pascal (1623–62) a French mathematician and physicist. Atmospheric pressure is about 1000 hPa (hectopascals); hecto means 100. A hectopascal is generally accepted to be equivalent to the old unit of a millibar (mbar), although this depends slightly on the definition of standard atmospheric pressure. Standard atmospheric pressure was originally 1013.25 hPa but since 1985 has been recommended to be a 1000 hPa. A change of 1 hPa in atmospheric pressure produces a change in elevation of $100/g\rho = 100/(9.81 * 1026)\,m = 9.93\,mm$ ~1 cm. For example, the pressure at the center of a severe tropical cyclone (hurricane) may be 100 hPa lower than the pressure of the surrounding atmosphere. This produces an isostatic increase in water level of 100 cm = 1 m. This adds to the surge created by the cyclone winds and associated alongshore currents. Figure 5.25 shows the isostatic change in surface elevation due to a cyclone with central pressure 100 hPa below ambient pressure. Notice that this elevation is strictly the isostatic elevation, and not the actual elevation that would be produced by a cyclone which includes major effects due to associated winds. Figure 5.26 illustrates the atmospheric pressure measured due to passage of a cyclone.

In older units atmospheric pressure was often quoted in inches of water and 1 inch of water is equivalent to 2.49 hPa. This is easily calculated by converting 0.993 mm of seawater to 0.39 inch of seawater and converting to the density of standard fresh water.

5.5 Further reading

See further reading from earlier chapters. Additional texts for this chapter are:

Cushman-Roisin, B. (1994). *Introduction to Geophysical Fluid Dynamics*. New York: Prentice Hall.

Emanuel, K. (2005). *Divine Wind: The History and Science of Hurricanes*. New York: Oxford University Press.

Lighthill, J. (2002). *Waves in Fluids*. Cambridge, UK: Cambridge University Press.

Neumann, G. and W. J. Pierson (1966). *Principles of Physical Oceanography*. Englewood Cliffs, NJ: Prentice Hall.

Newman, J. N. (1977). *Marine Hydrodynamics*. Cambridge, MA: MIT Press.

6

Modeling tides and long waves in coastal basins

6.1 Introduction

This chapter provides some background into our understanding and modeling of astronomical tides in coastal basins. We provide some simple models that illustrate some of the basic scientific ideas and methodology models of tides and long waves. We finish the chapter with some general comments on the way that we build spatial grids for numerical models.

6.2 Astronomical tides

6.2.1 Origin of tides

The astronomical tides of the ocean are primarily due to changes in the gravitational potential exerted on the large ocean basins by the Moon and Sun as the Earth rotates in their gravitational fields. These are called *astronomical tides* by oceanographers. The word *tide* in common parlance refers to any movement of ocean water or change in ocean state. Thus, we find the noun *tide* being used to describe phenomena in the coastal ocean, such as the state and direction of currents important to maritime operations, water height and also *red tides*, *rip tides*, etc., and there are many etymological links between the words *tide* and *time*. In oceanography *tide* is used in general to refer to any periodic changes in water level and the associated current; these may be due to astronomical forcing or to other processes, such as periodic winds. However, oceanographers often use the term *tides* as a synonym for *astronomical tides*, since these tides are a dominant feature of the world's oceans. There has been an attempt to tidy up, within earth sciences, the use of the word *tide*. For example, we now prefer to use the word *tsunami* to describe the surge in ocean elevation caused by underwater earthquakes, rather than the older (and still in common usage), *tidal wave* (*tsunami* literally translates as *harbor wave* in the Japanese language meaning that its effects are only visible on the coast).

Tides are not, of course, restricted to water on the Earth's surface. They are felt by any fluid on our planet such as the molten core of the Earth, and our continents move slightly in response to the changing gravitational fields of the Sun and Moon. All tides slowly damp these relative motions and planets become *tidally locked*, as seen in our Moon which keeps the same face towards us as it orbits around the Earth. Tides are also a vital part of stellar evolution and dynamics.

6.2.2 Newton's model and dynamical models of deep ocean tides

There is a very simple model of these tides based on an Earth totally covered with water, in which the ocean surface deforms instantaneously to the perturbed equipotential surface. This model was devised by Isaac Newton (1642–1727), mathematician and physicist, and one of the foremost intellects of modern science, and is called the *equilibrium theory of tides*. The model is clearly a quasi-static approximation because water requires a finite time to move under changes in gravitation potential and furthermore, the planet contains continents which break up the ocean into ocean basins. Nevertheless, Newton's model is very important as a means of understanding the basic physics of tides, and the predicted equilibrium tides are an important reference point for the real tides. Illustrations of Newton's model are given in most introductory texts on oceanography and will not be reproduced here, especially since we are not primarily concerned with the deep ocean, and tides in coastal basins are a consequence of those in the deep ocean, and not due directly to gravitational effects. Models of the real tidal motion in these basins are called *dynamic* tidal models.

6.2.3 Differences between tides in the deep ocean and coastal basins

The amplitude of the astronomical tide in an *enclosed ocean basin* evidently depends on the size of the basin. So, the variation in the water level at any point reflects the change in gravitational forcing *across* the basin. The *equilibrium tide* (Section 6.2.2) in the major ocean basins (which are some tens of thousand kilometers in size) has an amplitude *of order* 1 m, i.e., 1 m or a few meters but not 10 m, and so a (virtually isolated) basin such as the Mediterranean Sea (which is of size of order 1000 km) would be expected to have astronomical tides of about 0.1 m and inland water systems, such as one of the Great Lakes of North America, an astronomical tide of 0.01 m. The astronomical tides in the coastal basins considered in this book would be essentially zero if the basins were isolated from the continental shelf and deep ocean.

The astronomical tides under discussion so far are properly defined as *deep ocean tides*. They have the characteristic spatial scales of a 1000 km, and occur in the deep ocean as very long waves. When these waves arrive in the coastal zone, i.e., the shallow continental shelf, inshore waters and coastal basins, they produce *shallow water tides* which can be different in magnitude to the deep ocean tides. As a broad generalization, shallow water tides occur on spatial scales too small to be directly affected by the gravitational forcing of the Moon and Sun. Instead, they are a direct consequence of the change in water level at their *boundary* with the deep ocean. So, coastal models which include the astronomical tides are driven by the periodic oscillation of water level around their boundary (or boundaries), whereas tidal models of the deep ocean are driven by periodic gravitation forcing. Consequently, a coastal basin that is totally enclosed will have no astronomical tidal motions.

There is a significant boundary between the *deep ocean* and coastal region created by change of depth. Tidal waves in the deep ocean and coastal region have wavelengths much greater than water depth, and so propagate as *long waves*. This means that their velocity is given by $\sqrt{gh}$, where h is total depth and g is the acceleration due to gravity. So in moving from the deep ocean (h of order several thousand meters) to the coastal region (h of order 10 to 100 m) the speed drops by a factor of at least 10, and probably 100. During this change, the frequency evidently remains the same and so there is a corresponding reduction in wavelength. Although the magnitude of coastal tides is affected by friction, the dominant determinant of their magnitude is the size and topography of the coastal basin. So, in large coastal basins, the wavelength may become comparable with the size of the basin. This often occurs in *continental seas*. The term *sea* is usually reserved for large maritime waters that are bounded by land on at least two sides, although the noun is often replaced by *Gulf* or *Sound*, or even *Bay*. A *continental sea* is composed of continental shelf, i.e., it has depths of order 100 m or less, and is essentially a part of the continental plate that is submerged below sea level. In contrast, the Mediterranean Sea is a *marginal sea*, in that it is part of the deep ocean that is entrapped by land with depths of several thousand meters. Marginal seas are not coastal basins but do *contain* coastal basins; other good examples of marginal seas are the Gulf of Mexico and Gulf of California.

We shall consider some models of coastal basins that demonstrate the dependence of coastal tides on the ratio of wavelength to size of the basin. For special values of this ratio, *tidal resonances* occur, producing *macroscopic tides* with ranges up to 10 m. Coastal tides are also affected by friction and topographical restrictions. A common example of such a restriction is found in *coastal lagoons*: this term is usually used for inland basins with narrow connections to the open ocean. These connections often form tidal channels with tidal currents that are usually of order 0.1 to $0.2\,\text{m s}^{-1}$ in natural channels, and much higher ($1\,\text{m s}^{-1}$) in channels with artificial walls. In such basins the channel restricts water flow, and the tide inside the basin may be very small (*microscopic*, or *miniature*, tides of magnitude 0.1 m or less). Note that the tidal range in the deep ocean, and on most coasts bordering large ocean basins, is of order 1 or 2 m. A tide of less than 1 m can therefore reasonably be called small, with the term *micro-tidal* best reserved for tidal ranges of one or two tens of centimeters; tides with a range of 1 m often produce very significant flushing, mixing, and scour, and are a prime influence in the morphology of the world's coasts.

A very notable difference between the deep ocean and coastal region is the magnitude of tidal currents. Such currents are very small in the deep ocean, and models of deep ocean circulation usually neglect tides entirely. The magnitude of tidal currents can be estimated from the volume continuity equation developed in Chapter 2, which in one dimension is

$$\frac{\partial hu}{\partial x} = -\frac{\partial \eta}{\partial t} \tag{6.1}$$

where u is the current along the x axis, h is the depth of the basin, η is the surface elevation (assumed small relative to h in [6.1]), and t is time. In finite difference form (6.1) is

$$\frac{\Delta hu}{\Delta x} = -\frac{\Delta \eta}{\Delta t}. \tag{6.2}$$

Using $\Delta x = 1000$ km, $h = 5000$ m, $\Delta \eta = 1$ m, and $\Delta t = 12$ h, we get $u = 0.005$ m s^{-1}. The same estimate for the coastal region would have $\Delta x = 100$ km, $h = 10$ m, $\Delta \eta = 5$ m, and $\Delta t = 12$ h, which gives $u = 1$ m s^{-1}. In coastal basins tidal currents play a very important part in the dynamics of flushing and mixing.

6.2.4 Harmonic decomposition

An important (and perhaps the most distinguishing) feature of the astronomical tides in the deep ocean is that their forcing occurs at a unique set of frequencies that are determined by the astronomical motion of the Earth, Moon, and planets. These frequencies change only very slowly with time, or at least over periods of centuries (although there is ample evidence that the daily rotation of the Earth is slowing, and the period of the Moon increasing, and so the frequencies will have been different in the geological past). Provided that the ocean responds to the astronomical gravitation forcing in a *linear* fashion, these same harmonic frequencies will also be seen in the astronomical tidal motions and elevations of both the deep oceans and coastal basins. This is a fundamental property of a linear dynamical system such as a set of linear differential equations. Furthermore, and very importantly, there will be no component of the astronomical tidal response which cannot be described as a motion at one of these frequencies. Although the gravitational potential is forcing the Earth's deep oceans with a large series of discrete frequencies, the tides respond at this same set of discrete frequencies, i.e., no intermediate or additional frequencies are seen in the tidal response (provided the system is linear).

Once we include non-linear effects, the generalization of the last paragraph is no longer true. Non-linear processes change the simple relationship of *frequency out equals frequency in*, which we see in linear systems. We find new characteristic frequencies that are generated by non-linear effects in shallow water, and these are called *shallow water tides*. As a broad generalization, these new tides are usually small, and coastal basins have the same dominant tidal frequencies as the deep ocean, albeit at very different amplitudes.

Let us express these ideas mathematically. Suppose that the frequencies of astronomical gravitational forcing are denoted by a set of angular frequencies ω_1, ω_2, ω_3, ..., ω_n,.... . These are called the tidal harmonics. This list is not infinite in its extent, but it does extend to several hundred harmonics. So we usually arrange the list so that the highest frequencies occur at the beginning (up to twice-daily tides), with the frequencies getting smaller as we progress along the list, i.e., they have increasingly long periods. The members of the list near the beginning have much greater importance, in the sense that they have greater magnitude in the astronomical forcing, and therefore are major components of the response. Determining the frequencies in this list is a matter of studying the astronomical motion of the Earth and Moon, and is not a property of models of the deep ocean; we can use the same forcing for either Newton's model or dynamical models of the tides. Hence, the elevation η at any point in the ocean will have the form

$$\eta(t) = A_1 \cos(\omega_1 t + \phi_1) + A_2 \cos(\omega_2 t + \phi_2) + \cdots$$
$$= \sum_n A_n \cos(\omega_n t + \phi_n) \tag{6.3}$$

where the parameters A_n and ϕ_n are referred to as the tidal amplitude and phase angle of the *n*th harmonic (or *n*th *partial tide*). Both of these parameters are constant in time, but (in general) vary in space, i.e., they depend on where we are in the oceans. Equation (6.3) is the basic equation of *linear astronomical tidal theory*. If we take just one term in the equation, we see an oscillation in the surface elevation of the ocean that has angular frequency ω_n and phase ϕ_n, i.e., a simple sinusoidal variation in time with amplitude A_n. If we use Newton's model, the tidal harmonic due to the movement of the Sun is such that the elevation is positive when the Sun is directly overhead each day or directly beneath that point (12 h later).

The frequencies ω_n can be viewed as speeds in radians (or degrees if preferred) per unit time and are closely related to the basic relative speed of movement of the Sun, Moon, and Earth as shown in Table 6.1.

Thus referring to Table 6.1, the rotation of the Earth with respect to the fixed stars is $T + h = 15.04106864$ degrees h^{-1} (period = 23.93 h) and the change in the Moon's longitude per hour is $T + h - s = 14.49205211$ degrees h^{-1} (period 24.84 = h).

There are 23 tidal harmonics or *constituents* which are usually evident in ocean tides (of which 11 are usually most important and these are shown in Table 6.2). The notation M is used for tides predominantly due to the motion of the Moon, and S for those due to the Sun. The second letter in the name of the constituent is used to denote either the approximate frequency per day: 1 for daily, 2 for twice daily, or m for monthly and a for annual.

For each harmonic, a map can be produced with contours of all the points in the ocean that have the same amplitude A_n and these are called *co-amplitude* (or co-range) contours (or lines) for that harmonic. Points on these contours have the same tidal

Table 6.1. *Astronomical frequencies important to tidal harmonics*

Symbol	Process	Speed (degrees h^{-1})	Period
T	Rotation of the Earth on axis, with respect to the Sun	15	24 hours
h	Rotation of the Earth around the Sun	0.04106864	365.2425 days
s	Rotation of the Moon around the Earth	0.54901653	27.32 days
p	Moon's perigee (closest point to Earth) precession	0.00464183	8.85 years
N	Precession of the plane of the Moon's orbit	-0.00220641	18.6 years

Table 6.2. *Frequencies of the astronomical tidal harmonics or constituents*

Name	Derivation	Speed (degrees h^{-1})	Period
Semidiurnal constituents (about twice daily)			
M2	$2T - 2s + 2h$	28.984	12.42 hours
N2	$2T - 3s + 2h + p$	28.439	12.66 hours
S2	$2T$	30.000	12.00 hours
L2	$2T - s + 2h - p$	29.528	12.36 hours
nu2	$2T - 3s + 4h - p$	28.512	12.63 hours
K2	$2T + 2h$	30.082	11.97 hours
Diurnal (about daily)			
K1	$T + h$	15.041	23.93 hours
O1	$T - 2s + h$	13.943	25.82 hours
P1	$T - h$	14.958	24.07 hours
Monthly (about monthly)			
Mm	$s - p$	0.544	27.5735 days
Annual (once a year)			
Sa	h	0.041	365.24 days

amplitude (for the particular harmonic). Contours of constant phase ϕ_n are called the *co-tidal* lines for the particular harmonic, and represent points at which high and low water occur at the same time. Maps of co-tidal and co-range lines for the world oceans are available in standard texts and references.[1] Of special interest is the location of the *amphidromic points*, defined as a point in which the particular harmonic has zero amplitude. These usually lie at the center of *amphidromic systems*, which are caused by the rotation of the Earth and result in the waves associated with deep ocean tidal harmonics propagating in closed loops. Such amphidromic systems can also occur in

---

[1] For example Thurman and Trujillo (2003).

large coastal basins in which the basin size is of the same order, or larger, than the tidal wavelength.

6.2.5 *Tidal prediction*

Astronomical tidal prediction became a well-established science long before we had the computing power to establish numerical tidal models of the world's oceans. It is one of the most valuable of all services of oceanographers to the community and provides the times and heights of low and high water which are extremely important to all users of the coastal ocean. This has been especially true for communities living in regions of macro-tides, such as the north-west European shelf where much of the art of tidal prediction originated. The basic method used is simply to measure the tidal elevation over a period of at least 29 days, and then tidally decompose the resultant time series into the series (6.3). This method is called *harmonic analysis*. It determines the values of A_n and ϕ_n for the most important tidal harmonics. Once these are known, (6.3) can be used to obtain the value of η at any future time. The method is used at major ports that may be within coastal basins, and therefore subject to shallow water (non-linear) tidal components, and these are used in the analysis and prediction process.

The M2 constituent alone would give the astronomical tide if the tides due to the sun could be neglected, and if the moon orbited in a perfect circle in the plane of the earth's equator. For this reason, M2 is usually (but not always) the dominant tidal component. Table 6.3 presents the tidal constituents for Clearwater Beach, on the west coast of Florida. These tidal constituents are freely available to the public in many countries and are usually determined and updated by national governmental oceanographic organiza-tions. The amplitudes in Table 6.3 are in meters and the phases are in degrees, referenced to Coordinated Universal Time (UTC) or essentially Greenwich Mean Time (GMT). The *epoch* is the phase referenced to the equilibrium tide, i.e., the phase lag of the observed tidal constituent relative to the theoretical equilibrium tide, also known as phase lag. It is the angular retardation of the maximum of a constituent of the observed tide (or tidal current) behind the corresponding maximum of the same constituent of the theoretical equilibrium tide. As referred to the local equilibrium argument, its symbol is k. When referred to the corresponding Greenwich equilibrium argument, it is called the Greenwich epoch that has been modified to adjust to a particular time meridian for convenience in the prediction of tides, and represented by g or by k'. The relations between these epochs may be expressed by the following formula:

$$G = k + pL$$
$$g = k' = G - aS/15 \tag{6.4}$$

in which L is the longitude of the place and S is the longitude of the time meridian, these being taken as positive for west longitude and negative for east longitude; p is the

Table 6.3. *Observed major tidal constituents at Clearwater Beach, Florida, Gulf of Mexico (latitude: 27° 58.7' N longitude: 82° 49.9' W)*

Name	Amplitude (m)	Epoch (degrees)	Speed (degrees h^{-1})
M2	0.246	123.1	28.9841042
S2	0.096	141.0	30.0000000
N2	0.046	120.3	28.4397295
K1	0.158	12.4	15.0410686
O1	0.151	3.6	13.9430356
nu2	0.010	121.6	28.5125831
S1	0.013	93.8	15.0000000
J1	0.010	20.8	15.5854433
SSa	0.037	48.2	0.0821373
Sa	0.091	151.9	0.0410686
Q1	0.032	348.0	13.3986609
P1	0.053	12.5	14.9589314
K2	0.027	134.6	30.0821373

Source: National Oceanic and Atmospheric Administration (http://tidesandcurrents.noaa.gov).

number of constituent periods in the constituent day with $p = 0$ for all long-period constituents, $p = 1$ for diurnal constituents, $p = 2$ for semidiurnal constituents, and so on; and a is the hourly speed of the constituent, all angular measurements being expressed in degrees.

The existence of tidal harmonics allows us to produce very accurate tidal prediction tables for any point in the ocean for which we know all the tidal amplitudes A_n and all the phases ϕ_n simply by using (6.3). These predictions are usually made for the so-called standard ports and are essential for the maritime industry, since heavy ocean-going vessels may only be able to enter and leave some ports near high tide. So, we note that these predictions are not based on any sort of tidal *simulation* model (although in modern parlance they involve a *data synthesis model*).

To obtain values for the amplitudes and phases of the tidal harmonics, we need to measure the water elevation η at the point of interest over a sufficient time, and essentially invert (6.3) to find the series of amplitudes $A_1, A_2, \ldots, A_n \ldots$ and phases $\phi_1, \phi_2, \ldots, \phi_n \ldots$ The time series of η is usually obtained from an instrument that measures water elevation called either a *tide gauge* or *water level recorder*. The rate of sampling by this instrument needs to be sufficiently high to properly resolve the shortest period tidal harmonic, which is usually the M2 tide.

Harmonic tidal analysis can be applied to tidal currents in direct analogy to the theory that we have outlined for the tidal elevation. Currents are measured at a particular site over a minimum period of 29 days and the components in the west–east and south–north direction subjected to the same harmonic decomposition as we would use for tidal elevation. Currents in most coastal basins involve many processes apart from those due to astronomical tides and so tidal decomposition of currents will account for only a portion of the variability of the currents seen in the time series from the current meter. The same is also true of the time series of elevation in a coastal basin but usually to much less extent because water level variability in a coastal basin is mainly due to astronomical tides. The somewhat subtle point to note is that currents in coastal basins are fundamentally driven by *differences* in elevation across the basin. Nevertheless, tidal decomposition of currents is usually very valuable because of the unique frequency properties of tidal processes. Furthermore, in many basins tidal currents are dominant and one of the most important calibration and validation processes for numerical models is to compare measured and predicted amplitudes and phases for the harmonics of current. This process becomes less useful when non-linear effects cause tidal sub-harmonics but is nevertheless a recommended test of models.

6.2.6 Tidal models of one-dimensional basins

Before we consider simple tidal models of coastal basins, it is necessary to establish methods of writing numerical models of water motion in one- and two-dimensional basins. Models of the motion of a one-dimensional basin containing water of constant density require the equations of continuity of momentum and volume:

$$h\frac{\partial u}{\partial t} = -gh\frac{\partial \eta}{\partial x} \tag{6.5}$$

$$\frac{\partial \eta}{\partial t} = -\frac{\partial hu}{\partial x} \tag{6.6}$$

where we have ignored friction and the rotation of the Earth (Coriolis force), and also the non-linear advection terms and the non-linear terms in the volume continuity equation, i.e., $\eta \ll h$. For the moment, we will assume that the depth h is constant at $h = h_0$. In our numerical models we need to solve these two equations (6.5) and (6.6) in tandem, but for a frictionless irrotational basin it is possible to reduce the two equations to a single equation for either η or u, provided that the basin has constant depth h. To do this, we firstly differentiate (6.5) by x and (6.6) by t:

$$h_0\frac{\partial^2 u}{\partial t\partial x} = -gh_0\frac{\partial^2 \eta}{\partial x^2} \tag{6.7}$$

$$\frac{\partial^2 \eta}{\partial t^2} = -h_0\frac{\partial^2 u}{\partial t\partial x}. \tag{6.8}$$

If we add (6.7) and (6.8),

$$\frac{\partial^2 \eta}{\partial t^2} = c^2 \frac{\partial^2 \eta}{\partial x^2} \tag{6.9}$$

$$c^2 \equiv gh_0, \tag{6.10}$$

which is the one-dimensional wave equation, and in oceanographic terminology describes *long waves* (waves for which the water depth h is very much smaller than the wavelength). Any function of $x \pm ct$ is a solution of (6.9), and we need to find solutions that satisfy the boundary conditions at the ends of the basin.

Let us first cast the equations into dimensionless form. The advantage of this approach is that we minimize the number of physical parameters in the analytical or numerical model. If the basin is of length L and extends from $x = 0$ (the mouth) to $x = L$ (the head) we divide all times by L/c (which is the time for a long wave to travel the length of the basin), all distances x by L, and η by h. In these dimensionless units the speed becomes unity, and so

$$\frac{\partial^2 \eta}{\partial t^2} = \frac{\partial^2 \eta}{\partial x^2} \tag{6.11}$$

where (as usual) we are *redefining* x, t, and η as their dimensionless counterparts.

Assume that the basin is closed at its head, $x = 1$, so that $u(1)$ must vanish (current cannot pass through the closed end). As a first case, consider a basin that is also closed at $x = 0$, i.e., totally disconnected from the open ocean. In this case $u(0)$ is also zero. To find the equation for u we proceed by a similar method as for η and differentiate (6.5) by t and (6.6) by x which, in dimensionless form, is

$$\frac{\partial^2 u}{\partial t^2} = -\frac{\partial^2 \eta}{\partial t \partial x} \tag{6.12}$$

$$\frac{\partial^2 \eta}{\partial x \partial t} = -\frac{\partial^2 u}{\partial x^2}, \tag{6.13}$$

and adding (6.12) and (6.13) we obtain the (formally similar) wave equation for u:

$$\frac{\partial^2 u}{\partial t^2} = \frac{\partial^2 u}{\partial x^2}. \tag{6.14}$$

Analytically, there are two ways to view a solution of (6.14), in which u vanishes at $x = 0$ and $x = 1$. The first involves the so-called *progressive* waves which must suffer certain types of phase change on reflection at the boundaries in order to satisfy the boundary conditions. The second involves so-called *stationary* waves which are a sum of pairs of progressive waves, traveling in opposite directions, with their relative phase chosen so that the boundary conditions are satisfied at the ends of the basin. We treat these two methods in turn.

6.3 Long waves

6.3.1 Progressive waves

Consider an initial state of the basin in Section 6.8, in which the elevation η and current u are everywhere zero. Now impose a disturbance of the water level localized around some point in the basin. This will propagate as waves rather like the wavelets that spread out when a stone is dropped into a pond. When the waves reach the ends of the basin they will reflect in such a way that u remains zero at the ends (note that η is not in generally zero at the ends of the basin). The simplest way to visualize this reflection is to consider a wave of form

$$\eta = A \sin(\omega t - kx + \phi) \tag{6.15}$$

in which A is the amplitude, k is an arbitrary wave number, and ω the angular frequency. In order to satisfy the wave equation (6.11), ω and k must be related by the wave velocity, which in non-dimensional units is $\omega = \pm k$ (according to the direction of the wave), and in dimensionless units $\omega = \pm ck$. The corresponding form for u is

$$u = A \sin(\omega t - kx + \phi). \tag{6.16}$$

After reflection, k changes sign so as to represent a wave traveling in the opposite direction. The total phase of the wave must change so that u vanishes at the boundary. Hence, for the boundary at $x = 0$, the reflected wave has the following forms for η and u:

$$\eta_{\text{reflected}} = A \sin(\omega t + kx + \phi) \tag{6.17}$$

$$u_{\text{reflected}} = -A \sin(\omega t + kx + \phi) \tag{6.18}$$

and for $x = 1$ the reflected wave is:

$$\eta_{\text{reflected}} = A \sin(\omega t + kx + \phi - 2k) \tag{6.19}$$

$$u_{\text{reflected}} = -A \sin(\omega t + kx + \phi - 2k). \tag{6.20}$$

This process of continual reflection of waves eventually produces a set of so-called *standing waves*, and the study of these waves is the second method of solving the wave equation.

6.3.2 Standing waves

The standing wave method involves representing η and u as a sum of *sine* or *cosine* functions in kx which are multiplied by either sin ωt or cos ωt (according to our initial condition). This method is part of a general approach to solving linear partial differential equations called *separation of variables*, in which we look for solutions that consist of products of functions of time t and space x. Let us write the product as

$$u = X(x)T(t) \tag{6.21}$$

where X is a function of x only and T a function of t only. Hence, the wave equation for u becomes

$$\frac{\ddot{X}}{X} = \frac{\ddot{T}}{T} \tag{6.22}$$

where each dot indicates differentiation of the function with respect to its dependent variable (for X this is x and for T it is t). Since (6.22) must be true for all x and t, it follows that both fractions must be individually equal to the same constant:

$$\frac{\ddot{X}}{X} = -k^2$$
$$\frac{\ddot{T}}{T} = -\omega^2 \tag{6.23}$$

where $\omega^2 = k^2$. The solution for T is simply

$$T = \sin \omega t \tag{6.24}$$

and for X

$$X = A \sin kx + B \cos kx \tag{6.25}$$

where A and B are arbitrary constants that will be chosen to make the solution satisfy the boundary conditions at the ends of the basin. When the basin is closed at both ends, the spatial function for u must vanish at $x = 0$ and 1. Hence we put $B = 0$, which removes the cosine function in (6.25), and choose k so that $\sin kx$ is zero at $x = 0$ and 1, i.e.,

$$k_n = n\pi; \quad n = \pm 1, \pm 2, \pm 3, \ldots, . \tag{6.26}$$

Equation (6.26) corresponds to a wave that makes the return journey down the basin and back an integer number of times, in one period. The solution $n = 1$ is the fundamental mode of oscillation of the basin, and the higher values of n correspond to the higher *harmonics*. Each harmonic has a different frequency and wavelength linked by the propagation speed c. There is a close analogy to the vibration of a plucked string. The fundamental mode is often referred to as the *seiching* frequency of the basin, and represents the simplest back and forth motion (often seen in a bathtub). Such a motion is easily excited by a wind blowing over a coastal basin, and the frequency analysis of any instrument deployed in a coastal basin usually shows the fundamental seiching frequency. Seiching motion is a common phenomenon in numerical models and can be the cause of instability. The value of the seiching frequency depends only on the dimensions of the basin and the water depth (which controls the propagation speed c). Typical values of c are 3 to $5\,\mathrm{m\,s^{-1}}$, and for basins with dimensions of order 10 to 50 km we obtain seiching periods of order 1 h. In a two-dimensional basin, there are seiching motions in both directions and more complex two-dimensional modes. These involve the solution of the two-dimensional wave equation.

Finally (6.26) can be used in (6.25) to give the solution for u as

$$u_n = \sin(\omega_n t)\sin(k_n x) \tag{6.27}$$

where $\omega_n^2 = k_n^2$. Because the equation is linear, we can form a general solution of the equation by adding solutions (6.27) with all values of n:

$$u = \sum_{n=1}^{\infty} A_n u_n. \tag{6.28}$$

The solution (6.28) can be viewed as the sum of sets of two progressive waves moving in opposite directions. Differentiating (6.28) with respect to t and integrating with respect to x gives

$$\eta = \sum_{n=1}^{\infty} A_n \cos(\omega_n t)\cos(k_n x). \tag{6.29}$$

Because cosine is an even function of its argument, we have used only positive n. We note that the η modes do not vanish at the ends of the basin but involve zero change in volume, since the integral of the cosine terms with the allowed values of k is zero; this is a consequence of the velocity boundary condition which ensures the basin is closed so that there is no change in the volume of the basin.

When $t=0$, (6.29) gives

$$\eta(x,0) = \sum_{n=1}^{\infty} A_n \cos(k_n x) \tag{6.30}$$

and this determines the values of the coefficients A_n. To find a particular coefficient, say A_m, we multiply (6.30) by $\cos(k_m t)$ and integrate with respect to x from $x=0$ to 1. This produces an infinite sum of products and the integral of only one of these is non-zero, and that occurs for $n=m$, i.e.,

$$\int_0^1 \eta(x,0)\cos(k_m x)dx = \sum_n A_n \int_0^1 \cos(k_n x)\cos(k_m x)dx \tag{6.31}$$

$$A_m(\omega_n/k_n) = \frac{1}{\pi} \int_0^1 \eta(x,0) \cos(k_m x) \mathrm{d}x. \tag{6.32}$$

The number of modes needed to represent the basin depends on the initial form of η, and generally the more localized the initial disturbance in η, the greater the number of modes. For a highly localized disturbance, all modes are activated with equal amplitude. It is possible to activate only the fundamental mode ($n = 1$) simply by choosing an initial disturbance corresponding to that mode, although that is unlikely to occur in any practical problem.

The method of decomposing the initial disturbance into the normal modes and then representing the subsequent disturbance by a summation of those modes is a valuable technique. In practice with modern computers, it is probably easier to make a direct numerical solution of the original conservation equations (6.7) and (6.8). This also allows us to include friction and non-linear terms. Linear friction can be included in a wave equation, although it has not been included in our analysis.

6.4 One-dimensional hydrodynamic models

6.4.1 Finite differences

Our approach in this book is to include elementary computer codes. Where possible, MS Excel codes have been used, since this software is widely available. However, Excel can only easily be used for problems where there is one *independent* variable such as x or t. For the present case, we have both time and space as important components of the solution, so we are compelled to move to a higher language. The languages in common use in coastal models are Fortran and Matlab. The latter is much simpler for the novice programmer, and has very many advantages in high-performance computing environments so we will include some elementary Matlab codes. Matlab is a matrix (array) based language, and does have considerable sophistication in handling arrays which produces very compact code. We will not use those techniques, but instead write all code in terms of summations over arrays. Our reason is that such code is very readily translated into simple Fortran and is physically very transparent to the reader; it sacrifices sophistication for ease of understanding. We have not included code for output to graphics.

Before presenting code for a numerical model of the elevations and velocities inside a one-dimensional basin, we need to move to a finite difference form of the continuity equations. This process can involve some advanced techniques, and again we veer on the side of simplicity in order to produce models that illustrate the basic science of coastal basins. The interested reader can find more advanced finite difference systems in the scientific literature and texts specializing in numerical methods.[2]

---

[2] A very readable modern account can be found in Kantha and Clayson (2000).

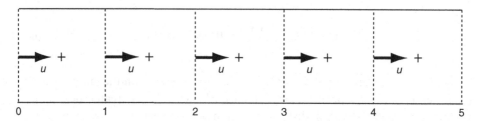

Figure 6.1 The finite difference system used for a one-dimensional basin showing five cells. The crosses are centers of cells at which elevation is determined. Arrows represent the current determined at points on cell faces corresponding to the bases of the arrows.

These methods reduce numerical errors and allow longer time steps in the subsequent solution via time stepping techniques; the codes in this book are merely illustrative.

The secret of the finite difference technique is the establishment of a spatial grid or set of cells in which the equations are to be solved. Let us suppose we use N cells extending from $x=0$ to $x=1$. The width of each cell is then

$$\Delta x = \frac{1}{N} \tag{6.33}$$

and so increasing N improves our accuracy. These cells are shown in Figure 6.1.

There are two types of points shown in the Figure 6.1: points on the faces of the cells, and points in the centers of the cells. We calculate the velocity u on the cell faces and the elevation at the centers of the cells (the crosses in Figure 6.1). The water depth is also specified at the centers of the cells. The reason for this difference in the spatial treatment of u and η is that u is determined by the gradient of pressure, i.e., the derivative of elevation, and η is determined from fluxes of volume across cell walls. So η represents the mean elevation in a cell, and u represents the mean current (volume flux) crossing a face. We number the cells $i=1$ to N. So η_i is the elevation in cell i. Our convention is that the velocity on the left-hand face of cell i is labeled u_i, or, in other words, the left-hand face belongs to the cell. It is often necessary to include an extra cell at one end, or both ends, of the array, i.e., outside the basin. A cell outside the boundary at $i=N$ is necessary if we need the velocity across that boundary. We use this artifice in the present case. An extra cell beyond the $i=1$ boundary is not, of course, required for u_i, but that cell is often added in order to represent the imposed tidal elevation of the boundary when the mouth is open; we shall see this technique in a subsequent example. The code below shows a closed basin, given an initial perturbation of the elevation corresponding to lifting the surface at one end; it involves a necessary depression at the other end so that the average water level stays at zero (since the total volume of water in the basin is kept constant). We call this the *bathtub model* since it simulates the familiar rocking back and forth of water in a closed basin.

6.4.2 Bathtub model

We are using non-dimensional units in which all elevations are scaled by the mean depth h, all velocities by the speed of waves $c = \sqrt{gh}$, and all times are scaled to the time for a wave to travel the length of the basin. Table 6.4 shows notation used for most models in this chapter.

Figure 6.2 shows the time series at the head of the basin for the model in Table 6.5 and Figure 6.3 shows the elevation as a function of distance along the basin. By analogy with the plucking of a guitar string, the modes that we excite depend very much on how we pluck or (in our case) how we initially disturb the basin. The modes that are excited continue indefinitely in the absence of friction. With linear friction the modes continue at the same frequencies but are damped. Even in the absence of friction, the models may never reproduce the original perturbation of the basin.

To run the model with non-linear friction, we need to return to the momentum continuity equation in non-dimensional form and include the frictional term

$$\frac{\partial u}{\partial t} = -\frac{\partial \eta}{\partial x} - C_d |u| u \tag{6.34}$$

where u is simply scaled by $\sqrt{gh}$. Hence, the code in Table 6.5 can be modified from

Table 6.4. *Notation used in models in Chapter 6; all variables are non-dimensional*

Symbol	Meaning	Value in demonstration version of model or means of calculation	Comments
N	Cells in model	100	Can be as large as user wishes
dx	Width of cell	$1/N$	$dx = 1/N$
dt	Time interval	$0.1*dx$	dt/dx called Courant number (less than 1 for stability)
$u(i)$	Velocity	$u(i) + dt*(-px(i) - u(i)/\tau)$	Add an extra cell to represent u at the $x = 1$ face; $u(1) = 0$ (basin is closed)
$px(i)$	Pressure gradient at edge of cell i for $i > 1$	$(\eta(i) - \eta(i-1))/dx$	
τ	Frictional damping time	10	We include simple linear friction in this model and let all velocities decay away with a characteristic time of τ
t_{max}	Run time of model	20	

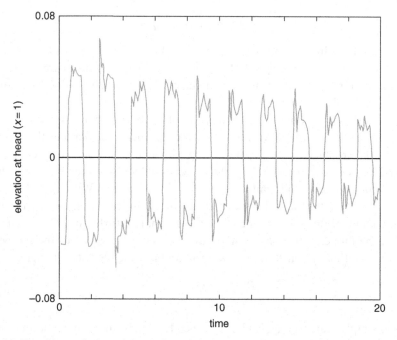

Figure 6.2 The time evolution of the surface elevation at the head of the basin computed by the model in Table 6.5. Notice the dominance of the fundamental model which has a period in non-dimensional time of 0.2, i.e., the time for a wave to move from the mouth to the head and back. The oscillation is slowly damping due to the use of a linear friction term with damping time of 10. This oscillation is called seiching and occurs in all natural basins: it is the demon that spills the coffee from a cup that is rather too full; notice that some of the spikes are higher than the original disturbance.

$$\text{for } i = 2 : N;$$
$$u(i) = u(i) + dt^*(-px(i) - u(i)/(tau^*height(i))); \tag{6.35}$$
$$\text{end};$$

to

$$u(i) = u(i) + dt^*(-px(i) - Cd^*abs(u(i)^*u(i))); \tag{6.36}$$

6.4.3 Non-linear surface waves

We can also include non-linear effects in the volume continuity equation for a basin of constant depth, in which case the volume continuity becomes

$$\frac{\partial \eta}{\partial t} = -\frac{\partial (1 + \eta) u}{\partial x} \tag{6.37}$$

Table 6.5. *Matlab code for bathtub model*

```
N = 100; dx = 1/N; dt = 0.1*dx; t = 0; tseries = t; tmax = 20; tau = 7.5;
for i = 1:N; u(i) = 0;ut(i) = 0; end; u(N + 1) = 0; %we set the basin to rest
at time t = 0;
for i = 1:N;
        if i < N/2; eta(i) = 0.05; else eta(i) = −0.05;end;
end;
while t < tmax;
        for i = 2: N; height(i) = 1 + eta(i);end;
        height(N + 1) = height(N); height(1) = 1;
        for i = 2:N; px(i) = (eta(i)-eta(i−1))/dx; end;
        for i = 2:N; u(i) = u(i) + dt*(−px(i)−u(i)/(tau*height(i))); end;
        for i = 1:N;
           eta(i) = eta(i)−dt*((height(i + 1)*u(i + 1)−height(i)*u(i))/dx);
           if eta(i) <−1;eta(i) =−1;end;
        end;
        t = t + dt
end;
```

in order to allow for the variation of the total height of the water column with elevation. The model grid that we have adopted determines η at the cell centers. In order to accommodate the non-linear terms in (6.37) we need to interpolate η to the edges of the cells after each time step. This involves an additional variable called height H

$$H \equiv 1 + \eta \qquad (6.38)$$

identified at the cell edges. The model code shown below also defines the volume flux

$$\text{flux} = Hu \qquad (6.39)$$

across the faces of the cells. The linear approximation that ignores the non-linear terms in the continuity equation involves $H \to h = 1$ (in our non-dimensional units). Note that there is also a minor change to the momentum equation:

$$\frac{\partial u}{\partial t} = -\frac{\partial \eta}{\partial x} - \frac{C_d}{H}|u|u. \qquad (6.40)$$

The code is shown in Table 6.6 (constant depth basin), in which we have also included non-linear friction. This simple model code is subject to possible instabilities and we add our usual caveat that the codes in this book do not including higher order, more complex numerical systems. Output from the model is shown in Figure 6.4.

The most striking effect of including the non-linear terms in the volume continuity equation is that the surface wave forms a steep edge, and this edge can be seen to

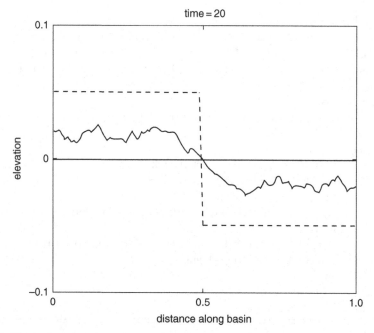

Figure 6.3 The surface elevation corresponding to Figure 6.2 at $t = 20$ (full line). The broken line shows the elevation at zero time. Notice the dominance of the fundamental mode with zero elevation at the center of the basin and equal and opposite elevations at its ends.

propagate into the shallower water. The reason for this tendency to form a steep edge is usually explained by the dependence of velocity on dependence, i.e.,

$$c = \sqrt{g(1 + \eta)} \tag{6.41}$$

so that part of the wave with positive surface elevation η moves faster than that with negative elevation. The result is that deeper water runs over the shallower water into the depression until the depression fills so that the slope of the elevation rises very suddenly, as can be seen in Figure 6.4. If we turn off the non-linear terms in the continuity equation by simply replacing flux$(i) = u(i) * H(i)$ with the linear form flux$(i) = u(i)$, the elevation profile changes to that shown in Figure 6.5. This change involves ignoring the effect of the elevation on the volume flux (so that there is no *interaction* between elevation and current, although of course the volume flux determines the rate of change of elevation). Notice that the slope of the elevation in Figure 6.5 remains constant and is much lower than in the previous figure.

6.4.4 Shoaling

The propagation of long waves is sensitive to total water depth. This result is true whether shoaling is due to a variation of the depth of the basin or to a wave whose

Table 6.6. *Matlab code for the bathtub model with non-linear terms in the volume continuity equation and non-linear friction*

```
N = 100; dx = 1/N; dt = 0.001*dx; t = 0; %we start the model at time t = 0;
for i = 1:N; u(i) = 0; ut(i) = 0; end; u(N + 1) = 0; %we set the basin to rest at time t = 0;
for i = 1:N; eta_initial(i) = 0.1*(N/2−i)/(N/2); eta(i) = eta_initial(i); end;
tmax = 8; Cd = 1;
while t < tmax;
        for i = 2:N; H(i) = 1 + 0.5*(eta(i) + eta(i−1)); flux(i) = u(i)*H(i); end;
        H(N + 1) = H(N); flux(N + 1) = 0;
        for i = 2:N; px(i) = H(i)*(eta(i)−eta(i−1))/dx; end;
        for I = 2:N; u(i) = u(i) + dt*(−px(i)−Cd*abs(u(i))*u(i))/H(i); end;
        for i = 1:N;
           eta(i) = eta(i)−dt*(flux(i + 1)-flux(i))/dx;
           if eta(i) <−1; eta(i) =−1; end;
        end;
        t = t + dt
end;
```

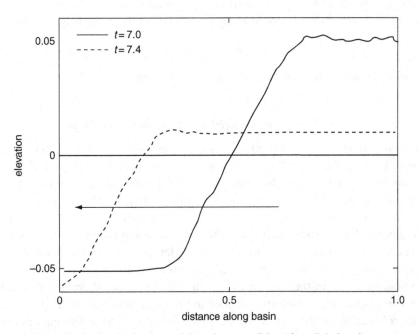

Figure 6.4 Elevation in the bathtub model at times $t = 7.0$ and $t = 7.4$ showing a steep wave propagating from right to left.

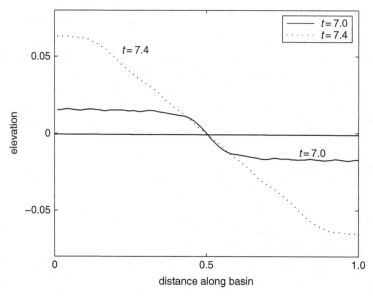

Figure 6.5 Elevation in the bathtub model at times $t = 7.0$ and $t = 7.4$ without non-linear terms.

height is comparable to water depth. This effect on waves is very familiar near the beach and is the basis for the terrible effects of a tsunami. We illustrate this process by considering a basin whose depth shoals by some fraction γ over the length of the basin:

$$h = 1 - \gamma x. \qquad (6.42)$$

The code is shown in Table 6.7 and the results in Figures 6.6 to 6.8.

6.4.5 A simple tidal model of a one-dimensional basin

Tides in coastal basins are *not* produced by the action of the gravitational field of the Sun and Moon on the basin, but are instead due to the action of the tides of the deep ocean on the coastal region; the tides in the deep ocean are themselves due to the gravity of Sun and Moon. Thus, for example, even a basin of the size of the Mediterranean Sea has no significant tides due to the action of the Moon and Sun: it has some small tides with small effects from the Straits of Gibraltar. So, in order to construct a tidal model of a basin we do not consider the action of the Sun and Moon, but merely apply a boundary condition in which the water elevation varies according to the tidal elevation in the exterior ocean. In one dimension, there is a choice as to whether this boundary condition is applied in the first cell of the model ($i = 1$), or an *exterior cell* is added at the mouth. We adopt this latter method and simply specify the elevation in this exterior cell ($i = 0$). Note that $u(1)$ is non-zero (in contrast to the models of the last sections) because the basin is now open at the mouth. The value of

Table 6.7. *Matlab code for bathtub model with shoaling*

```
N = 100; dx = 1/N; dt = 0.001*dx; t = 0; %start the model at time t = 0;
for i = 1:N; u(i) = 0; ut(i) = 0; end; u(N + 1) = 0; %set the basin to rest at time t = 0;
for i = 1:N; eta_initial(i) = 0.1*(N/2−i)/(N/2); eta(i) = eta_initial(i); end;
tmax = 8; Cd = 1; gamma = 0.9;
while t < tmax;
     for i = 2:N; H(i) = 1−gamma*dx*(i−1) + 0.5*(eta(i) + eta(i−1)); flux(i) = u(i)*H(i);
end;
     H(N + 1) = H(N); flux(N + 1) = 0;
     for i = 2:N; px(i) = H(i)*(eta(i)−eta(i−1))/dx; end;
     for I = 2:N; u(i) = u(i) + dt*(−px(i)−Cd*abs(u(i))*u(i))/H(i); end;
     for i = 1:N;
        eta(i) = eta(i)−dt*(flux(i + 1)−flux(i))/dx;
        if eta(i) <−1; eta(i) =−1; end;
     end;
     t = t + dt
end;
```

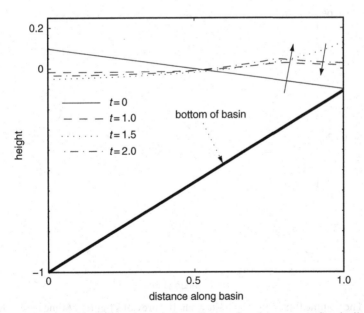

Figure 6.6 Elevation in a shoaling basin at various times *t* after the water surface is tilted upward at the left-hand end ($x = 0$) at $t = 0$. The figure shows the bottom of the basin and the shoaling towards the right-hand end $x = 1$. Notice the steepening of the wave as it shoals.

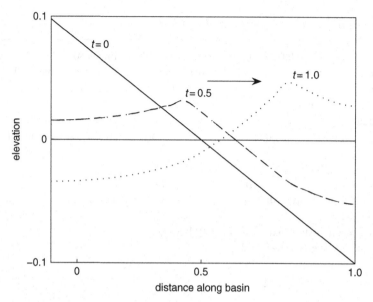

Figure 6.7 The initial behavior of the wave shown in the Figure 6.6 Notice that the pressure gradient is initially greater in the deeper water on the left of the basin, so that the surface moves more quickly than in the shallower water on the right.

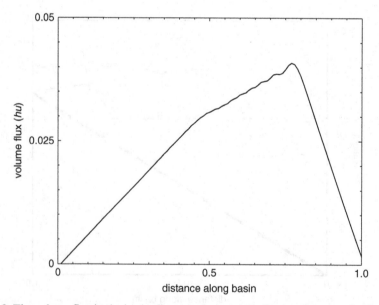

Figure 6.8 The volume flux in the basin shown in the previous figures at time $t = 1$. The slope of the curve with respect to x may be used in the volume continuity equation to show that the regions on the left with positive slope have a surface which is falling with time, and for those on the right with negative slope, the surface is rising.

Table 6.8. *Matlab code for a simple one-dimensional tidal model of a coastal basin*

```
N = 100; dx = 1/N;dt = 0.1 *dx; amp= 0.05; initial_eta = 0;
for n = 1:N; u(n) = 0;eta(n) = initial_eta; end; u(N + 1) = 0; tau = 1;
tidal_period = 5; t = 0; amp = 1;
while t < 30;
        for n = 2:N; px(n) = (eta(n)−eta(n−1))/dx; end;
        px(1) = (eta(1)−amp*sin(2*pi*t/tidal_period))/dx;
        for n = 1:N; u(n) = u(n) + dt*(−px(n)−u(n)/tau); end;
        for n = 1:N; ux(n) = (u(n + 1)−u(n))/dx; end;
        for n = 1:N; eta(n) = eta(n)−dt*ux(n); sum = sum + eta(n); end;
        t = t + dt; tinterval = tinterval + dt;
end;
```

$u(1)$ has to be calculated and this is done using the pressure gradient at the $u(1)$ face, which is derived from the difference between $\eta(1)$ and the boundary elevation in the extra cell ($i = 0$).

We can specify the boundary elevation in various forms. One method is to use the observed elevation as measured from a tide gauge (more accurately called a water depth recorder) and such data files may be readily available, although there is usually some decision on the part of the modeler as to whether the spatial location is suitable. For an astronomical tidal model, the boundary elevation uses the predicted astronomical tidal elevations based on known harmonic constants. The simplest model uses a sinusoidally varying boundary elevation and this corresponds to a single tidal harmonic. This method is used in the code shown in Table 6.8.

If the model is linear, i.e., there are no non-linear terms (such as those associated with advection, friction or the continuity equation), the basin will settle into an oscillation at exactly the same frequency as the boundary elevation. This is a property of the linear differential equations on which the model is based. After the boundary elevation forcing is first turned on, there will be oscillations at other frequencies known as *transients* but these will decay away due to friction. These transient oscillations involve the *natural modes* of the basin discussed earlier, while the oscillation at the frequency of the boundary elevation corresponds to the *forced* oscillation of the basin. This is a common problem in physics and analogous to the forced oscillation of a simple (linearly damped) harmonic oscillator. The oscillator has a natural *resonance* frequency. If the *forcing frequency* is much lower than the natural frequency of the oscillator, the response is simply locked to the forcing providing an immediate reaction to all changes in forcing. If the forcing frequency is very much higher than the natural frequency, the oscillator does not react at all. When the forcing frequency corresponds to the natural frequency of the oscillator, the response involves large amplitude oscillations limited only by damping. This analogy applies exactly to a

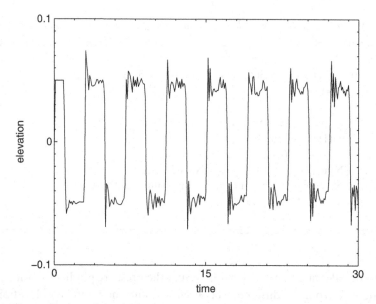

Figure 6.9 The oscillation of the surface elevation η of a coastal basin, with an initial elevation of 0.05 (in dimensionless units), connected to an ocean that has fixed surface elevation $\eta = 0$. Notice the fundamental oscillation with period 4 (in dimensionless units so that the wave can travel four times the length of the basin in one period) and sets of higher modes with shorter periods (higher frequencies) that exactly fit inside the fundamental, i.e., periods of 2, 1, 0.5, and lower.

basin, except that there are a series of natural oscillation modes (rather than just one as for a simple harmonic oscillator) as we saw in the previous sections. The modes are affected by the opening of the mouth of the basin: the fundamental mode has a frequency corresponding to a period which is sufficient for two journeys of the wave from the mouth to head, i.e., the basin is a *quarter wave* oscillator (the length of the basin is a quarter of the wavelength of a progressive wave) as opposed to the closed basin which is a *half wave* oscillator. The closed basin has a fundamental period of 2 (in our dimensionless units) and the open basin a fundamental period of 4. The same behavior is found in many one-dimensional systems in physics and engineering; examples are found in oscillators as diverse as steel railway lines and microwave guides. Figure 6.9 shows the oscillation of the basin produced by the model in Table 6.8 with amp $= 0$ and initial_eta $= 0.05$.

If we run the model with an initial elevation of zero and an ocean boundary elevation of period 4 (tidal_period $= 4$) and amplitude amp $= 0.05$ and tau $= 1$, Figure 6.10 shows the elevation at the mouth and head of the basin and also at a point midway along the basin ($x = 0.5$). It is evident that there is considerable amplification as x increases and the elevation is approaching the fundamental quarter wave mode. That mode has zero amplitude at the mouth and maximum at the head. Notice

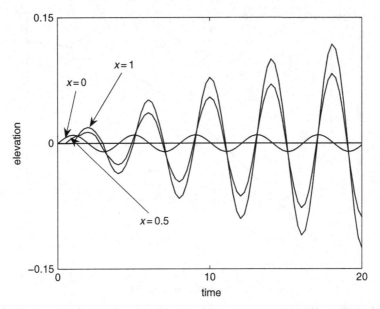

Figure 6.10 Surface elevations along an open basin due to an oscillating ocean elevation at the boundary ($x=0$) of $0.01\sin(2\pi t/4)$ with linear damping time of 10. Notice the time delay of $t=0.5$ and $t=1$ before the tidal elevation arrives at $x=0.5$ and $x=1$ (respectively).

also that there is a phase lag down the basin. The amplitude at the head is close to 0.18 (an amplification of 18); the value increases as we reduce the damping time tau. Also note that the forced oscillation reaches its steady amplitude, i.e., the transits die away, with a characteristic time of tau $=10$. Figure 6.11 shows the surface elevation as a function of distance down the basin at $t=50$, and we note that it is close to a quarter wave mode.

We repeat in Figure 6.12 a run of the same model with an ocean elevation of period 5 and amplitude of 0.05 and the same damping time of tau $=10$. Notice that the elevation shows much less variation with x and the amplification is much smaller. The elevation in the basin is not a horizontal surface, but it changes only by a maximum of 0.04 over the length of the basin compared with 0.18 for a tidal period of 4. As the period becomes increasingly long, the surface becomes closer to a horizontal surface.

The model that we have presented does predict the occurrence of macro-tides in continental seas. This is basically due to resonance (or near resonance) of (usually) the M2 semidiurnal (twice daily) tide with the dynamics of the basin. The period of M2 is about 12.4 h, and so resonance requires a basin in which a wave can travel its length in a quarter period, i.e., about 3 h. Continental seas have typical depths from 50 to 100 m, giving wave speeds (6.10) from 20 to $30\,\mathrm{m\,s^{-1}}$, i.e., the basin length must be 200 to 350 km. Characteristic basin lengths on the north-west European Shelf, the Bay of Fundy in Canada, and also on the north-west shelf of Australia fall within this range

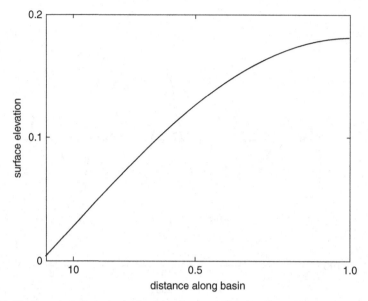

Figure 6.11 Surface elevation along the open basin shown in Figure 6.10 at $t = 50$, due to an oscillating ocean elevation at the boundary $(x = 0)$ of $0.01\sin(2\pi t/4)$ with linear damping time of 10. Notice that the surface elevation is close to a quarter sine wave and also the formation of a macro-tide due to quarter wave resonance.

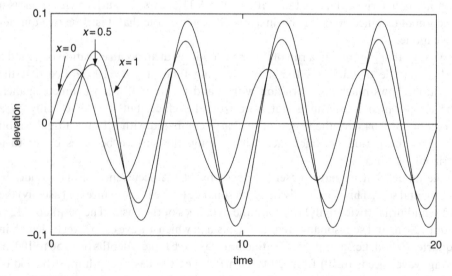

Figure 6.12 Surface elevation as a function of time along an open basin due to an oscillating elevation at the boundary $(x = 0)$ of $0.05\sin(2\pi t/5)$ with linear damping time of 10.

and so as predicted we find macro-tides. A resonance with the diurnal (once daily) tide O1 (which has a period of about 24 h) requires basins of about twice this size (since waves have about twice the time in which to travel the length of the basin). Macro-tides involve large currents, and for the model run in this section and wave speed $c = 22 \, \text{m s}^{-1}$ the currents are of order $0.4 \, \text{m s}^{-1}$; note that the tidal amplitude at the head in 50 m of water would be 9 m. These large currents imply a considerable transport of energy from the mouth to the head, and the currents themselves are driven by the pressure (elevation) gradient evident in the figures of this section.

For basins of depth 1 to 10 m the wave speed is 3 to $10 \, \text{m s}^{-1}$, and so in a semidiurnal quarter wave period of about 3 h a wave travels 30 to 100 km. However, in such basins friction is one of more orders of magnitude more effective and so the tau $= 10$ in the previous model runs would be replaced by tau $= 0.1$ or less, and the tidal amplitude would actually *attenuate* (rather than grow) down the length of the basin. The critical value for a transition from amplification to attenuation is near tau $= 0.5$. Many basins with typical depths of 1 to 10 m are smaller than the quarter wavelength discussed so far, which means that the period of the boundary forcing is rather greater than 4. In many small basins, the period of tidal forcing is very high. As an example, we can run the model of this section with a period of 20. In this case the elevation has the form of an almost horizontal surface, with the tidal amplitude essentially unchanged through the basin and almost no lag in phase. When this occurs, the tidal velocity decreases linearly from the mouth to the head, i.e.,

$$u = u_0(1 - x) \tag{6.43}$$

and this is very easy to prove from the continuity equation for volume (6.6): since $\partial\eta/\partial t$ is independent of x, it follows that $\partial u/\partial x$ is also independent of x so that since u vanishes at the head $x = 1$, we obtain (6.43).

6.5 Two-dimensional models

6.5.1 Non-rotating Earth

We can easily extend the bathtub model to two dimensions, and Table 6.9 shows the code for a basin on a non-rotating Earth. This simply involves changing the one-dimensional grid into a two-dimensional grid with rectangular cells of side dx by dy (the special case $dy = dx$ gives square cells). There are M cells in the y direction. This means that y extends from 0 to some maximum value y_{max}. The component of velocity in the y direction is called v, and all variables are now functions of i and j. The velocity component v for cell (i, j) is measured at the lower side of each cell as shown in Figure 6.13. This is called a *staggered Arakawa C grid*, named after Akio Arakawa (who was responsible for the invention of the finite difference grids necessary to retain stability in numerical models of the atmosphere and ocean). There are two other major

Table 6.9. *Matlab code for two-dimensional model of a coastal basin without rotation of the Earth*

```
N = 100; M = 100; dx = 1/N; dy = dx; t = 0; tmax = 2.5; dt = 0.001; tau = 4;
for j = 1:M;
        for i = 1:N;u(i, j) = 0; v(i, j) = 0; eta(i, j) = 0.1*exp((-i^2 – j^2)/(N^2 + M^2)); end;
end;
for j = 1:M + 1; u(N + 1, j) = 0; end; for i = 1: N + 1; v(i, M + 1) = 0; end;
while t < tmax;
        for j = 1: M;
          for i = 2: N; px(i, j) = (eta(i, j)–eta(i–1, j))/dx; end;
          for i = 1: N; u(i, j) = u(i, j) + dt*(–px(i, j)–u(i, j)/tau); end;
        end;
        for i = 1: N;
          for j = 2: M;
            py(i, j) = (eta(i,j )–eta(i, j–1))/dy; v(i, j) = v(i, j) + dt*(–py(i,j)–v(i, j)/tau);
          end;
        end;
        for j = 1:M
            for i = 1:N;
                eta(i,j) = eta(i,j)–dt*((u(i + 1,j)–u(i,j))/dx + (v(i,j + 1)–v(i,j))/dy);
            end
        end;
t = t + dt
end;
```

types of grid called *A* and *B* grids in which the currents are measured at the centers of the cell (the *A* grid) and at its corners (the *B* grid). The grids are used for different types of model.

Matlab is a matrix language. A matrix (array) has elements written as *aij* where *i* is conventionally the number of the *row*, and *j* the number of the column. So if we think of the array as representing our grid, the second index represents the distance across the page and the first the distance along the side of the page. This is contrary to the usual convention of representing functions of *x* and *y*, such as η as $\eta(x, y)$ where the *x* axis is drawn across the page and *y* along the side of the page. When we use an array such as eta in Matlab, we must remember that if cell *x*, *y* is represented by indices *i*, *j*, then $\eta(x, y)$ becomes eta(*j*, *i*). Fortunately, when Matlab plots functions such as eta it does so with the low values of *j* at the bottom of the figure. In the code in Table 6.9 the basin is given an initial symmetrical perturbation in elevation, and this symmetry is retained as the water in the basin moves back and forth. Output from the model presented in Table 6.9 is presented in Figure 6.14.

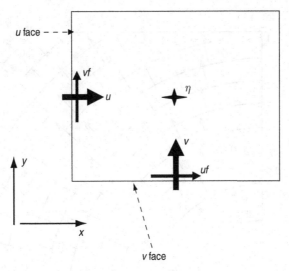

Figure 6.13 Configuration of points and vectors in finite difference cell. Here u and v are the components of velocity in the x and y directions. The left side of cell is called the u *face*, since u is measured on this face as shown by the arrow. The lower face is similarly the v *face*. Vectors uf and vf are current vectors in the x and y directions (parallel to u and v), but are measured on the v and u faces respectively. This is a device which aids representation of the Coriolis force.

6.5.2 Rotating Earth

On a rotating Earth it is necessary to add the Coriolis force discussed in Chapter 4. This force acts to the right of the velocity of a moving particle (to the left in the southern hemisphere). In dimensional form the momentum equations (without friction) for u and v now have the form

$$h\frac{\partial u}{\partial t} = -gh\frac{\partial \eta}{\partial x} + fhv \tag{6.44}$$

$$h\frac{\partial v}{\partial t} = -gh\frac{\partial \eta}{\partial y} - fhu \tag{6.45}$$

where f is the Coriolis parameter having the units of inverse time (f is positive in the northern hemisphere and negative in the southern hemisphere). It is convenient to write f in terms of the inertial period $t_{\text{intertial}}$ defined as

$$t_{\text{inertial}} = \frac{2\pi}{f}. \tag{6.46}$$

Hence, f is essentially the rate of turning of the velocity vector in radians per unit time. So, in non-dimensional form, (6.44) and (6.45) become

(a)

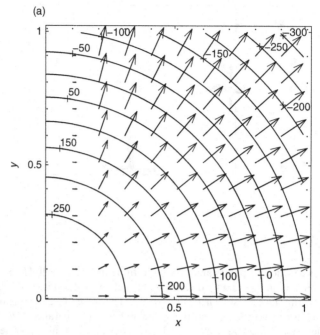

Figure 6.14 Output from the model presented in Table 6.9. The figure shows contours of elevation η and the arrows indicate direction of the current. Contours of $\eta' = 10^4\eta$ are shown where η is the dimensionless elevation and the numbers on the contours are $\eta' - 700$. The basin is closed and initially the surface is elevated by $\eta' = 1000$ at the bottom left-hand corner with the elevation decreasing to $0.04\,\eta' = 400$ at the top right hand corner. (a) Initial flow after one time step. (b) Surface elevation and current vectors at $t = 0.5$. Notice that the current vectors have a similar orientation to those in panel (a), but the contours of elevation have shifted due to the filling of the areas where the surface is lower. (c) Elevation and current vectors at time $t = 2.5$. Note that the currents do not necessarily follow the pressure gradients, because the slope of water elevation and current in wave motion are not necessarily in phase.

$$\frac{\partial u}{\partial t} = -\frac{\partial \eta}{\partial x} + f'v \qquad\qquad (6.47)$$

$$\frac{\partial v}{\partial t} = -\frac{\partial \eta}{\partial x} - f'u \qquad\qquad (6.48)$$

where

$$f' = \frac{2\pi}{t_{\text{inertial}}/t_{\text{basin}}} \qquad\qquad (6.49)$$

in which t_{basin} is the fundamental scaling time which we are using in this chapter, i.e., the time for a long wave to travel the length of the basin. Hence f' is the angle (in radians) through which the velocity vector turns in time t_{basin}. The Matlab code for a

(b)

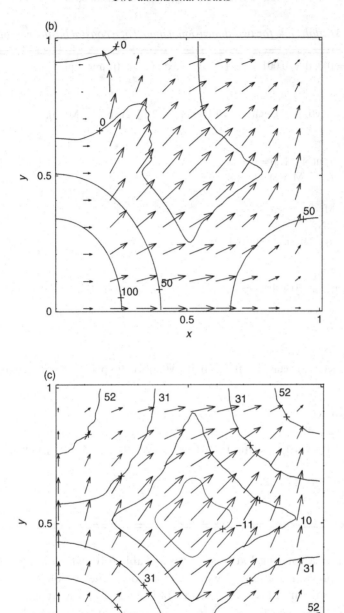

(c)

Figure 6.14 (cont.)

Table 6.10. *Matlab code for two-dimensional model of a coastal basin on a rotating Earth*

```
N = 100; M = 100; dt = 0.001; dx = 1/N; dy = dx; tau = 4; fprime = pi;
for j = 1:M;
    for i = 1:N;
        u(i, j) = 0;v(i, j) = 0; eta(i, j) = 0.1*exp((−i^2 − j^2)/(N^2 + M^2));
    end;
end;
for j = 1: M + 1; u(N + 1, j) = 0; end;
for i = 1: N + 1; v(i, M + 1) = 0; end;
while t < 0.5;
    for j = 1: M;
        for i = 2:N;
            px(i, j) = (eta(i, j)−eta(i−1, j))/dx;
        end;
        for i = 2:N;
            u(i, j) = u(i, j) + dt*(−px(i, j) + fprime*v(i, j)−u(i, j)/tau);
        end;
end;
for i = 1: N;
    for j = 2: M;
    py(i, j) = (eta(i, j)−eta(i, j−1))/dy; v(i, j) = v(i, j) + dt*(−py(i, j)−fprime*u(i, j)−v(i, j)/tau);
    end;
    end;
    for j = 1: M
        for i = 1: N;
        eta(i, j) = eta (i, j)−dt*((u(i + 1, j)−u(i, j))/dx + (v(i, j + 1)−v(i, j))/dy);
        end
    end;
    t = t + dt
end;
```

closed basin on a rotating Earth is shown in Table 6.10, which is equivalent to the model of the previous section but with the inclusion of the Coriolis force. This code is not optimized in terms of the full potential of Matlab, and is written in a simple array form to allow the reader to convert to any equivalent array form such as Fortran. It also uses a simple finite differencing method, designed only as introductory code.

Figure 6.15 shows a run of the model at time $t = 0.5$. It is evident that the current vectors show *Ekman rotation* (see Chapter 5) of $\pi/2$ radians (90°) in the time $t = 0.5$, which is consistent with our choice of $f' = \pi$. Note that rotation is not exactly $\pi/2$ because there are other forces acting on the water column (pressure and friction). If we perform a spectral analysis on the time series of any point in the basin, we will see, additionally to all the normal modes of oscillation of the basin, a clear signal at the

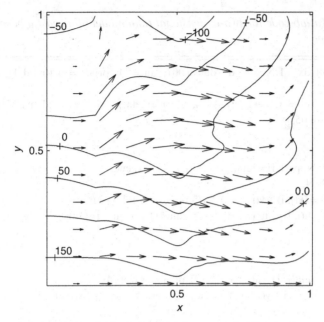

Figure 6.15 The closed basin model in Table 6.10 for a rotating Earth after a time $t=0.5$ for $f'=\pi$. Notice that the current vectors have moved through an angle of about $\pi/2$ which is $f't$. Compare Figure 6.14 for a non-rotating Earth.

inertial frequency. This is evident in all times series obtained from current meters and water level recorders deployed in coastal basins in which the inertial oscillator is not *over-damped*. To determine the inertial frequency, or period, it merely necessary to use (6.46):

$$t_{\text{inertial}} = \frac{12*3600}{\sin(\text{latitude})} \tag{6.50}$$

i.e., the period is 12 h at the poles and diverges at the equator. For example, a current meter deployed at latitude $30°$ would show inertial oscillations with a period of 24 h (provided damping is not too great). Such oscillations are interestingly very close (but not identical) to the diurnal tidal periods (see Table 6.1), and importantly, the coincidence of the M2 tidal period and inertial period occurs at a latitude somewhat less than that of the poles.

6.5.3 Tidal model with rotation

We can extend the models of the previous sections to provide a two-dimensional tidal model. The tidal boundary is applied at the mouth, i.e., $x=0$. In two dimensions, we can make a decision as to whether the mouth extends the entire length of the $x=0$

Table 6.11. *Matlab code for two-dimensional tidal model of a coastal basin on a rotating Earth*

```
N = 100; M = 50; dx = 1/N; dy = dx; dt = 0.001; tau = 20; fprime = pi/10; tidal_period = 5;
a(1:M) = 0;
for j = 1:M + 1; for i = 1:N + 1; u(j, i) = 0; v(j, i) = 0;eta(j, i) = 0;end; end; a(1:M/5) = 0.05;
a(4*M/5:M) = -0.05;
while t < 2.5;
    for j = 1: M;
        for i = 2: N; px(j, i) = (eta(j, i)-eta(j, i-1))/dx; end;
        for i = 2: N;
            vf(j, i) = (v(j, i) + v(j + 1, i) + v(j, i-1) + v(j + 1, i-1))/4;
            u(j, i) = u(j, i) + dt*(-px(j, i) + fprime*vf(j, i)-u(j, i)/tau);
        end;
end;
for j = 1: 1: M;
        px(j, 1) = (eta(j, 1) - a(j)*sin(2*pi*t/tidal_period))/dx;
        vf(j, 1) = (v(j, 1) + v(j + 1, 1))/2; u(j, 1) = u(j, 1) + dt*(-px(j, 1)-u(j, 1)/tau);
end;
for i = 2: N;
        for j= 2: M;
            py(j, i) = (eta(j, i)-eta(j-1, i))/dy;
            uf(j, i) = (u(j, i) + u(j, i + 1) + u(j-1, i) + u(j-1, i + 1))/4;
            v(j, i) = v(j, i) + dt*(-py(j, i)-fprime*uf(j, i)-v(j, i)/tau);
        end;
end;
for j = 1: M
        for i = 1: N;
            eta(j, i) = eta(j, i)-dt*((u(j, i + 1)-u(j, i))/dx + (v(j + 1,i)-v(j, i))/dy);
        end
    end;
        for i = 1:N + 1; u(M + 1, i) = u(M, i);u(M + 1,i) = 0; end; for j = 1:M + 1; v(j, N + 1) =v (j, N);
v(j, N + 1) = 0; end;
        t = t + dt
end;
```

boundary, i.e., from $y = 0$ to some maximum value of y, denoted by y_{max}, or over any part of the boundary. In the code in Table 6.11 we have chosen to put the open mouth along the whole boundary; this is illustrated in Figure 6.16. The grid shown in this figure is a staggered Arakawa C grid which we illustrated in Figure 6.13.

Figure 6.17 shows that the effect of the Coriolis force is to make the current vectors tend to run parallel to the contours of elevation; we have deliberately accentuated that effect by reducing friction below values realistic for a coastal basin, and the mouth of

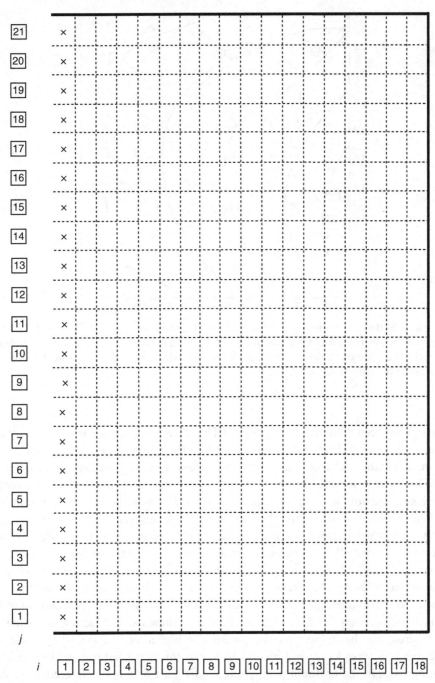

Figure 6.16 An illustration of the tidal boundary for a two-dimensional basin model with $N = 18$ and $M = 21$. The cell centers marked x have tidal elevation boundary conditions. The solid bold lines denote the closed boundaries of the basin and the broken lines are the boundaries between cells. Refer to Figure 6.13 for the details of variables within each cell. Currents u and v are measured perpendicular to the faces of the cells (currents uf and vf are measured along the faces of the cells). Surface elevation η is measured at the center of each cell.

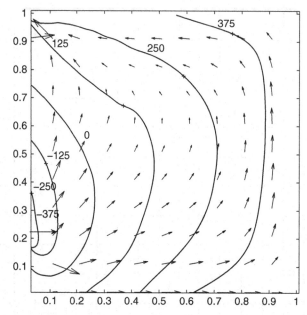

Figure 6.17 Surface elevation contours (isobars) of $10^4 \eta$ and current vectors at a time $t = 2.5$ for the model code in Table 6.11 for a tidal model of a coastal basin on a rotating Earth (northern hemisphere) connected to the ocean through two tidal channels each occupying 20% of the width of the basin at its edges.

the basin has been reduced in lateral extent The current in the basin is in approximate geostrophic balance (Section 5.7), and notice that the high pressure (surface elevation) region is to the right of the direction current vectors (northern hemisphere). The Ekman number for Figure 6.17 is $1/(50\pi)$. Figure 6.18 shows the same basin with higher friction, and we have chosen $\tau = 20$, $f' = \pi/10$, i.e., an Ekman number of $1/(2\pi)$. Figure 6.18 shows the elevation contours (isobars) and current vectors at a time $t = 2.5$, which corresponds to the ocean at mean water level. In Figure 6.18, we have applied a spatially uniform elevation along the whole of the y edge of the basin as shown in Table 6.12. Care is required with this type of boundary condition in a rotating basin using the present simple geometry. Figure 6.18 shows that at higher Ekman number, the velocity vectors are closer to being perpendicular to the isobars, i.e., they are following the pressure gradients. The isobars are slightly tilted across the basin, so that the pressure is higher on the right-hand side of the current vectors. As we will see in Chapter 8, this is the natural circulation in a rotating basin in the northern hemisphere, and Figure 6.18 is the antecedent of the classical estuarine circulation which is counterclockwise in the northern hemisphere. The reason for this sense of circulation is that the current swings to its right, and therefore on the ebb tide the current tends to run along the upper side of the figure,

Table 6.12. *Modification of the Matlab code in Table 6.11 in order to apply a spatially uniform elevation along the whole of the y edge of the basin*

```
for j = 1: 1: M;
        px(j, 1) = (eta(j, 1) − a(j)*sin(2*pi*t/tidal_period))/dx;
        vf(j, 1) = (v(j, 1) + v(j + 1, 1))/2; u(j, 1) = u(j, 1) + dt*(−px(j, 1)−u(j, 1)/tau);
end;
```

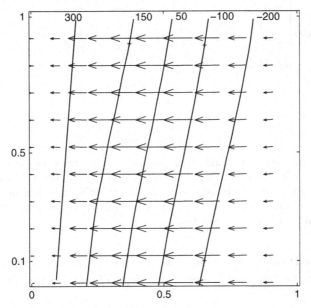

Figure 6.18 Surface elevation contours (isobars) of $10^4 \eta$ and current vectors at a time $t = 2.1$ just after the elevation boundary condition has started to increase from zero on its second cycle.

while on the flood it swings against the lower side. In the southern hemisphere, the circulation has the opposite sense (i.e., is clockwise).

The solution of the wave equation in the presence of the Coriolis force is an important problem in geophysical fluid dynamics. When solved near a coast, the plain waves that we encountered in Chapter 4 develop an exponential decrease away from the coast. The wave is then called a *Kelvin wave*. If we consider a straight open coast, the only length scale that appears is c/f where $c = \sqrt{gh}$; this is basically the distance moved by a gravitation wave in the inertial time $1/f$. This is called the *Rossby deformation radius*. In terms of the basin we are studying in this section, a wave takes unit (dimensionless) time to travel the length of the basin, and so in time $1/f'$ (for $f' = \pi/10$) the wave travels $R = 10/\pi$ lengths. Hence, the deformation radius R is much greater than the width of the basin. A run of the model with the unreasonably high value of $f' = \pi$ and a basin of width 2 is shown in Figure 6.19. This has $R = 1/\pi$, and

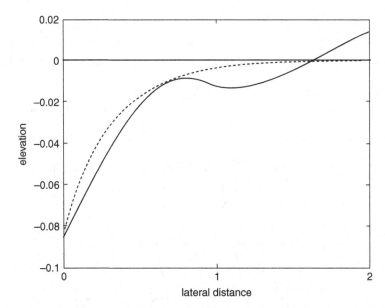

Figure 6.19 Lateral variation of elevation (full line) of the basin in the model shown in Table 6.12 at time $t = 2.5$ when $f' = \pi$. The dimensionless Rossby radius is $1/\pi = 0.32$, and the dotted line shows an exponential Rossby decrease (corresponding to the Kelvin wave that would occur if the basin were infinitely wide); there is reasonable agreement until the center of the basin when the effect of the other coast is evident.

would correspond to an unreasonably wide basin several thousand kilometers wide for a depth of 50 m, and hundreds of kilometers at a depth of 5 m (which would be dominated by friction).

6.6 Model speed and the cube rule

Traditionally numerical models of coastal basins used simple rectangular grids, of the type illustrated in Figures 6.13 and 6.16, and these are both extremely practical and simplest for coding purposes. One of the problems of using such grids is that they do not provide the greatest economy of computing time versus spatial resolution. This is an important consideration for two reasons. The first is that the number of cells in a model increases as the inverse square of the side-length of the cell; so doubling the resolution, i.e., halving the side-length of the cells, increases their number by a factor of four. If the time step can be keep constant, this means that the computer run time of the model increases by a factor of four since the computer program needs to perform the same numerical operations on four times as many cells.

In reality, the time step in numerical models must also be reduced as the size of the cells is reduced. The reason is that the controlling factor for the maximal time step allowed by stability is the time for a surface wave gravity wave of speed c to propagate across the cell i.e. $\Delta x/c$. We define the ratio of our time step Δt to this propagation time as the *Courant number C*,

$$C \equiv \frac{c\Delta t}{\Delta x}. \tag{6.51}$$

If the advective speeds in the basin are comparable with c, as they may be in shallow basins, then c should be modified to include particle advection speeds.

The maximum value of the Courant number allowed by stability is dependent on the way that the model integrates the dynamical equations of the basin. The simplest method is so-called *explicit* integration and this is easily illustrated by the simple differential equation

$$\frac{dy}{dt} = 1 - y \tag{6.52}$$

in which we wish to find y as a function of t starting with $y=0$ at $t=0$. The analytic solution of the equation is simply obtained by integration:

$$\int_0^y \frac{dy}{1-y} = \int_0^t dt \tag{6.53}$$

which gives

$$y = 1 - e^{-t}. \tag{6.54}$$

If we integrate the equation numerically, an explicit method would use the finite difference approximation

$$\Delta y = \left.\frac{dy}{dt}\right|_t \Delta t = [1 - y(t)]\Delta t. \tag{6.55}$$

This method does involve an error which can easily be seen graphically in Figure 6.20. The explicit method essentially approximates the quantity $\Delta y/\Delta t$ by the slope of the $y(t)$ curve at the initial point time t whereas the true value should be based on the slope of the chord which joins the values of y at t and $t+\Delta t$. Since we know the exact solution we can easily find the error from the explicit method,

$$\Delta y|_{\text{explicit}} - \Delta y|_{\text{exact}} = e^{-t}\Delta t - \left(-e^{-(t+\Delta t)} - e^{-t}\right) \tag{6.56}$$

and assuming that $\Delta t \ll 1$ the error is

$$\Delta y|_{\text{explicit}} - \Delta y|_{\text{exact}} \sim \frac{(\Delta t)^2}{2} e^{-t}. \tag{6.57}$$

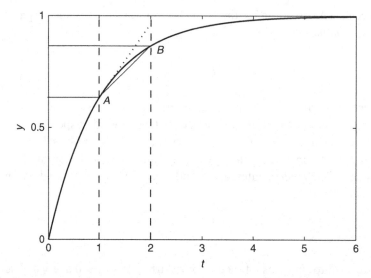

Figure 6.20 The curve represents the function $y(t)$ in Section 6.6. Two values of t have been chosen and corresponding values of y indicated by the intercepts with the $y(t)$ curve and marked A and B. The figure also shows the tangential slope of the curve at A and the chord joining A and B. Numerical integration of the differential equation, whose solution is $y(t)$, involves finding the point B by integrating the differential equation from A to B. This is properly done using some approximation to the slope of the chord but an explicit method approximates this slope by the slope of the tangent. For clarity, values of t are here illustrated further apart than would be used in a practical integration procedure.

It may be possible to use a better approximation for $\Delta y/\Delta t$ to reduce this error. There is in mathematics the *mean value theorem* which states effectively the value of $\Delta y/\Delta t$ is equal to the derivative dy/dt somewhere in the interval t to $t + \Delta t$. A possible route is to assume that $\Delta y/\Delta t$ is the average of the derivatives of y at t and $t + \Delta t$. If we define

$$y_1 = y(t)$$
$$y_2 = y(t + dt)$$
(6.58)

then using (6.52) we would obtain

$$\frac{\Delta y}{\Delta t} = 1 - \frac{y_1 + y_2}{2}$$
(6.59)

and hence

$$y_2 = y_1 - \left(\frac{y_1 + y_2}{2} - 1\right)\Delta t$$
(6.60)

which gives

$$y_2\left(1 + \frac{\Delta t}{2}\right) = y_1\left(1 - \frac{\Delta t}{2}\right) + \Delta t \tag{6.61}$$

and leads to

$$\Delta y = \frac{\Delta t}{1 + \Delta t/2}(1 - y_1) \tag{6.62}$$

which for $\Delta t \ll 2$ leads to

$$\Delta y = \Delta t(1 - \Delta t/2)e^{-t} \tag{6.63}$$

which is correct to second order. This method differs from the explicit solution in that it treats the derivative as being only an implied quantity within the differential equation. Such *implicit* methods allow much longer time steps in numerical models. They do involve much more complicated coding since the equivalent of y_1 and y_2 in hydrodynamic models are arrays on the model grid. Implicit models use a variety of methods to optimize the approximations to the mean rate of change over the time step and many models use *semi-implicit* techniques. The use of implicit methods allows the model to remain stable at longer time steps assuming that the numerical scheme is able to accurately solve the equations of motion.

For *explicit models*, the Courant number in (6.51) must be less than unity for numerical stability. This is known as the Courant, Friedrichs, and Levy (CFL) criterion.[3] It is a severe restriction on the maximum time step which can be used in the model, i.e.,

$$\Delta t < \frac{\Delta x}{c} \tag{6.64}$$

and limits the speed of the model. This restriction can be overcome by use of *implicit* models which allow values of the Courant number larger than unity and in shallow basins typically of order 5 to 10. The time step Δt must be the same in all cells of the model and such that in the cell with maximum $c/\Delta x$, C is less than the maximum allowed value. Hence if we have a shallow basin with typical depths of a few meters but a navigation channel with depths up to $20\,\mathrm{m}$, it is the wave speed in the channel where c is maximal that controls the time step everywhere in the model. Implicit models avoid this problem slightly by much greater numerical care in the time stepping method and work with very much higher values of Courant number closer to 10, i.e., $C < 10$.

---

[3] See Courant, Friedrichs, and Levy (1928).

A technique used by some models is to extract the two-dimensional dynamics of the coastal in what is called the *external mode*. Essentially this involves one solving the two-dimensional depth-averaged equations at the time step Δt dictated by the CFL criterion and only solving the baroclinic, or *internal mode*, at some multiple of Δt. This is called *mode splitting* and was used by one of the earliest models to have widespread use by non-specialist models: the Princeton Ocean Model (POM)[4] available on the internet.

However no matter what care we take to increase the allowed value of C it remains true that in a particular basin, Δt is proportional to the value of Δx. This means that the run time of a model becomes proportional to the inverse cube of Δx; i.e., the computational speed of the model increases as the cube of cell size. So, if we try to increase our spatial resolution by a factor of 10, for example go from 500 m to 50 m, our run time increases by a factor of 1000. In practice this means that for a given simulation time of the model, and practical limit on the run time, the minimum size of the cells is determined to within about a factor of about 25%; for example changing cell size from 80 m to 100 m almost doubles the run time. The converse of this *cube rule* relating cell size to computational speed is that increasing the simulation time (at given run time) is comparatively inexpensive in terms of spatial resolution: for example if we go from a 1-month simulation to a 1-year simulation, the cell size need only be increased by only a factor of $\sqrt[3]{12} = 2.3$.

6.7 Horizontal grids

Our general discussion of model grids has been based on the structured rectangular mesh. An example of such a grid with square cells of side 500 m is given in Figure 6.21 for Tomales Bay on the west coast of Florida. Such grids are the simplest to use and are capable of representing all of the dynamics of coastal basins. The disadvantage of rectangular grids is that they do not optimize the speed with which a model can run given the constraints of the cube rule discussed in the last section. The answer to the dilemma posed by the cube rule for model speed is to use our spatial grid more economically and only have small cells were we need them. This question is fundamentally bound up with the role of coastal topography and bathymetry in the behavior of coastal basins and on larger spatial scales in the coastal region in general. There are several ways to reduce the number of cells and these are presented in the next sections.

6.7.1 Nesting

The first method is to use a grid that contains cells of varying size. The most popular approach is to use a finite number of cell sizes that are arranged so that they fit inside

---

[4] Blumberg and Mellor (1987).

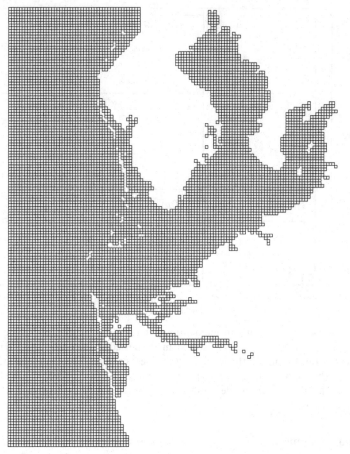

Figure 6.21 A grid consisting of square cells of side 500 m used in Tampa Bay on the west coast of Florida.

one another in a modular form; for example, we might start with cells of size 810 m square and then decide that we need more resolution in certain parts of the basin. We therefore introduce a second cell size of 270 m and we can fit exactly nine 270 m cells inside a 810 m cell. So, we take a continuous *rectangular block* of 810 m cells and nine of the 270 m cells inside each of the 810 m cells in the rectangle. This is called *nesting*. There are some choices as to how the nesting numerical analysis works and it can be either *one way* or *two way*. In the former method, the large cells are merely used as boundary conditions for the smaller nested cells but the smaller cells do not affect behavior in the larger cells. The two way method means that the smaller cells do affect the larger cells and this is clearly much more difficult in a coding sense. Inside the smaller cells, we can go ahead and fit yet smaller cells and for our present example these might be 90 m square so that we can fit exactly nine cells inside a 270 m cell. This is illustrated in Figure 6.22. The same rules apply and we need to nest the 90 m cells

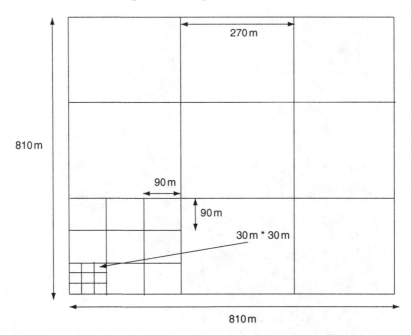

Figure 6.22 Illustration of nesting inside a square cell of side 810 m. Cells of side 270 m are nested inside one of the cells and cells of side 90 m are nested inside one of the 270 m cells. The figure only shows one 810 m cell (for clarity) but these would be expected to extend across the entire coastal basin. In this example, high spatial resolution is only needed near the bottom left-hand corner of one cell. To run a model with 30 m resolution over the entire basin would involve 9^2 times as many cells.

inside a rectangular mesh of 270 m cells. We can continue this process but each time we introduce smaller nested cells we reduce the time step *everywhere* in the model. The process of nesting illustrated in Figure 6.22 is an example par excellence because high resolution is needed in only one small area of the basin and so we reduce the number of required cells by a factor of 81. Unfortunately, the usefulness of the nesting procedure is limited by the need to maintain rectangular edges on the nested area and also it is very difficult to allow nested areas to touch one another. Consequently, the reduction of cell numbers is often rather limited.

6.7.2 Curvilinear coordinates

Another way of introducing cells or varying size is to make a spatial transform in the coordinates that describe the basin and then use the transformation to focus the resolutions. This stirs mathematical memories of the methods that we used in Section 2.9 but is now used a pre-processor of the full three-dimensional hydro-dynamic equations of motion for a coastal basin. The method has been used to produce very small cells near to a coastal outfall based on the expectation that there is a *near field* region in the immediate vicinity of the outfall that shows smaller-scale spatial variability. The most useful transformation is the logarithmic transformation taken about the point of entry of the outfall to the basin and in this case the size of the cells increases logarithmically with distance from the point. Another method is to use so-called orthogonal curvilinear coordinates, of which polar and spherical coordinates are well-known examples. In these systems, the contours of constant coordinates intersect one another at right angles although they are curves and not straight lines as for Cartesian coordinates. Curvilinear coordinates can be mapped onto a rectangular grid system by the analogue of the conformal mapping techniques that we saw in Chapter 2. This mapping essentially straightens out all the complexities of the coastline of the basin so that we can solve for the hydrodynamics of the basin in very much simpler geometry. Furthermore curvilinear grids have variable spatial resolution over the basin (typically varying by an order of magnitude) and therefore (like nested grids) can use high resolution only in regions of the basin where it is required and so economize on the total number of cells in the model. The reason for this variability of resolution can be seen in the simple case of polar coordinates, i.e., the use of a set of concentric circles. As the circles become smaller so the area between fixed radii becomes smaller. Furthermore the radii of the circles can be made closer to one another as needed near to region where high resolution is required. Curvilinear coordinates are ideally suited to complex coastlines with narrow inlets and islands where they can produce savings in the total number of cells of two orders of magnitude. The use of curvilinear coordinates is dependent on grid generation software which provides the necessary mapping coefficients. The method is illustrated in Figure 6.23 which shows a curvilinear grid for part of the coast of Zeeland (Netherlands) created by the model DELFT 3D.[5] In this method the cells sides of the cells may be either quasi-orthogonal or non-orthogonal but the basic aim remains the same: to bend the axes of the grid as that they run along the major axes of the topographic features of the basin or coastline. The method works especially well for rivers and canals. Curvilinear coordinates can follow smooth but complex topography although branching structures can present challenges.

6.7.3 Flexible meshes

A *flexible*, or *unstructured, mesh* is illustrated in Figure 6.24 for Tampa Bay on the west coast of Florida. It is developed by the DHI model MIKE3.[6] It consists of a set of triangles and has the great advantage over a fixed rectangular grid of allowing the cells to have variable size; each triangle abuts three others with different lengths. The

---

[5] WLIDelft Hydraulics, Rotterdamsweg, Delft, The Netherlands.
[6] DHI Water and Environment, Agern Allée 5, 2970 Hørsholm, Denmark.

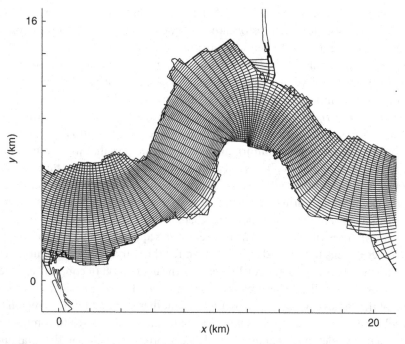

Figure 6.23 Curvilinear coordinates used for a coastal basin on the coast of Zeeland developed by the model DELFT 3D (WLIDelft Hydraulics).

method has very great utility in providing cells that cover an almost continuous spectrum of cell sizes that range from cells of some tens of meters increasing without limit: Figure 6.25 shows the same flexible mesh as in Figure 6.24 but extending outwards to much larger spatial scales (in this case over the Gulf of Mexico). The model in Figure 6.25 has two external boundaries, one with the Atlantic Ocean and the other with the Caribbean Sea.

The important, and vital, ingredient of a flexible mesh is that the cells can vary in size to accommodate the scale of the local bathymetry and coastal features. Notice that in Figure 6.24 the grid has a high density of small cells in the navigation channels of Tampa Bay as well as cells that resolve coastal and structural features of the bay. Flexible mesh grids can be automatically generated by programs that recognize coastal features but may require manual intervention by the modeler to assist the grid resolution to follow the bathymetric variability; such a grid is called a *bathymetry following flexible mesh*. This attribute is a feature which is possibly unique to the flexible mesh since bifurcating and convoluted bathymetric contours are harder to follow with curvilinear coordinates. Note that the flexible mesh, and all the grids that we are discussing, are subject to limitations imposed by the Courant number; a flexible mesh only serves to reduce the number of cells and does not affect the time step. So a flexible mesh has a time step dictated by the cell with the largest value of $c/\Delta x$ but will

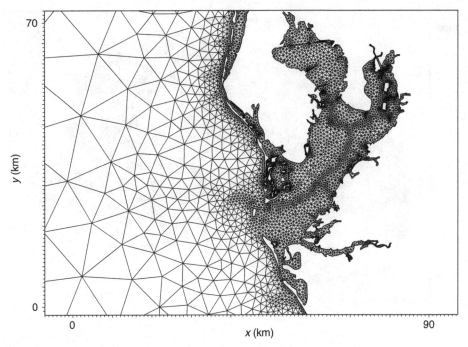

Figure 6.24 The DHI bathymetry following flexible mesh (DHI Water and Environment) for Tampa Bay on the west coast of Florida.

nevertheless have many less cells than a rectangular mesh with all cells the same size as the smallest cell in the flexible mesh.

6.8 Vertical structure of model grids

There is a choice as to the type of vertical spacing used for the grid. These choices divide broadly into two types. This first is the so-called *z layers* in which layers are used which have constant thickness across the basin; the layers may not be all of the same thickness but their thickness does *not* vary with (*x, y*). This is illustrated in the middle and bottom panels of Figure 6.26. There are many ways of choosing the thickness of individual layers, and for example, modelers may choose to have thinner layers at the top and bottom in order to better resolves boundary layers. The *z* layer method lends itself well to *wetting/drying* because an individual *z* layer may simply be dry in one particular spatial cell (and wet in neighboring cells).

The other type of vertical spacing is referred to as *σ* (sigma) layers or *σ* coordinates. This name simply comes from the transformation

$$\sigma = \frac{z}{h} \qquad (6.65)$$

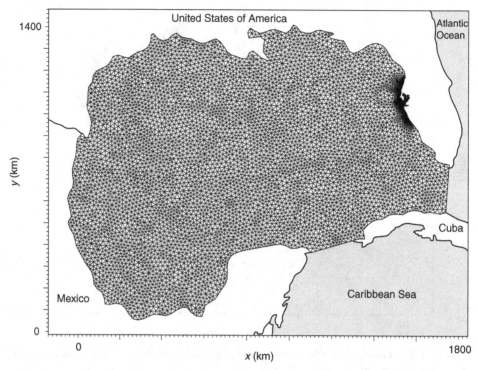

Figure 6.25 The DHI flexible mesh (DHI Water and Environment) extended from Tampa Bay (on the west coast of Florida; see Figure 6.24) to the Gulf of Mexico.

which is a common way of creating a non-dimensional height in the water column. As (6.65) implies, we make a σ transformation before coding the model and then use a vertical spacing uniform in σ. The layers of constant thickness in σ space have variable thickness in real space and so we always have the same number of layers below a horizontally level datum surface; they have varying thickness in proportion to depth h as shown in the top panel of Figure 6.26. The advantage of σ coordinates is that they maintain the same number of layers at all points in the basin and this is especially valuable in regions of rapidly varying depth. They can also have closer spacing (relative to the height of the water column) near to the bottom in order to better resolve the bottom boundary layer (Figure 6.27). Provided the model has a *free surface* (as for most models of coastal basins), i.e., the surface moves up and down under tidal and other influences, the σ coordinates expand as the water level rises and compact as the level falls (Figure 6.28). There are reservations from modelers as to the veracity of results from both types of coordinate systems in special circumstances and the user is best advised to be wary of effects which could be associated with the choice of coordinates especially in very shallow water and particularly in the presence of strong bottom curvature. Some of these effects are associated with numerical diffusion effects.

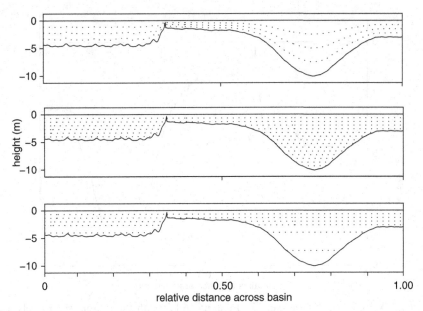

Figure 6.26 Illustrations of the vertical structure of numerical models in common use. The top panel shows the sigma coordinate (σ coordinate) system in which the thickness of the vertical layers is scaled to the depth of the basin. The interfaces between the layers are not horizontal but have the same shape as the bottom of the basin. This allows much greater vertical resolution in the shallower parts of the basin. The middle and bottom panels show a z coordinate system in which the interfaces between the layers are horizontal. The thickness of the layers can either be all equal (middle panel) or more closely spaced near the surface (bottom panel). It is also common practice with both σ and z coordinates to reduce spacing near the bottom and surface to resolve the log layer (not shown).

There are noted examples of σ coordinates in very shallow basins failing to represent properly the stabilizing effects of vertical density gradients and tending to remove stratification. It should be emphasized that all finite difference systems tend to produce numerical diffusion in the horizontal and vertical mixing simply because they use horizontal and vertical cells of finite size. A z layer *mixes* over a range of z corresponding to its thickness and disperses over a finite horizontal area (in that z interval). A σ layer does the same thing but mixes over a range of σ over the horizontal area of the cell and therefore moves particles in the vertical without regard for vertical stability imposed by the density gradient. This is only a significant effect if there is significant bottom curvature relative to depth on the spatial scale of the cells and can be a problem in shallow basins.

The vertical structure of the model determines stability criteria associated with vertical diffusion or mixing. Generally these produce restrictions that are less severe than those imposed by the CFL criterion discussed above (in Section 6.6) although that may not be true in very shallow water.

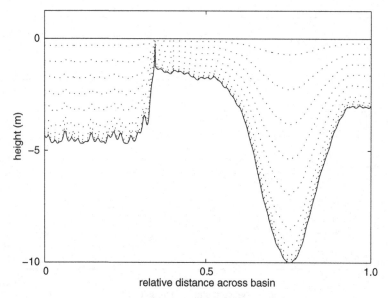

Figure 6.27 A repeat of the top panel of Figure 6.26 but with the bottom six σ levels scaled logarithmically to improve vertical resolution of a bottom log layer.

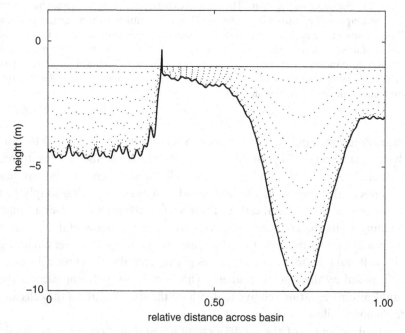

Figure 6.28 A repeat of the top panel of Figure 6.26 but with a tidal decrease in water level to −1 m. Notice the way that the σ levels are squeezed together to accommodate the reduction in the height of the water column and further that this compaction is greatest in the shallowest parts of the basin that have remained wet.

6.9 Further reading

See further reading from earlier chapters. Additional texts for this chapter are:

Abbott, M. B. (1994). *Coastal, Estuarial and Harbour Engineer's Reference Book*. London: E. & F. Spon.

Boon, J. (2004). *Secrets of The Tide: Tide and Tidal Current Analysis and Applications, Storm Surges and Sea Level Trends*. Chichester, UK: Ellis Horwood.

Brenner, S. C. and L. R. Scott (2005). *The Mathematical Theory of Finite Element Methods*. New York: Springer.

Cartwright, D. E. (2001). *Tides: A Scientific History*. Cambridge, UK: Cambridge University Press.

Haidvogel, D. B. and A. Beckmann (1999). *Numerical Ocean Circulation Modeling*. Singapore: World Scientific Publishing Co.

Open University. (2000). *Waves, Tides and Shallow-Water Processes*. Oxford, UK: Butterworth-Heinemann.

Pugh, D. (2004). *Changing Sea Levels: Effects of Tides, Weather and Climate*. Cambridge, UK: Cambridge University Press.

Randall, J. L. and M. J. Ablowitz (2002). *Finite Volume Methods for Hyperbolic Problems*. Cambridge, UK: Cambridge University Press.

Versteeg, H. and W. Malalasekra (1996). *An Introduction to Computational Fluid Dynamics: The Finite Volume Method Approach*. New York: Prentice Hall.

7

Mixing in coastal basins

7.1 Introduction

This chapter considers various aspects of mixing in coastal basins, with particular emphasis on the comparative roles of turbulence and molecular processes. Some of the applications will be pursued in more detail in later chapters, and here we mainly discuss vertical mixing processes. The key elements of mixing theory as presented in this book are Fick's law and idea of the coefficient of mixing. Our focus in this chapter is largely with mixing in unstratified basins and this forms the basis for our later understanding of the way in which stratification reduces vertical mixing. The only exception to this order of presentation is that we do briefly consider in this chapter the effects of the thermocline at the bottom of a surface wind mixed layer.

7.2 Theory of mixing

7.2.1 Nature of mixing

Mixing is a crucial component of the dynamics of coastal basins, and involves all quantities that are of interest in models of coastal basins such as momentum, energy, salt, heat, dissolved nutrients and chemicals (such as oxygen), organic and non-organic particulate material, and so on. There are a number of specialized terms that are used for different types of mixing. The mixing of momentum is also called friction or *viscosity*. The word *diffusion* is synonymous with mixing, and we will often use it instead of mixing. Mixing is one of the most important topics in oceanography, and plays a major role in the basic dynamics of the oceans and estuaries. It also affects all life forms that live in the ocean. The ideas that we will develop in this chapter apply equally well to lakes and rivers, and also to the atmosphere although with possibly very different spatial and temporal scales. So, much of the material of this chapter forms a part of the elementary introduction to the more general topic of *mixing in the environment*, and we could talk about *environmental fluids* rather than coastal basins. We shall see that the way in which quantities mix depends on the *spatial scale* at which we work, so that the strength of mixing in a cup of coffee is very different from that in a coastal basin. Mixing can occur either in the vertical where it is described as *vertical*

228

mixing or in the horizontal where we describe it as *horizontal dispersion*. Dispersion simply means spreading, and we find the same word used for dispersion of a wave packet due to the change of propagation speed with frequency.

Let us start with an example of vertical mixing of heat. Mixing heat in the vertical changes the vertical profile of temperature. This is illustrated in Figure 7.1, which has been obtained from a very simple model of vertical mixing, which is the exact analogue of the one-dimensional model of horizontal dispersion that we met in Chapter 3. It relies on the definition of a vertical coefficient of mixing similar to that we which also met in Chapter 3, and which will be explored in more detail later in this present chapter. The water column in Figure 7.1 has been modeled on the assumption that there is no heat flow across the surface, i.e., no net gain or loss of heat from any of the processes that we discussed in Chapter 3. Similarly, there is no heat exchange at the bottom of the basin which is a very common assumption in coastal basins (although there are many interesting cases in which this is not true). So the heat content of the basin is simply rearranged vertically with time, based on the rule that heat flows from warmer to colder regions. This means that the total heat is conserved, and so the depth-averaged temperature remains constant. The final profile of temperature represents a situation in which the basin has become *vertically mixed* or *well mixed*. The boundary conditions of zero heat flux at the surface and bottom are evident in Figure 7.1 as zero gradients of temperature T with respect to the vertical coordinate z.

This type of vertical mixing of heat usually occurs in water columns of height from a fraction of a meter to tens of meters. On such scales, fluids do not mix naturally in the vertical, but require some *external agency* to mix them. These external agencies are said to *stir* the fluid. We can add energy to the system by *stirring*, and a common example is simply stirring a cup of coffee to induce mixing of the coffee and cream. At much smaller spatial scales, there are different mixing processes that do not require such external agencies but rely on the natural thermal energy to produce mixing.

There are two important processes by which fluids move in coastal basins and throughout the ocean. They are called *advection* and *diffusion*. The distinction between these two processes is most evident in the horizontal. Advection is associated with an ordered, or *average*, motion of particles. Mixing, or diffusion, is due to a *random* motion of particles. In our initial discussion of mixing we will ignore advection totally, and consider the motion of all particles to be random with zero average motion. For example, the vertical mixing of heat in Figure 7.1 is due to diffusion, and there is no contribution from advection. Advection can certainly move heat and for example a warm current moves heat from warmer regions, but this is not considered to be mixing. In the vertical, mixing is the dominant process for moving heat, although advection can sometimes be important as in the upwelling or downwelling of water masses. The random motion which gives rise to mixing is a natural property of fluids; at very small spatial scales it is due to thermal motion and at larger scales involves some form of stirring. Hence, all mixing is conceptualized as individual packets of

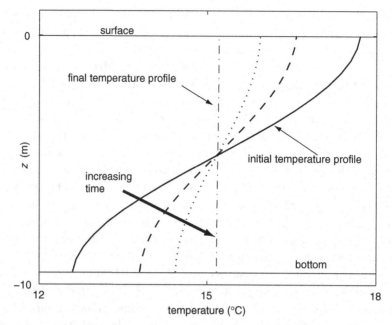

Figure 7.1 Schematic of the vertical mixing of heat which, in the absence of other processes, tends to equalize the temperature at all points of the water column. Illustration shows a basin of depth 9.5 m with an initial temperature profile ranging from 17.7 °C at the surface, to 12.6 °C at the bottom of the basin. Mixing would tend to create a vertically mixed water column at a uniform temperature of 15.15 °C.

fluid interchanging places in the fluid, because in the absence of any mean motion particles cannot move without being replaced by others. For example, we might imagine packets moving as shown in Figure 7.2.

The interchange of particles in Figure 7.2 will tend to reduce the vertical temperature gradient, since colder water moves upward and warmer water downward. Note that the total heat and mean temperature remain constant. Mixing always *conserves* the quantity being mixed within the system under study. Notice that the mixing of heat is quite different from the *upwelling* of cold bottom ocean water, or *downwelling* of warm surface water, which are both due to advection and *not* mixing. All types of physical, chemical, and biological quantities can be mixed. In the first example in Figures 7.1 and 7.2 we could have discussed horizontal momentum (current velocity) rather than temperature. Movement of packets of the water column involve movement of heat (tending to equalize temperature), salt (tending to equalize salinity), and momentum (tending to equalize velocity). It may also involve movement of dissolved oxygen and nutrients, tending to equalize concentrations of oxygen and nutrients. The rate of mixing depends in general on the strength of the random motion and on the quantity being mixed, although in certain circumstances this rate is the same for all quantities.

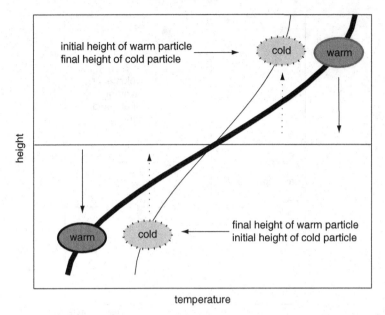

initial height of warm particle
final height of cold particle

cold   warm

warm   cold

final height of warm particle
initial height of cold particle

height

temperature

Figure 7.2 Schematic of the vertical mixing of heat due to particles interchanging position in the water column. The vertical temperature profile is shown by thicker full line at some initial time. Two hypothetical particles are shown: one warm and the other cold, with the warm particle initially near the top of the water column and the cold particle initially near the bottom of the water column. The particles then interchange positions and the temperature profile changes to that represented by the thinner line.

7.2.2 Fluctuations and diffusion

Fluids consist of a very large number of particles which have the ability to move relative to one another in an almost infinity of ways. We say a fluid has a very large number of *degrees of freedom*. So, to describe properly the motion of a fluid we would need to divide it into a huge number of small volumes, or masses, and assign a velocity vector in three dimensions at every time to each of these small masses. In a sense, this is exactly what we would do if we set up a grid for a numerical model of a coastal basin that had infinitesimally small cells and time steps. In reality, we can only use finite cells and finite time steps. Within each of these cells and time steps, our model is essentially performing an average over the space in each cell, and over the time that elapses during a time step. Let us start with the example of the component of velocity in the x direction (and suppose that we have a one-dimensional world). Our assumption is that there exists a function $u(x, t)$ which exactly describes the variation of u with x and t. Our model performs an average of u over both x (over a cell) and t (over a time step). This average means an integral over a small volume of space and time and is indicated by the multiple integral

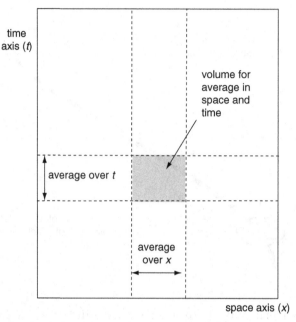

Figure 7.3 Schematic of volume of integration in space and time used to determine the mean components of velocity.

$$\bar{f}(x,y,z,t) = \int\limits_{t-\delta t/2}^{t+\delta t/2} \left(\int\limits_{x-\delta x/2}^{x+\delta x/2} f(x,y,z,t)\mathrm{d}x \right) \mathrm{d}t \tag{7.1}$$

where the spatial average is taken from $x - \delta x/2$ to $x + \delta x/2$ and the time average from $t - \delta t/2$ to $t + \delta t/2$. This is illustrated in Figure 7.3.

In three dimensions this average becomes a fourfold integration over x, y, z, and t. Using the same notation as in (7.1), the average involves

$$\bar{f}(x,y,z,t) \equiv \underbrace{\int\int\int\int f\mathrm{d}x\mathrm{d}y\mathrm{d}z\mathrm{d}t.}_{\text{space－time－volume}} \tag{7.2}$$

In (7.2), $f(x, y, z, t)$ represents any of the properties of the fluid. The space and time volume need not of course have a regular shape as in the figure. In typical numerical models of coastal basins, the space and time volume has dimensions of 10 to 100 m in the horizontal, 1 to 10 m in the vertical, and seconds to minutes in time. We also find it convenient occasionally to write the average using triangular brackets, especially when the quantity being averaged consists of a number of terms, i.e.,

$$\langle f(x,y,z,t)\rangle \equiv \bar{f}(x,y,z,t). \tag{7.3}$$

Let us write the components of the velocity (u, v, w) in the fluid in the (x, y, z) directions as

$$u = \bar{u} + u'$$
$$v = \bar{v} + v' \qquad (7.4)$$
$$w = \bar{w} + w'$$

where the over-bar represents an average in time and space as in (7.2), and the prime indicates the difference between the actual value of the component and its mean value. A relation similar to (7.4) can be expressed for any quantity with concentration $c(x, y, z, t)$ in the water column, i.e.,

$$c = \bar{c} + c', \qquad (7.5)$$

and we refer to c as representing a *species* of dissolved chemical, biological entity, or particulate material. Detailed numerical models are used to determine the mean values of current, i.e., $\bar{u}, \bar{v}, \bar{w}$, and of species concentration $\bar{c}$.

The primed quantities in (7.4) and (7.5) are referred to as *fluctuations*. It is important to prove that the mean value of any of the fluctuations taken over the volume of space-time used in defining the fluctuations is zero. So, for example, from the first of (7.4)

$$\langle u \rangle = \langle \bar{u} + u' \rangle \qquad (7.6)$$

and we note that the mean on the right-hand side of (7.6) is the sum of the means of the two terms inside the angular brackets (since it involves the integral of their sum) and hence, (7.6) becomes

$$\langle u \rangle = \langle \bar{u} \rangle + \langle u' \rangle. \qquad (7.7)$$

The first term on the right of (7.7) is the average of a constant, i.e., $\bar{u}$ is not a function of position inside the averaging volume and so it follows that

$$\langle u' \rangle = 0 \qquad (7.8)$$

and the same is true of the fluctuations of v, w, and c, and any quantity in the system, i.e., the fluctuations are *fluctuations about the mean*.

The space-time interval over which the average is taken is dependent on the scales used in the model, and this affects the value of the mean and the size of the fluctuations. The fluctuations are necessarily defined on smaller spatial and temporal scales than the mean quantities. Writing (7.4) is essentially a formality, since we will not be able to calculate the fluctuations. But as we shall see, these fluctuations are the basis of mixing, and we can express the rates of mixing (formally) in terms of averages over the products of the fluctuations. A difference between the mean quantities and the fluctuations depends on the spatial and temporal scales over which we define the averages. This means that there is no fundamental difference between the movement of quantities in a coastal basin by advection (which involves the average components of velocity and species concentration) and by mixing (which involves the fluctuations).

In order to try to model the fluctuations we could, of course, build a numerical model at smaller spatial and temporal scales, but these will always necessarily be at *finite* scales. So, there will always be fluctuations on smaller scales than those at which we can model. In coastal basins we terminate the numerical modeling process at the sort of scales that we have mentioned above, and then use other methods to represent averages of products of the fluctuations.

The advective flux density of the species with concentration c in the x direction is just $\bar{u}\bar{c}$ as discussed in Chapter 2. We now consider the flux density that would be obtained if we really knew $u(x, y, z, t)$ and $c(x, y, z, t)$ and not just their averages, i.e., we consider the *true* flux uc. Using (7.5) and (7.4) we can write this flux as $(\bar{u} + u')(\bar{c} + c')$. It is evident that this flux fluctuates in the same way as the components of current and concentration, but we shall see that it has a mean value (averaged over the same space-time volume as used for u and c alone) that is different from $\bar{u}\bar{c}$. So, in order to use the flux in models of coastal basins, it is necessary to take an average of this *total flux*, i.e.,

$$\langle(\bar{u} + u')(\bar{c} + c')\rangle = \bar{u}.\bar{c} + \langle u'\bar{c}\rangle + \langle\bar{u}c'\rangle + \langle u'c'\rangle. \tag{7.9}$$

The first term on the right of (7.9) is the advective flux discussed earlier. The second and third terms can be written as $\bar{c}\langle u'\rangle$ and $\bar{u}\langle c'\rangle$, since the mean values are constant over the integration volume and time interval in (7.1). But since the mean of any fluctuation is zero, these terms vanish and so without approximation

$$\langle uc\rangle = \bar{u}.\bar{c} + \langle u'c'\rangle \tag{7.10}$$

which shows that the true flux consists of the advective flux plus a term $\langle u'c'\rangle$, which we call the *mixing flux* or *diffusive flux*. To be more specific $\langle u'c'\rangle$ is the diffusive flux of the species represented by concentration c. Similarly, the diffusive flux of all other quantities can be written as shown in Table 7.1. As an example, Figure 7.4 illustrates the way that the fluctuations in w and u lead to the vertical flux of horizontal momentum in the x direction. The means of the fluctuations in w and u are zero, but the mean of their product $u'w'$ is non-zero with $\langle u'w'\rangle/\sqrt{\langle u'^2\rangle\langle w'^2\rangle}$.having a value of -0.06. The unpublished data in figure 7.4 were collected by using an acoustic Doppler velocimeter (ADV) at the coral reef mesocosm of Biosphere 2.[1]

7.2.3 Turbulence and scales of motion

One of the most important distinctions in the flow of fluids is that between *laminar flow* and *turbulent flow*. If we pour oil very carefully from a can into a car engine, we

[1] For details of the work at Biosphere 2 see Atkinson, Falter, and Hearn (2001).

Table 7.1. *Notation used for some diffusive fluxes in coastal basins*

Diffusive flux	Description of flux
$\langle w's' \rangle$	Vertical diffusive flux of salt
$\rho C_p \langle w'T' \rangle$	Vertical diffusive flux of heat
$\rho \langle w'u' \rangle$	Vertical diffusive flux of horizontal momentum
$\langle s'u' \rangle$	Horizontal diffusive flux of salt
$\rho \langle v'u' \rangle$	Diffusive flux in x direction of y momentum

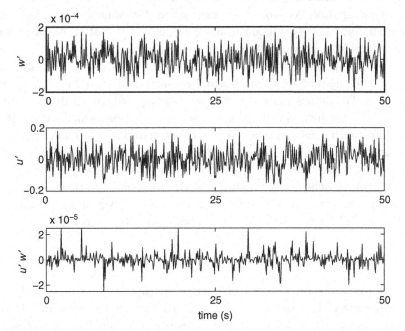

Figure 7.4 Upper panels show time series of fluctuations in w and u (vertical and horizontal components of velocity) and the bottom panel is the corresponding time series of the product $u'w'$. All velocities are in $m\,s^{-1}$.

notice that the flow is even and regular. A speck of dust on the surface of the oil can be seen to move downwards in a smooth path. This is known as *laminar flow*; the flow takes place like plates moving smoothly over one another. By contrast, consider the flow of water over a waterfall. A particle of water is seen to move in a very irregular fashion being tossed and turned as it falls. This is called *turbulent flow*. The difference between laminar and turbulent flow is controlled by the speed of motion in relation to the *viscosity* of the fluid, and by the *physical dimensions* of the flow. Oil is a very viscous fluid, which means that it tends to stick together as a fluid and all parts of the fluid tend to move together as a whole. This is due to strong molecular forces within the fluid which hold it together. By contrast, water has very *low viscosity*: compare the tasks of stirring a bath of water and a bath of oil (or perhaps syrup) with a spoon. Generally, the individual parts of a body of water only move in a laminar fashion if the flow is very slow. Usually, water flows in a turbulent manner and this is true of most of the oceans. If water was more viscous, we would probably not exist in our present form since our bodies are based on water. Also, we would not be able to move boats through water with such ease if water had the viscosity of oil.

A good example of turbulent motion in water is seen in a swimming pool filled with swimmers. Arriving early in the morning when the pool has been unused for hours, one sees that the water is perfectly still. If you swim very gently, you can almost maintain laminar flow. As more swimmers arrive, the water becomes increasingly turbulent and increasingly difficult to move through. The motion of water is more likely to be turbulent as the size of the body of water increases. Hence, a very slow flow of water from a tap might look to be laminar, while water flowing at the same speed over a waterfall might be turbulent. When the water is turbulent it is obeying exactly the same laws of dynamics as for laminar flow. The only difference is that the motion is much more complicated, and the *momentum advective terms* which we discussed in Chapter 4 are much greater than the forces of *friction* (or viscosity). Each particle of a fluid is capable (in principle) of behaving totally independently of all others, and is only prevented from so doing by friction (or viscosity). The ratio of these forces is called the *Reynolds number* (Re).[2] When this number is small the flow is laminar, and when it is high the flow is turbulent. The value of Re is usually greater than unity for all coastal basins and for the oceans. The definition of the Reynolds number is

$$\mathrm{Re} \equiv \frac{Lv}{\nu} \tag{7.11}$$

where ν is the viscosity, L is the characteristic length of the flow, and v is its characteristic speed. It is important to emphasize that the viscosity in (7.11) is the molecular viscosity and not the effective viscosity that results from turbulence. The value of molecular viscosity for water is about $10^{-6}\,\mathrm{m}^2\,\mathrm{s}^{-1}$, and so for a flow with speed $v = 0.1\,\mathrm{m\,s}^{-1}$, $R > 1$ unless L is very small, i.e., L is less than $10\,\mu\mathrm{m}$. Generally we are not interested in analyzing the motion of each individual water particle, because the details of that motion tell us very little. Instead, we are more interested in the average motion of all the particles (advection) and the turbulence, or statistical distribution of the motion about this mean (which controls mixing). Turbulence has an effect that is different from that of the average motion, in that it produces mixing and spreading of quantities such as heat, salinity, and pollutants.

We have a choice as to the *volume* over which we average the motion, and the *time* for which we take the average. This choice is largely governed by the size of the system we are considering, and the time over which currents may vary. Two examples might

---

[2] After the pioneering work of Osborne Reynolds (1842–1912), first published in 1883.

be the seasonal change in currents in the Pacific Ocean, and the daily variability of currents in a coastal basin. In the first case, we might divide the ocean up into cells each of surface area of 100 km * 100 km and average over each of these cells for a period of a week at a time. In the second case, we might average over much shorter periods of about 30 min (because we are interested in variation through a day) and use cells of 500 m * 500 m (because coastal basins are so much smaller). All motion below those scales is said to be *turbulence*. When we talk about the *scales of motion* or *space and time scales*, we mean the size of boxes over which these averages are performed and the time for which the average is taken. These scales are equally important to our measurements of turbulence. So, the difference between advection and diffusion in a basin depends on the space and time scales used to define advection. Consider a coastal embayment of dimensions a few kilometers on the west coast of California. The coastal basin will make very little contribution to the net advection in a model of the North Pacific Ocean. However, a detailed model of that basin will contain advective currents at subscales of the basin, which averaged over the whole basin sum nearly to zero. In the model of the whole North Pacific Ocean these individual advective currents contribute to the local turbulence, and represent the coastal boundary zone.

7.2.4 A simple wind mixing event

Much of mixing theory is concerned with *boundary layers* or *boundary zones* simply because the stirring which produces turbulence usually occurs near surfaces, or because surfaces absorb momentum or heat, nutrients, and other substances. This process is well illustrated by *wind mixing*. Consider a shallow basin in which the temperature is higher at the surface than the bottom, as in Figure 7.1. We saw in this previous example that *complete mixing* would produce a water column in which the temperature is the same at all heights. Suppose that we take the unmixed water column and apply a wind stress to the surface. We recall from Chapter 5 that wind stress produces currents. Those are an advective effect and not of direct interest here. Wind also has another effect of creating turbulence near the surface. This process is due to *stirring of the surface by waves*. What happens is that the wind creates a *surface boundary layer* in which there is strong mixing, and outside of which the temperature profile remains unchanged from the initial form. This is illustrated in Figure 7.5, which is an idealized version of the vertical profile of temperature that results from wind stirring. The thickness of this surface mixed layer will increase with time, provided that there are no other influences other than the wind stirring.

Notice that in Figure 7.5, there is a discontinuous (sudden) change in temperature with height at the bottom of the *surface mixed layer*. This sharp change, called the *thermocline*, is really smoothed out as illustrated in Figure 7.6. The *thermocline* forms the boundary between two layers, or strata, and we say that the water column is

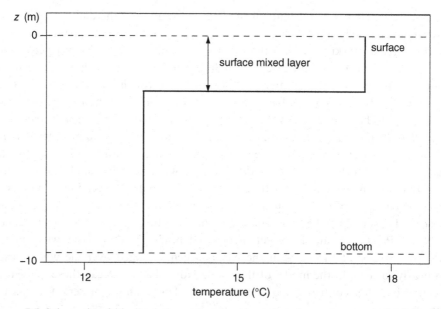

Figure 7.5 Schematic of idealized thermal surface boundary layer due to wind mixing with infinitely sharp thermocline; compare Figure 7.6.

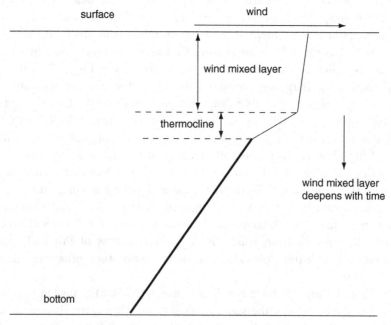

Figure 7.6 Schematic of thermally stratified water column showing a thermocline of finite width. The discontinuities in temperature gradient are typical of a layered model.

stratified. Stratification is a very common situation in the water column of coastal basins, and involves a less dense layer of water overlaying a denser layer. The difference in density between the two layers can be due to temperature or salinity (usually both). The reason that we can identify two strata, is that there is a very sharp jump in the temperature at the interface. If we measure wind stirring events in the field we do not see discontinuous changes in temperature with height, but the change is nevertheless very sudden. As an example, we might measure temperature profiles in a lake during a calm period followed by the sudden onset of wind. We will probably find some minor variations in temperature in the surface layer and (as we said) a sharp, but not truly discontinuous, change in temperature to the bottom layer.

Note also that the constant temperature inside the mixed layer is an idealization; in reality the temperature will change slowly from top to bottom of the *mixed layer*. The reason for the sudden jump in temperature at the thermocline is that the total heat within the mixed layer is exactly the same as the heat that was in that part of the water column prior to mixing. This is an example within the surface layer of the conservation of heat that we mentioned previously. So, the temperature of the mixed layer is the same as the average temperature of that part of the water, and this is clearly greater than the temperature at the top of the unmixed region.

Figure 7.6 shows the thermocline deepening with time. The process that keeps the mixed layer *well mixed* during deepening is heat constantly mixing (or conducting) from the warmer regions to the colder water just entering the mixed layer at the bottom. Heat conduction is the vertical movement of heat through mixing. Notice that there is no movement of volume (no vertical advection) because particles of equal volume merely interchange places. There is, however, a vertical movement of heat because the upper particles are warmer. Hence, we get a vertical heat flux but no vertical volume flux. If the thermal expansion is linear in temperature, there is no effective change in mean density. The surface layer is turbulent, while the water below is much less turbulent. The thickness of the surface layer increases with time as shown in Figure 7.6. This proceeds until it fills the whole water column, and its temperature drops until it reaches the temperature that was the mean of the original water column.

There are actually two types of mixing processes at work in this example. One of them is mixing of water *within* the surface layer. The other is mixing between this layer and the water below called *entrainment* which involves water from beneath the mixed layer being captured by the mixed layer. This water is then mixed within the layer. It is this entrainment which causes the bottom of the mixed layer to deepen with time. At first, the layer is very thin but it gradually thickens by entrainment, until it engulfs the entire water column. Mixing across the interface (between the mixed layer and calm water) is much slower than the mixing within the mixed layer. An imaginary particle in the bottom layer (but near the thermocline) would be very slowly captured by the mixed layer and then rapidly moved around in the mixed layer. The reason for the slowness of this capture process arises from the temperature difference between the surface water and bottom water, so that there is a density difference across the interface. As a result, an interchange of denser and less dense particles, which raises mass across the interface, has a tendency to reverse due to the natural tendency of the heavier particles to sink to the bottom; instead of mixing we get internal waves.

We shall see that entrainment (and mixing) changes the *potential energy* of the water column. This energy comes from the wind and is easily calculated from the change in the profile of temperature with height. Note that the wind stress causes a downward flux of momentum. Associated with this process is a loss of energy to the water column in the form of turbulence. So, while momentum is conserved, the transfer of momentum (by eddy viscosity) from water of higher to lower speed involves a loss of energy to turbulence. The mixed state has higher potential energy than the unmixed state, and this is the reason that mixing of a stratified fluid does not occur unless some external agency stirs the water and adds the necessary energy. Entrainment into the mixed layer involves an immediate increase of potential energy, and this limits the speed of entrainment.

7.2.5 *Rate of entrainment*

It is possible to calculate the rate of descent of the surface mixed layer. This requires knowledge of the rate at which the wind puts energy into the system, and must match the rate of increase of potential energy due to the thickening of the layer. This also depends on the change in density across the thermocline. For a wind speed of $10 \, \text{m s}^{-1}$, such as might be experienced in a strong sea breeze, a mixed layer of thickness a few meters will form in about 1 day. The speed of descent of the bottom of the mixed layer depends on the wind strength and the variation of density with height in the water column. If the water is very highly stratified, i.e., a large density difference exists between the top and bottom water, it will take a very strong wind to mix the water column. Furthermore, the thicker the mixed layer, the less effective is the wind because the turbulence is distributed over the larger volume of the mixed layer and is therefore more dilute. In the deep ocean basins, at low and middle latitudes, there exists a pycnocline at depths of several hundred meters which is very stable and called the *permanent pycnocline;* it is further stabilized by the advection of cold bottom water from higher latitudes.

The rate of entrainment gives an upward volume flux into the mixed layer (and an equal downward flux); from this we can determine the fluxes of all other quantities such as mass, heat, salt, etc. To characterize entrainment we merely need to specify the vertical *entrainment velocity*, which is just the volume flux of bottom water per unit area of interface (there is an equal flux of top water downwards and so no net advection). We denote this entrainment velocity by w_e (see Figure 7.7), and clearly it determines the rate of descent of the bottom of the mixed layer,

$$w_e \equiv \frac{dh_l}{dt}, \qquad (7.12)$$

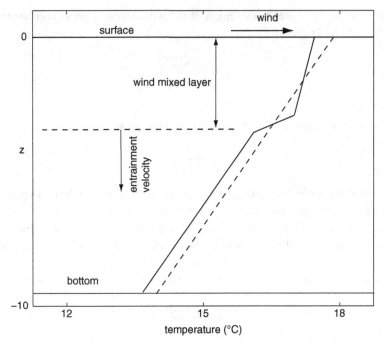

Figure 7.7 Illustration of the deepening of the wind mixed layer formed when a wind blows over a basin with an initially uniform vertical temperature gradient. The linear temperature variations would be obtained from a layered model.

where h_1 is the thickness of the mixed layer. The entrainment must ultimately be related to $u*$ which is the surface wind frictional velocity related to the surface stress τ_w by

$$u_* \equiv \sqrt{\frac{\tau_w}{\rho}} \tag{7.13}$$

where ρ is the density of the surface water. The surface wind stress τ_w is conveniently determined from a drag equation of form

$$\tau_w = \rho_a C_w W^2 \tag{7.14}$$

where W is the wind speed, ρ_a the density of air (about $1.2 \, \text{kg m}^{-3}$), and C_w a drag coefficient. Equation (7.14) is one we use extensively in our models of coastal basins. It is derived from the theory of drag on a surface, which we discuss later. The form of (7.14) essentially defines the drag coefficient C_w, and its value must be derived from observations. A typical value is $C_w \sim 0.001$. The interaction involves the roughness or wave state of the surface, and it is usually found that C_w increases with wind speed and is much higher in storm conditions (say $W > 20 \, \text{m s}^{-1}$) than under more gentle winds.

The drag speed u_* measures the strength of the wind stress in more convenient form, and (7.13) and (7.14) give

$$u_* = \sqrt{\frac{\rho_a}{\rho} C_w} \, W \qquad (7.15)$$

so $u_* \sim W/1000$.

We define the entrainment coefficient as the ratio of w_e to u_*.

$$c_e \equiv \frac{w_e}{u_*}. \qquad (7.16)$$

The entrainment coefficient c_e obviously depends on both u_* and the density difference between the two layers (the greater the difference, the more energy required to lift water from the lower to upper layer). This dependence is modeled as a function of a number denoted by r called the *effective bulk Richardson number* (more about the Richardson number in Chapter 9), and denoting the difference between the average density of the mixed layer and the density at the bottom of the thermocline as $\Delta\rho$,

$$r \equiv \frac{g h_1 \Delta\rho}{\rho u_*^2} \qquad (7.17)$$

where h_1 is the thickness of the mixed layer. So a high value of r indicates large density difference and implies reduced entrainment. Classic experiments by Kato and Phillips (in a laboratory tank)[3] show

$$c_e \sim \frac{2.5}{r} \qquad (7.18)$$

giving from (7.16) to (7.18)

$$w_e = \frac{2.5}{g h_1 \Delta\rho/\rho} \left(\frac{\rho_a}{\rho} C_w \right)^{3/2} W^3 \qquad (7.19)$$

and so using values for constants as mentioned above,

$$w_e = \frac{A W^3}{h_1 \Delta\rho/\rho} \qquad (7.20)$$

and $A \sim 3 * 10^{-10} \, \text{s}^2 \, \text{m}^{-1}$. For example, consider a typical situation in a coastal basin, $h_1 \sim 10 \, \text{m}$, $\Delta\rho/\rho \sim 10^{-3}$, and $W = 10 \, \text{m s}^{-1}$ so $u_* \sim -0.01 - \text{m s}^{-1}$ giving $r \sim 1000$, so $c_e \sim 0.003$ and $w_e \sim 3 * 10^{-5} \, \text{m s}^{-1}$. Thus, the thickening of the mixed layer would proceed at a rate of 0.1 m per hour. Equation (7.20) shows the expected reduction in entrainment as density difference (and hence Richardson number r) increases. Notice

---

[3] Kato and Phillips (1969).

that w_e decreases as h_1 increases due to the decrease of the velocity shear. Note also that the density difference can be calculated from the difference in temperature (and/or salinity) by a formula of type

$$\Delta\rho/\rho = \alpha\Delta T \qquad (7.21)$$

where α is the coefficient of thermal expansion which has a value of about $2 \times 10^{-4}\,°C^{-1}$.

7.2.6 Layered and one-dimensional models

The tendency of the water column to form relatively sharp horizontal interfaces has lead to the development of *layered models* in which discrete layers each have individual properties such as temperature. The layers exchange quantities such as heat and salt across the interface. In such models, the number of layers is decided by the user to best represent vertical exchange processes in the water column. Individual layers can change thickness due to entrainment across interfaces. This type of model is very useful in applications such as those discussed in the last section and especially value for simple analytical solutions.[4] The models can generalized so that the thickness of the layers varied across the basin. With increases in computer power such models have tended to give way to *one-dimensional* models in the vertical. Modelers tend to make decisions as to whether the behavior of a coastal basin is controlled primarily by processes in the vertical or horizontal. One of the most important parameters that control the relative importance of the two types of process is the ratio of the depth of the basin to its horizontal dimensional. Many inland lakes and especially reservoirs have ratios that emphasize vertical processes so that layered and one-dimensional models have proven very accurate. Layered and one-dimensional models are very suitable to studies of biological and chemical properties in the water column and often used in studies of water quality in coastal basins and inland water bodies.[5]

7.2.7 Mixing coefficient

In spite of our success in defining the mixing flux in (7.10), there is less elegance in finding a way of determining this flux from other properties of the water within the coastal basin. The approach that we use here (and the approach that is common throughout models of coastal basins) is to define a coefficient of mixing based on Fick's law of diffusion. This *mixing coefficient* (or *diffusivity*) is denoted by K (upper case) often with a subscript z (for vertical) or either x or y (for horizontal), although the symbol A is sometimes adopted. Other subscripts and superscripts are also attached to denote the quantity being mixed; although at spatial scales greater than a fraction of a

---

[4] See especially Officer (1976).
[5] For example, DYRESM developed by the Centre for Water Research at the University of Western Australia: Patterson, Hamblin, and Imberger (1984).

meter, in unstratified coastal basins, K is largely independent of the quantity being mixed. The coefficient of mixing of momentum (velocity) is called the coefficient of *viscosity* or *eddy viscosity*, and may be denoted by a different symbol. The dimensions of K are length squared per unit time, i.e., $m^2\,s^{-1}$, and is defined as the flux of the quantity per unit gradient of its concentration. For example, the vertical flux of heat H_z (per unit area per unit time) associated with a vertical temperature gradient of dT/dz is

$$H_z = -K_z \frac{dT}{dz}. \tag{7.22}$$

So mixing increases with the magnitude of K. Mixing is being modeled as proportional to the gradient of the concentration of the quantity being mixed. This is called *Fick's law of diffusion*, and is borrowed from the diffusion of ions (and is usually considered to have some validity in the turbulent conditions of coastal basins).

The general rule is that the diffusive flux of any quantity of concentration c is written as

$$\langle c'w' \rangle = -K_z \frac{\partial c}{\partial z} \tag{7.23}$$

$$\langle u'c' \rangle = -K_x \frac{\partial c}{\partial x} \tag{7.24}$$

$$\langle v'c' \rangle = -K_y \frac{\partial c}{\partial y} \tag{7.25}$$

where we have dropped the over-bar on the average value of c. We have seen already that the advective flux in the three directions is uc, vc, and wc (respectively) so that the total flux (advective plus diffusive) is

$$J_z = -K_z \frac{\partial c}{\partial z} \tag{7.26}$$

$$J_x = vc - K_x \frac{\partial c}{\partial x} \tag{7.27}$$

$$J_y = vc - K_y \frac{\partial c}{\partial y} \tag{7.28}$$

where we have assumed that there is no vertical advective flux and dropped the over-bar on the mean of c and u, v. If we generalize the continuity equation in one dimension to

$$\frac{\partial c}{\partial t} = -\frac{\partial J_x}{\partial x}, \tag{7.29}$$

i.e., replace the purely advective term by the total flux, and substitute (7.26) in (7.29) we obtain

$$\frac{\partial c}{\partial t} = -\frac{\partial uc}{\partial x} + \frac{\partial}{\partial x} K_x \frac{\partial c}{\partial x} \qquad (7.30)$$

for a one-dimensional model in the x direction (and similarly in any other direction). More generally, we should take the divergence of all components of J. The coefficient of diffusion K is often modeled by a constant, i.e., it is assumed that c varies more rapidly with x than does K and so (7.30) simplifies to

$$\frac{\partial c}{\partial t} = -\frac{\partial uc}{\partial x} + K_x \frac{\partial^2 c}{\partial x^2}. \qquad (7.31)$$

Note that (7.30) and (7.31) apply to a particular height z. We often take the depth average of (7.31), i.e., integrate with respect to z from the bottom $(z = -h)$ to surface (at $z = \eta$):

$$\frac{\partial c}{\partial t} = -u\frac{\partial uc}{\partial x} + K_x \frac{\partial^2 c}{\partial x^2}. \qquad (7.32)$$

7.2.8 Physical interpretation of Fick's law

Fick's law can be easily rationalized by using the concept of *mean free path*. This model considers that particles in a fluid move a certain distance as *free* particles, denoted by λ and called the mean free path. After traveling a distance l a particle interacts with other particles, and so comes into equilibrium with the fluid. During the free path, a particle carries with it memory of the bulk properties of the fluid at the start of the free path. In Figure 7.8 we show two planes normal to the z axis and separated by the mean free path. There is no advection between the planes in Figure 7.8, but there are fluxes due to the random motion of the particles. Suppose we consider a quantity with concentration $c(z)$. Let us consider that there is a pair of equal and opposite velocities, w' and $-w'$, which drive the two fluxes of the quantity $f(z)$ and $f(z+l)$ as shown in the figure. These fluxes are not equal, because c is different on the two planes. The fluxes are

$$\begin{aligned} f_1 &= w'c(z) \\ f_2 &= -w'c(z+l) \end{aligned} \qquad (7.33)$$

and so the net flux in the positive z direction is

$$f = f_1 - f_2 = -w'[c(z+l) - c(l)] \qquad (7.34)$$

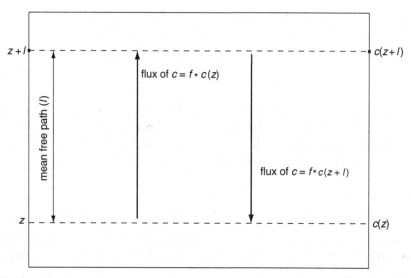

Figure 7.8 Schematic of particles moving vertically between two horizontal planes separated by their mean free path l. The fluxes of particle number are equal in the upward and downward directions, but there is a net flux of the quantity measured by the concentration c because its value is different on the two planes.

and since λ is very small compared to the spatial scales on which we study processes in coastal basins, (7.34) simplifies to

$$f = -w'l\frac{\partial c}{\partial z} \tag{7.35}$$

which is identical to (7.23) with

$$K_z = w'l. \tag{7.36}$$

This type of approach works well for diffusion due to the thermal motion of gases, but is merely illustrative for turbulent fluids.

7.3 Vertical mixing time

In Section 7.2, we saw that mixing by surface winds involves the creation of a surface mixed layer within which the surface turbulence is confined, but we can also envisage situations in which the turbulence, and hence mixing coefficient, extends throughout the water column. Dimensionally, the time to totally mix a water column of height h is inversely proportional to this value of K_z and must have a form

$$t_z \sim \frac{h^2}{K_z}. \tag{7.37}$$

For a non-stratified water column it is common to assume

$$t_z = 0.3 \frac{h^2}{K_z}. \tag{7.38}$$

In the horizontal, we often start with a small patch of the quantity that mixes, or disperses, with time so that the patch gets bigger, i.e., its surface area increases with time. An example might be a spill of buoyant material on the surface of a coastal basin. In this case $4\pi K_x$ gives the rate of increase of area with time (to within a constant). Thus if K_x is $10\,\mathrm{m^2\,s^{-1}}$, the area increases at $125\,\mathrm{m^2\,s^{-1}}$. Note that dispersion is much greater in the horizontal than in the vertical and (as we shall see later) it is driven by turbulence over much larger spatial scales; spatial scales are necessarily much larger in the horizontal than the vertical in coastal basins. When considering mixing in the horizontal (usually called *dispersion*), it is to be emphasized that advective processes due to buoyancy forces may occur, which also spread the patch. A good example is the spreading of a buoyant plume from a river. Since the area increases proportionally to time, the radius increases as the square root of time. Hence, the rate of spreading along a line decreases as the patch spreads. This is a general property of diffusion.

7.4 Examples of mixing

Consider a coastal basin of area $100\,\mathrm{km^2}$. Suppose that a patch of nutrients enters the basin with an initial area is $10\,\mathrm{m^2}$ and the coefficient of dispersion of the nutrients is $10\,\mathrm{m^2\,s^{-1}}$. After 3 days the area of the patch is $10 + 10 * 4\pi * 3 * 3600 * 24 = 2.6 * 10^6 = 5.4\,\mathrm{km^2}$, which is less than the area of the coastal basin. Assuming the patch is circular, the radius after 3 days is $1300\,\mathrm{m}$, so the spatial scale of the patch varies from a few meters to hundreds of meters. If the nutrient patch occurred, for example, in the Pacific Ocean off the Hawaiian Islands with an initial area of $10\,\mathrm{km^2}$, much greater spreading would occur because the coefficient of dispersion is probably as high as $500\,\mathrm{m^2\,s^{-1}}$. After 3 days the area would become $10 * -10^6 + 4\pi * 500 * 3 * 3600 * 24 = 16 * 10^8 = 1600\,\mathrm{km^2}$ which corresponds to a radius of $23\,\mathrm{km}$, so that spatial scale is a few kilometers. The dispersion coefficient is greater in this example, because the patch experiences turbulence on a much greater spatial scale, i.e., a scale of kilometers as opposed to 10 to 100 meters in the first example. Also the turbulence off Hawaii is greater (at any spatial scale) than in a typical coastal basin because of the much greater currents. In particular, the North Equatorial Current passes through the Hawaiian Islands causing large-scale eddies to form on their leeward side. The patches in the above examples may be affected by advection. Typical currents in coastal basins are $0.1\,\mathrm{m\,s^{-1}}$, so the center of the patch would probably move $0.1 * 3 * 3600 * 24 = 26\,\mathrm{km}$, which is large compared to its spreading, and a typical value for the North Equatorial Current is $0.5\,\mathrm{m\,s^{-1}}$. So, the patch moves $130\,\mathrm{km}$, which is again much larger than dispersion.

Let us consider another example. In the North Sea, which is on the European continental shelf, tidal currents are very large with maximum values of $1\,\mathrm{m\,s^{-1}}$ but residual (long-term) currents of order $0.01\,\mathrm{m\,s^{-1}}$. The tides create considerable turbulence, and hence dispersion, with a typical dispersion coefficient of $100\,\mathrm{m^2\,s^{-1}}$. Dispersion of a patch over a diurnal tidal cycle (about 12 h) produces a change in area of $4\pi * 100 * 12 * 3600 = 54\,\mathrm{km^2}$, giving a radius of 4.1 km. The advective term will take the patch a distance of 0.5 (mean tidal current) $* 6 * 3600 = 10.8\,\mathrm{km}$, but then return it to nearly the same place. The residual current of $0.01\,\mathrm{m\,s^{-1}}$ will ensure that the patch does not return to quite the same position, there being a net movement of $0.01 * 12 * 3600 = 432\,\mathrm{m}$. So dispersion is dominant.

To illustrate mixing in the vertical, consider a patch of heat introduced into a water column of total height 10 m. The patch initially occupies the water at $z = -5\,\mathrm{m}$ with negligible vertical spread. If the vertical diffusion coefficient of mixing for heat is 10^{-4} $\mathrm{m^2\,s^{-1}}$, the vertical spread after 1 day is $2\sqrt{0.0001 * 3600 * 24} = 6$ m and the vertical mixing time t_z is $0.3h^2/K_z = 1.2$ days.

7.5 Mixing processes and spatial scale

7.5.1 Molecular mixing

The molecules which make up a fluid (gas or liquid) have *random thermal motion*. If very fine particles are added to a gas, it is possible to see quite abrupt and sudden motion of these particles due to the random motion of the molecules. This is called *Brownian motion*. This random motion produces mixing, dispersion, and diffusion. It is referred to as *molecular motion*. The energy of this random thermal motion is contained in the kinetic energy of the molecules and is the same, per molecule, for any fluid (gas or liquid) at the same temperature. Consequently, as the mass of the molecule increases, so the speed of this thermal motion decreases. In fluids, the random motion is also controlled by collisions between molecules, and the mean free path of a molecule is smaller in a fluid than a gas (because of the greater density). Molecular motion of molecules manifests itself as microscopic variations in density. That variation gets larger as the spatial scale gets smaller. So, if you measure the density of a gas, or fluid, within a very small cell, the answer can fluctuate according to the number of molecules that happen to be in that cell at that time; these fluctuations get smaller as the cell gets larger, just because the average number of molecules gets larger. Such fluctuations cause a scattering of light that passes through air; smaller wavelengths experience greater variations in density and are scattered more, which is the reason that the sky is blue and sunsets are red (called Rayleigh scattering).

We use the symbol D for a molecular mixing coefficient and v for the molecular viscosity. It is greatly reduced by inter-molecular collisions which reduce the mean free path, and is therefore much smaller in a liquid than in a gas. A release of gas at one side of a room is soon detected at the other side and hence, we soon smell a gas leak. Apart from advection by thermal currents, the dispersion of the gas leak around the room is caused by molecular diffusion. If we fill the room with water and release a gas, say oxygen, it would take a very long time to reach the other side by molecular diffusion. For example, molecular diffusion takes some years to spread oxygen from one side of a 5 m pond to the other but spreads a distance of a few millimeters in tens of minutes. This illustrates the important point that molecular mixing is very weak on the spatial scales of coastal basins, but nevertheless mixes particles over very small distances very quickly.

7.5.2 Turbulent energy cascade

We visualize turbulence in terms of random turbulent eddies, or vortices with no net motion (all net motion has been subsumed into the advective term). Although there are many complications to our understanding of turbulence, we adopt a picture here in which stirring produces eddies at small scales and away from the direct zone of energy production. These eddies do not have an infinite lifetime, but decay by losing their energy through molecular viscous (*frictional*) forces. Individual eddies have velocity shear, which produces frictional forces that oppose the motion. This decay of eddies essentially reduces their size and energy, and occurs in a very interesting way. Imagine vortices of a whole range of sizes in the fluid. Large vortices tend to be created by the stirring processes that feed turbulent energy into the fluid. These vortices do not just collapse due to friction, but decay by passing their energy to smaller vortices. This is called an *energy cascade* and was first proposed by Lewis Fry Richardson.[6] No energy is lost in the actual Richardson cascade but when the vortices become small enough, the velocity gradient is sufficiently high for molecular viscosity to convert the motion to heat. This occurs at the *Kolmogorov* length scale. Richardson proposed that turbulence has a fractal structure (Chapter 11), with the property that turbulence is *self-similar* at all spatial scales (above what we would now call the Kolmogorov scale). The cascade process only occurs away from the zone of production. Near that zone eddies actually grow in size, and are then often stabilized by buoyancy to form fossilized turbulence. This can be seen for example in the condensation trails (called *contrails*) from jet aircraft engines in the stratified atmosphere which remain for many hours and are visible in satellite images of parts of the Earth that have high levels of jet aircraft traffic.

There is a minimum size for eddies beyond which molecular viscosity damps the eddy away totally. The size is typically about 10 mm, but depends on how rapidly energy is being input into the turbulence from winds, tides, and other stirring processes. This quantity is known as the *turbulence dissipation rate* and normally denoted by ε.

---

[6] Mathematician, physicist, and psychologist (1881–1953) who made fundamental contributions to the science of weather forecasting through the solution of differential equations and was interested in turbulence. He also created the *Richardson plot* of the length of a coastline against the spatial scale used in its measure which we discuss further in Chapter 11.

It is important to understand the meaning of the dissipation rate. All energy that enters the water column through any type of forcing works its way down the Richardson cascade, and is *dissipated* as heat due to the effects of molecular viscosity. It is measured in watts per kg (W kg^{-1}), i.e., J kg^{-1} s^{-1}, or equivalently m^2 s^{-3} (some texts quote ε per unit volume which is in our terms $\rho\,\varepsilon$). As ε increases, the coastal basin becomes more *energetic*. If ν is the molecular viscosity (units m^2 s^{-1}), then the minimum length scale for eddies is

$$L_\nu = 2\pi \left(\frac{\nu^3}{\varepsilon}\right)^{1/4} \tag{7.39}$$

which is called the *Kolmogorov length*.[7] The factor of 2π in (7.39) belies the fact that the Kolmogorov length is based on a scaling argument but the factor is commonly used as an aid or calibration.

A typical value of ε for a wind of strength of $15\,\mathrm{m\,s^{-1}}$ is $10^{-6}\,\mathrm{W\,kg^{-1}}$ which gives $L_\nu = 2\pi * 10^{-3} = 6\,\mathrm{mm}$ if we use $\nu = 10^{-6}\,\mathrm{m^2\,s^{-1}}$. Clearly, the greater the input of energy the smaller will be the scales at which turbulence can exist. Also, the smaller the molecular viscosity, the smaller will be Kolmogorov length. For a very energetic system on top of a coral reef, ε might be as high as $10^{-3}\,\mathrm{W\,kg^{-1}}$, giving a Kolmogorov length of 1 mm (which is smaller than the previous example by a factor of about six).

7.5.3 Diffusive boundary layer

Mixing at small scales is controlled by molecular motion, and the mixing of momentum is controlled by molecular viscosity. Molecular viscosity and molecular mixing are very weak processes in the ocean and so only effective over small distances. The reason is that the distance d over which a point patch of particles with diffusivity D spreads in a time t is $d = \sqrt{Dt}$. So that the *effective* distance over which a diffusive process is felt is proportional to the square root of D. We have seen examples in Section 7.4 with turbulence spreading material over many kilometers in hours. But for molecular viscosity $D \sim 10^{-6}\,\mathrm{m^2\,s^{-1}}$, and so the diffusion length $d \sim 0.06\,\mathrm{m}$ in 1 h. So, molecular viscosity is able to damp out instabilities at very small spatial scales. On scales below millimeters, molecular viscosity is a major force. Creatures that exist on these scales have small mass, so to them the fluid seems highly viscous.

The coefficients controlling the mixing of momentum (viscosity), heat, salt, and nutrients by molecular processes are all different. This is in contrast to the mixing

---

[7] Andrei Nikolaevich Kolmogorov (born April 25, 1903 in Tambov, Russia and died October 20, 1987 in Moscow). Kolmogorov (often spelt Kolmogoroff) was one of the great mathematicians of the twentieth century. His contributions spread over a vast variety of mathematical fields. The promulgation of his ideas on turbulence in the mid twentieth century is due to an Australian scientist George Batchelor, who arrived in Cambridge, England to work with G. I. Taylor and discovered a seminal paper by Kolmogorov (published in 1941), and wrote a masterly review of the work in 1947. Batchelor went on to develop the theory in terms of its application to heat, salt, and other species of the water column.

Table 7.2. *Typical values of molecular viscosity used as examples in this book*

Quantity	Diffusion coefficient ($m^2 s^{-1}$)
Momentum (viscosity)	10^{-6}
Heat	$1.5 * 10^{-7}$
Salt and nutrients	$1.5 * 10^{-9}$

of these quantities by turbulence, in which these coefficients are the same (provided that the water is not stratified). These coefficients are referred to as the molecular diffusivity for momentum, heat, salt, nutrients, etc. We show values in Table 7.2.

Notice that the coefficient of diffusion for heat is lower than that for momentum (viscosity) by a factor of six, with the coefficients for salt and nutrients being a further factor of 100 (two orders of magnitude) smaller. The lower diffusivity of salt compared with heat is responsible for a well-known double-diffusive instability. The effect of these lowered diffusivities is to permit turbulent fluctuations in these quantities to exist on scales smaller than the Kolmogorov scale. These minimum length scales are called the *Batchelor length* scales

$$L_D = 2\pi \left(\frac{\nu D^2}{\varepsilon} \right)^{1/4} \tag{7.40}$$

where D is the diffusivity of the quantity. For heat, salt and nutrients $L_D < L_\nu$, so fluctuations exist on much smaller length scales. The Batchelor length scale for momentum is identical to the Kolmogorov length but for nutrients is a factor of 40 smaller.

Some *submerged aquatic vegetation* (SAV) can take nutrients directly from seawater. Imagine the surface of submerged aquatic vegetation which is absorbing nutrients from the surrounding water through its surface. If there were no mixing of nutrients through the water column, it would simply absorb nutrients until the water immediately outside that surface was totally depleted of nutrients. In reality, there is mixing of nutrients and this essentially controls the rate at which nutrients can be absorbed by the surface. The main constraint on the delivery of nutrients to the surface comes in the last few millimeters, when molecular diffusion alone is responsible for their transport. If we go a centimeter away from the surface, we find comparatively high nutrient concentrations, but very close to the surface the nutrient concentration is depleted.

The depleted layer is called the *diffusive boundary layer* (Figure 7.9). Its thickness is controlled by the Batchelor scale for nutrients. The flow of nutrients to the surface is faster if the thickness is smaller, e.g., if the vegetation grows in a high-energy environment such as the intertidal region or the top of a reef. The diffusive boundary layer is discussed more fully in Chapter 11.

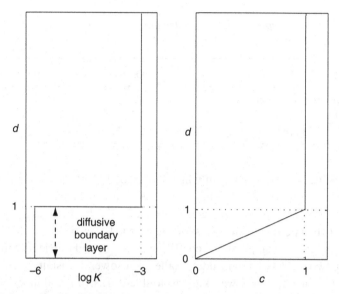

Figure 7.9 Schematic of diffusive boundary layer. The left-hand panel illustrates a discontinuous change in a nutrient mixing coefficient K from $10^{-6} \, \mathrm{m^2 \, s^{-1}}$ to $10^{-3} \, \mathrm{m^2 \, s^{-1}}$. The axis labeled d refers to the distance from the nutrient uptake surface relative to the thickness of the diffusive boundary layer. The right-hand panel shows the steady state variation of nutrient concentration c with d assuming that it vanishes at $d = 0$. The value of c is measured relative to its ambient value in the water column, i.e., $c \to 1$, for $d \gg 1$. The slope of c with d is three orders of magnitude lower when $d > 1$ than for $d < 1$.

7.5.4 Sinking of a small sphere (Stokes' law)

A classic experiment that provides a very good way of measuring molecular viscosity is the free fall of a sphere in a viscous fluid. It is also widely used in sediment analysis to measure the diameter of sediment grains. The sphere needs to be negatively buoyant, i.e., the density of the sphere must be greater than that of the fluid. This is called the Stokes experiment after research in 1850 by George Stokes. We simply release a sphere at the top of a column of fluid and watch it fall. Initially, the sphere accelerates under its net weight, i.e., the negative buoyancy force. The acceleration then decreases due to friction, i.e., viscosity, and the velocity reaches a steady value in which the net weight is exactly balanced by the viscous forces on its surface:

$$F = -\frac{4}{3}\pi r^3 g \left(\rho_{\text{sphere}} - \rho_{\text{fluid}} \right) \tag{7.41}$$

where the negative indicates that the buoyancy force acts downwards in the negative z direction; r is the radius of sphere. The total viscous force is

$$J = -6\pi \rho_{\text{fluid}} \nu w r \tag{7.42}$$

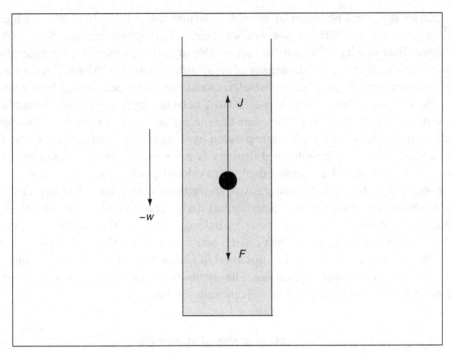

Figure 7.10 Schematic of a sphere falling through a vertical water column in the Stokes' law experiment: F is the negative buoyancy force, J is the frictional drag and $-w$ is the steady state fall velocity.

where w is the velocity of the sphere ($w < 0$). In the steady state the total force is zero, i.e.,

$$F + J = 0 \tag{7.43}$$

and so (7.41) to (7.43) give

$$w = -\frac{gd^2}{18\nu}\left(\frac{\rho_{\text{sphere}}}{\rho_{\text{fluid}}} - 1\right) \tag{7.44}$$

and note that w is negative which is consistent with the fall velocity being directed downwards and note that F is negative (directed downwards) and J is positive (directed upwards) (Figure 7.10).

For a sphere of density $3000\,\text{kg m}^{-3}$ and diameter $2\,\mu\text{m}$ (a grain of fine silt) in water (viscosity $\nu \sim 1*10^{-6}\,\text{m}^2\,\text{s}^{-1}$), $w = -0.04\,\text{m s}^{-1}$. If we replace the water with olive oil, at $20\,°\text{C}$ the viscosity is about $81\ 10^{-3}\,\text{m}^2\,\text{s}^{-1}$, and so for the same sphere $w = -0.5\ 10^{-6}\,\text{m s}^{-1}$.

Stokes' law is often used to assess the speed of fall of plankton (which are slightly negatively buoyant). For plankton of density of $1050\,\text{kg m}^{-3}$, seawater with a density of $1026\,kg\,m^{-3}$, and plankton of diameter $100\,\mu m$, $w = -1.25 * 10^{-4}\,m\,s^{-1}$ which is $10\,m$ per day. As cells become smaller their sinking rate decreases. Some cells can lower their density sufficiently to become almost positively buoyant, but generally cells are negatively buoyant; the density of cytoplasm is 1030 to $1100\,kg\,m^{-3}$ (cells may have harder material such as calcite which is much denser). So, cells tend to be small so that their slight negative buoyancy is almost balanced by friction, i.e., molecular viscosity, in order for them to stay near the surface and receive sunlight. Although such particles are moved in the water column by large turbulent eddies, the conventional thinking is that turbulence diffusivity is not effective on such small spatial scales, and that molecular viscosity damps out velocity gradients at such scales.

Stokes' law (7.44) is used in the study of the settlement of sediments (Chapter 12), and has been subject to extensive experimental study. Above a certain critical size, the falling sphere does start to create its own turbulence field and although this is damped out away from the region of production, it is active near the particle and does serve to slow the speed of descend as will be discussed in Chapter 12. This leads to a modified form of Stokes' law at higher grain size. The viscous force also depends on the shape of the particle which leads to other modifications to Stokes' law.

7.6 Vertical mixing of momentum

The vertical mixing of momentum by turbulence is controlled by the same diffusivity K_z as for other quantities. For momentum, we call this diffusivity by the special name of *viscosity*. Consider a water column with higher velocity at the surface than at the bottom, and let us assume that no other forcing is present on the surface or bottom. The momentum (velocity) will simply mix downward by the vertical interchange of particles. The mean momentum will be conserved in these mixing events (although energy is lost to turbulence). This process of downward mixing of momentum determines the form of the velocity profile. The process is akin to the surface input of (say) heat, and the downward mixing of that quantity. In the case of heat, there is a flux of heat from the water to the seabed. For momentum, that same type of transfer occurs through bottom friction. In a steady state, the rate of input at the surface exactly matches the loss to the bottom, and the downward momentum flux would be the same at all heights in the water column. Momentum input also comes from other sources such as pressure gradients. The form of the velocity profile (whether steady state or time dependent) involves a balance between momentum production, mixing and loss.

The vertical flux of horizontal momentum in the x direction, ρu, is denoted by $\rho \langle u'w' \rangle$ (note that positive momentum flux implies an upward flux) and given by Fick's law as

$$\langle u'w' \rangle = -K_z \frac{\partial u}{\partial z} \tag{7.45}$$

so that momentum moves from a region of high momentum (velocity) to lower momentum (velocity). The dimensions of momentum flux are clearly equivalent to momentum per unit area per unit time, and can also be written as force per unit area. This is identical to the shear stress in the x direction, except for a change of sign, because a positive shear stress represents a downward flux of momentum, i.e., for positive shear stress u must increase with z:

$$\tau_x = \rho K_z \frac{\partial u}{\partial z}. \tag{7.46}$$

It is essential to know K_z in order to find the variation of velocity with height z. Most of the early three-dimensional models of coastal basins were based on a K_z that is independent of z.

7.7 The logarithmic layer

7.7.1 Von Kármán's experiments

The logarithmic layer (usually abbreviated to *log layer*) represents one of the cornerstones of our understanding of the creation of turbulence and our modeling of mixing. The present section provides an introduction to the log layer and more details are given in section More details of the log layer are given in Section 7.11. Consider the flow of water over the bottom of a basin. If the speed of flow is sufficiently high (and this is usually the situation in a coastal basin), the flow will be turbulent. The bottom exerts an important influence on the flow including a stress on the water column, and if we know that stress τ_x we can determine the quantity, $\rho K_z \, \partial u / \partial z$, at the bottom, i.e., $z = -h$. Although τ_x varies through the water column, there is clearly a region close to the bottom where this variation is only a small fraction of its value at the bottom. Let us assume that within this *constant stress region* of z, a steady state exists. It is convenient to define the distance of a point from the bottom as z', i.e.,

$$z' \equiv z + h \tag{7.47}$$

so that z' (like z) increases upward. In a model based on turbulent diffusivity we would need to know K_z in order to find the velocity shear du/dz', but alternatively we can measure the velocity shear in the laboratory or in the field. If we can determine the shear stress, it is possible to obtain information on the magnitude and form of K_z. Let us assume that the only turbulence in the water column is created by the effect of the bottom friction acting on the water column.

Theodore von Kármán performed a set of classical experiments on turbulent flow near a rough wall and found that the velocity shear du/dz' (the rate of change of velocity with height) is not constant, but instead it has the form

$$\frac{\partial u}{\partial z'} \sim \frac{1}{z'}. \tag{7.48}$$

Equation (7.48) shows the velocity shear increasing indefinitely as z' approaches zero, so there must a minimum value of z' for which (7.48) is valid. Von Kármán's experiments used a pressure-driven flow along a tube. He was able to determine the stress at the wall by establishing a steady state flow, and matching the total frictional force on the wall to the difference in the height of the water surface at the ends of the tube. He established a very remarkable result. For each flow he calculated the stress τ_x derived *shear velocity u_** via

$$u_*^2 = \frac{\tau}{\rho}$$ (7.49)

and then plotted the velocity shear against $1/z'$ and found the very important result

$$\frac{\partial u}{\partial z'} = \frac{u_*}{\lambda z'}$$ (7.50)

where λ is called the *von Kármán constant* which has an observed value of 0.41; it is an experimental constant based on measuring the velocity profile across a fluid in a pipe. This is a very general result in turbulent fluids and also called the *law of the wall*. It applies to the velocity of a coastal basin near to a surface such as the bottom, or top, of the water column. It also applies to the wind velocity over the ground or ocean where, in general, z' is the distance from the surface. We shall see later the form of (7.50) is due to the way that turbulence is produced by the wall.

Since the velocity shear is positive in (7.50), the velocity must be increasing with distance from the wall. Equation (7.50) shows that the velocity shear, i.e., the gradient of u, decreases with distance from the wall; it follows that the velocity increases very rapidly near the wall and then more slowly as we move away from the wall. To calculate the velocity, we need to integrate the velocity shear, i.e.,

$$u = \int \left(\frac{du}{dz'}\right) dz'.$$ (7.51)

Now using the law of the wall (7.50),

$$u = \frac{u_*}{\lambda} \int \frac{1}{z'} dz' = \frac{u_*}{\lambda} \log z' + \text{constant}.$$ (7.52)

Note that $\log(z')$ tends to minus infinity as z' tends to zero, but the law of the wall only applies above some minimum value of z'. Let us just assume that u is zero when $z' = z_0$ (more details are given in Section 7.11). This is very reasonable, since we expect that the velocity will vanish very close to the bottom; the precise value of z' at which u vanishes merely changes the value of the additive constant in (7.52). With this assumption, we can now determine the constant of integration which just amounts to writing

$$u = \frac{u_*}{\lambda} \log(z'/z_0).$$ (7.53)

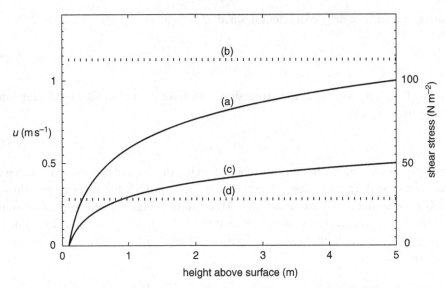

Figure 7.11 Curves (a) and (c) show the variation of the current u (left-hand scale) in a logarithmic boundary layer as a function of height above the surface. For curve (a) the current is set at $1\,\mathrm{m\,s^{-1}}$ at a height $5\,\mathrm{m}$ above the surface, and for curve (c) the current is set at $0.5\,\mathrm{m\,s^{-1}}$ at that height. In both cases, current is assumed to vanish at a height of $0.1\,\mathrm{m}$ above the surface The dotted lines (b) and (d) represent the shear stress corresponding to curves (a) and (c) respectively (right-hand scale). Notice the factor of four increase in shear stress for a factor of two increase in current.

Because u varies logarithmically with z', the region is called a *logarithmic boundary layer*. Note that log as used here means the natural logarithm, i.e., $\log_e$, which is sometimes written as ln in other texts. The variation of u with z' is a function of z'/z_0 but scaling inside a log function only adds a constant, and so z_0 is not a spatial scaling (it merely produces an additive constant). Figure 7.11 illustrates logarithmic current variations for two profiles in which the current at a fixed height above the surface has values of 0.5 and $1\,\mathrm{m\,s^{-1}}$. The figure also shows the shear stress (broken lines) for each current profile (which is independent of height above the surface).

7.7.2 Quadratic friction

One of the interesting properties of a logarithmic boundary layer is that (7.53) shows that u at any z' is proportional to u_* and since $u_*^2 = \tau/\rho$, the shear stress is quadratic in the current, i.e.,

$$\tau = \rho C(z')u^2(z') \tag{7.54}$$

where C is a surface drag coefficient defined as

$$C \equiv \left[\frac{\lambda}{\log(z'/z_0)} \right]^2. \tag{7.55}$$

This is a familiar result from the expression, which we use for surface wind drag and bottom friction, i.e.,

$$\tau = \rho C u^2_{surface}, \tag{7.56}$$

the only difference being that the drag coefficient is a function of the point at which u is measured. The drag coefficient C increases as we approach the surface, and this is simply a consequence of the decrease in u. Many authors take the approach of defining a quadratic friction law based on the value of u at the *edge of the log layer*. In practice, this means that u is referenced at some fraction of the height of the water column h, say γh, in which case

$$C \equiv \left[\frac{\lambda}{\log(h) - \log(z_0) + \log(\gamma)} \right]^2 \tag{7.57}$$

Equation (7.57) illustrates the dependence of C on z_0, and without pre-empting the more detailed discussion of Section 7.11, it is clear that z_0 increases as the surface becomes rougher in the sense that the current speed reaches zero further away from the surface. This increase in roughness does, as expected, increase the value of C. For example, if $\gamma h/z_0 = 5000$ and we increase z_0 by an order of magnitude, C increases from 0.0023 to 0.0044.

Equation (7.53) gives for the average current for that part of the water column that extends from $z' = z_0$ to $z' = l$,

$$\bar{u} = \frac{u_* z_0}{\lambda l} \int_1^{l/z_0} \log(p) \mathrm{d}p \quad p \equiv z'/z_0 \tag{7.58}$$

and evaluating the integral

$$\bar{u} \sim \frac{u_*}{\lambda} (\log l/z_0 - 1) \quad l/z_0 \gg 1. \tag{7.59}$$

We notice, of course, that there is a very weak divergence of the average current as z_0 vanishes. If we define a drag coefficient in terms of the this average current we find

$$C_\mathrm{d} = \left(\frac{\lambda}{\log l/z_0 - 1} \right)^2 \tag{7.60}$$

which is essentially the same as (7.55). Figure 7.11 shows the bottom shear stress for two logarithmic current profiles which have their associated shear stress (constant through the water column in both cases) determined by a boundary condition fixing the current at a height of 5 m above the surface. Notice that the shear stresses (broken linens) shown in the figure differ by a factor of four for a change in current speed of a factor of two. The value of z_0 in the figure is 0.1 m. Using (7.55) with $z' = 5$ m gives a drag coefficient of $C_d = 0.011$. Equation (7.54) gives a shear stress of 11.29 and 2.82 N m^{-2} which agrees with the values shown in the figure.

7.7.3 Viscosity near the wall

Let us write the usual expression for shear stress

$$T_x = \rho K_z \frac{\partial u}{\partial z'} \tag{7.61}$$

and remembering that τ has the constant value ρu_*^2 for the law of the wall, the velocity shear (7.50) gives

$$K_z \frac{u_*}{\lambda z'} = u_*^2 \tag{7.62}$$

which effectively determines the dependence of diffusivity on z':

$$K_z(z') = \lambda u_* z' \tag{7.63}$$

indicating that K_z is proportional to both u_* and distance z' from the surface, which can be interpreted as the reason that the velocity shear decreases with z' under conditions of constant shear stress. The dependence of K_z on u_* is the very root of the quadratic law for bottom friction and is a consequence of the generation of turbulent kinetic energy. The eddy viscosity K_z arises from the turbulence created by the stirring of the fluid by the surface stress. That turbulence increases with scale, and so K_z increases with z'. Equation (7.61) is the only weak assumption in determining this spatial dependence, i.e., the assumption of an eddy viscosity, but the result nevertheless has phenomenological validity.

It must be emphasized that the form (7.63) applies only to the constant stress region near to a surface. It is based on an observational result, and in its present form does not have any theoretical basis; that is, it has not been derived from a model of turbulence near a surface (although such theoretical treatments are available). There are some simple ways of extending the result to provide a model of the behavior of K_z throughout the water column. One of these concerns the combination of logarithmic layers at the surface and bottom. In that case, it is suggested that K_z initially increases with distance above the bottom, but with a slope dK_z/dz which starts to decrease with height towards the middle of the water column. This decrease with height is effectively a linear increase with distance measured downwards below the surface, i.e., we could combine two forms of (7.63) for the bottom and surface boundary layers:

$$K_z(z') \sim \frac{\lambda}{\sqrt{\rho}} \begin{cases} \sqrt{\tau_b}(z+h) & z \sim -h \\ \sqrt{\tau_w}(\eta-z) & z \sim \eta \end{cases}.$$ (7.64)

where τ_b and τ_w are the bottom and surface shear stresses (respectively), and the quantities in the round parentheses are distance from the bottom and surface respectively. It is possible to make the two forms in (7.64) part of a continuous function of z by various methods, such that K_z that has the appropriate form at either end of the water column. However, this is very much a construct, and it is possible that in shallow coastal basins the combined effects of mixing by both surface and bottom stress should be considered, i.e.,

$$K_z(z') \sim \frac{\lambda}{\sqrt{\rho}} [\sqrt{\tau_b}(z+h) + \sqrt{\tau_w}(\eta-z)].$$ (7.65)

Note that the assumption of a form for the height dependence of K_z will control the profile of current $u(z)$. Another modification that is used in the literature is to correct the logarithmic profile for a linear variation in shear stress through the water column. For example, a steady state pressure-driven flow with zero surface stress has

$$\tau(z) = \tau_b \frac{\eta - z}{\eta + h}$$ (7.66)

and *if* the linear form for K_z applies throughout the water column

$$\frac{du}{dz'} = -\frac{u_*}{\lambda h} \frac{z' - h}{z'}$$ (7.67)

where we have taken the water surface as $z = 0$, hence,

$$u = \frac{u_*}{\lambda} \left(\log \frac{z'}{z_0} - \frac{z' - z_0}{h} \right)$$ (7.68)

which is illustrated in Figure 7.12.

7.7.4 Calculating the vertical eddy viscosity

Determining the vertical eddy viscosity K_z is of prime importance to three-dimensional models of the hydrodynamics of coastal basins. Early models assumed that K_z is independent of height z in the water column, and a number of formulations existed that are mainly based on (7.63) evaluated at the center of the water column (these are discussed in this section). Later models have used the more detailed form of (7.63) and

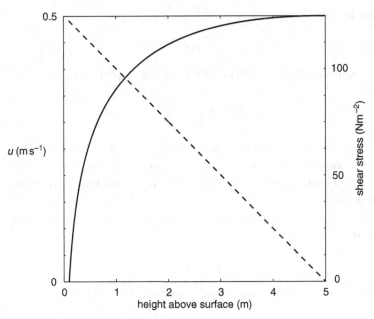

Figure 7.12 Full line is a repeat of Figure 7.11 with the velocity profile for (c) recalculated for a shear stress (broken line) that reduces linearly to zero at height 5 m above the surface. This is a prediction of the velocity profile over a rough bottom surface under conditions of zero shear stress at the water surface, i.e., zero wind stress.

these have now been superseded by the use of so-called *turbulence closure* techniques discussed briefly in Section 7.9.

If we are to assume a constant eddy viscosity for the whole water column we might take z' in (7.63) as $h/2$, i.e.,

$$K_z = \lambda u_* h/2. \tag{7.69}$$

It is at first sight a surprise to find the average value of K_z increasing with the total height of the water column, The reason lies in the increase of eddy viscosity with distance from a surface and is ascribed physically to the increase of scale of the eddies. Some of our basic ideas on mixing in basins and rivers are based on the pioneering work of Hugo B. Fischer,[8] which indicates that if we generalize (7.69) to

$$K_z = \gamma u_* h \tag{7.70}$$

then (7.69) corresponds to $\gamma = 0.21$, but experiments indicate a considerably smaller value $\gamma \sim 0.07$. It is smaller by a factor of three probably because the logarithmic layer does not stretch fully into the whole water column.

---

[8] American engineer (1937–83), professor of civil engineering at the University of California, Berkeley.

For wind forcing

$$\tau = \rho_a C_w W^2 \tag{7.71}$$

where W is wind speed and ρ_a the density of air and $C_w \sim 0.001$, and using $\gamma \sim 0.07$

$$K_z = AWh \quad A = \gamma \sqrt{\frac{\rho_a}{\rho}} C \approx 8 * 10^{-5} \tag{7.72}$$

and so for example if $W = 10\,\mathrm{m\,s^{-1}}$, and $h = 10\,\mathrm{m}$, we get $K_z = 0.008\,\mathrm{m^2\,s^{-1}}$. As another example, we can use (7.72) to find the vertical mixing time due to a $5\,\mathrm{m\,s^{-1}}$ wind blowing over a basin of depth $50\,\mathrm{m}$: $K_z = 2\ 10^{-2}\,\mathrm{m^2\ s^{-1}}$ giving a vertical mixing time, $h^2/\pi^2 K_z$ of 3.5 h.

For tidal flows we can similarly determine an *eddy diffusivity* for turbulence created by bottom friction. In this case

$$\tau = \rho C_d u^2, \tag{7.73}$$

and taking a bottom drag coefficient value for sand $C_d = 0.0025$,

$$K_z = Buh \quad B = \gamma \sqrt{C_d} \approx 0.003 \tag{7.74}$$

and so for example, for a tidal flow of $u = 0.1\,\mathrm{m\,s^{-1}}$, $h = 10\,\mathrm{m}$, we get $K_z = 0.003\,\mathrm{m^2\,s^{-1}}$. Macro-tidal waters might have $u = 1.0\,\mathrm{m\,s^{-1}}$ and if $h = 100\,\mathrm{m}$, we get $K_z = 0.3\,\mathrm{m^2\,s^{-1}}$. As another example, we can find the tidal vertical mixing time due to a current of $0.5\,\mathrm{m\,s^{-1}}$ in water of depth $50\,\mathrm{m}$: $K_z = 7.5 * 10^{-2}\,\mathrm{m^2\,s^{-1}}$ giving a vertical mixing time of 0.9 h.

7.7.5 Mixing length

We have previously considered three types of dependence of the eddy viscosity K_z on height z in the water column. The first is simply to assume a constant value, and this is useful as a good first approximation in initial models of coastal basins. The second is to assume a linear increase of K_z with distance from the bottom of the coastal basin due to the logarithmic boundary layer. The latter is essential to the proper determination of the current profile near to the bottom, and incorporates in a very fundamental way the process by which K_z is generated by the turbulence created by flow over the rough surface. In some coastal basins, the log layer may occupy only a small fraction of the water column, and so resolving the current profile in the log layer may not be essential to many aspects of the dynamics of the basin. However, processes close to the bottom or in principle, on the bottom itself, do require the model to resolve the log layer. Many numerical models increase their vertical resolution near the bottom to achieve that aim. An example of such a method is make logarithmic transformation of z', i.e.,

$$z'' \equiv \log(z'/z_0), \tag{7.75}$$

and so (7.53) becomes

$$u = \frac{u_* z''}{\lambda} \qquad (7.76)$$

which means that a constant spacing of vertical cells in z'' would produce cells of increasing thickness as z' increases. This type of log transformation can also be used in the horizontal to increase spatial resolution near a coast, and there is some evidence that the same type of logarithmic boundary layer does occur just beyond the roughness elements provided by twists and turns of an undulating coastline (as discussed in Chapter 11).

Apart from the log layer as a region of intrinsic interest, the great advantage of utilizing the parameterization associated with this proscription for eddy viscosity is that it allows K_z in the water column to be determined *ab initio* from the velocity of flow and *roughness* of the surface. This is of very fundamental value to the dynamics of coastal basins, and is widely used as an easy way of finding K_z for wind and tidal forcing (this was fully discussed in Section 7.7.4). However, the method is not necessarily accurate at points of the water column that are not in the direct vicinity of the surface in time dependent flows, where the shear stress is not usually constant. So, the third method of modeling K_z is to assume a so-called *mixing length*, so that K_z does not depend just on z', but has a dependent on the vertical shear of horizontal velocity. The physical reasoning behind this method is that the velocity shear has a controlling influence in the creation of turbulent energy.

The mixing length was first introduced by Ludwig Prandtl.[9] As we saw in Section 7.2, the mixing coefficient is the product of a velocity associated with random motion w' (as may be due to turbulence) and a length which characterizes mean free path l, i.e., (7.36) gave

$$K_z = w' l. \qquad (7.77)$$

If we return to our simple conceptual picture in Figure 7.8, particles move with the velocity w' between planes parallel to the x axis, with currents of $u(z)$ and $u(z+l)$. So a particle with turbulent velocity w' in the z direction has a variation of current in the x direction of $l du(z)/dz$ in the x direction. Prandtl suggested that the fluctuations in w are of the same order as the variation in u. His reasoning was that $w \ll u$. We discussed this inequality in Section 4.3 for coastal basins, and the same is true of the air flows considered by Prandtl. As a consequence, we can reasonably suppose that it is variations in u that are responsible for fluctuations in w. This can be visualized as eddies that rotate in the vertical plane with their velocity components having similar

---

[9] German physicist (1875–1953), who developed fundamental principles of subsonic aerodynamics and was well known for the Prandtl number (the ratio of momentum diffusivity to thermal diffusivity).

magnitude along the x and z axes. The eddies are fed by fluctuations in u. Hence, Prandtl proposed that

$$w' \sim l\left|\frac{\partial u}{\partial z}\right| \tag{7.78}$$

where l is now called the *mixing length*. Using (7.77) and (7.78) (as an equality),

$$K_z = l^2\left|\frac{\partial u}{\partial z}\right|. \tag{7.79}$$

If we substitute (7.50), the observed variation of u with z' near a wall, in (7.79) we obtain

$$l(z') = \lambda z' \tag{7.80}$$

or

$$K_z = (\lambda z')^2\left|\frac{\partial u}{\partial z}\right| \tag{7.81}$$

and with (7.61) we find

$$\tau_x = \rho(\lambda z')^2\left|\frac{\partial u}{\partial z}\right|\frac{\partial u}{\partial z} \tag{7.82}$$

which can be written

$$|\langle w'u'\rangle| = \left(\lambda z'\frac{\partial u}{\partial z}\right)^2 \tag{7.83}$$

as we would expect from our initial assumption of $w' \sim u'$ together with (7.78) and (7.80).

If we take (7.82) as our starting point in the study of the behavior of fluid near a wall, and consider a region of constant shear stress, we find

$$\left|\frac{du}{dz'}\right|\frac{du}{dz'} = \left(\frac{u_*}{\lambda z'}\right)^2 \tag{7.84}$$

and this is immediately seen as predicting the observed log law as contained in (7.50). This is a good example of observations inspiring a theoretical model that does naturally reproduce those observations. But of course (7.81) has much greater value than merely reproducing the log law, since we have sufficient confidence to apply it to a wide range of situations in a coastal basin. We will explore some of these applications in later sections.

7.8 Friction and energy

7.8.1 Energy input to turbulence

As shown in Section 7.6, the vertical flux of horizontal momentum manifests itself as a horizontal stress τ:

$$\tau = -\rho\langle w'u'\rangle \tag{7.85}$$

where

$$\tau = \rho K_z \frac{\partial u}{\partial z}. \tag{7.86}$$

The shear stress is essentially a frictional or drag force between neighboring elements in the water column and the velocity shear $\partial u/\partial z$ is a measure of the velocity difference between those elements. If we return to the momentum continuity equation, and assume that there is no pressure gradient or Coriolis force,

$$\rho \frac{\partial u}{\partial t} = \frac{\partial \tau}{\partial z} \tag{7.87}$$

so that the vertical change in τ determines the acceleration of each element. In the steady state τ must be constant, which basically just means that the stresses on the upper and lower faces of any element are equal giving zero net force. However, unless $\partial u/\partial z$ is zero, frictional forces are present and energy is lost from the mean motion. This lost energy creates turbulence. This is an interesting situation because K_z itself is a manifestation of turbulence, and it is turbulence that allows stress or momentum flux to penetrate through the water column.

7.8.2 Energy and shear stress

The momentum flux always moves from high to low velocity, so while momentum is conserved we find that energy is always lost from the mean motion. To prove this, just consider two particles of mass m with velocities u_1 and u_2 ($u_1 > u_2$), and then transfer a lower velocity. Denote the original energy by E_{initial} and the final energy by E_{final}:

$$E_{\text{initial}} = \frac{m}{2}\left(u_1^2 + u_2^2\right) \tag{7.88}$$

$$E_{\text{final}} = \frac{m}{2}\left[(u_1 - \delta u)^2 + (u_2 + \delta u)^2\right] \tag{7.89}$$

Equations (7.88) and (7.89) give to first order in δu

$$E_{\text{final}} = E_{\text{initial}} - m\delta u(u_1 - u_2) \tag{7.90}$$

and so $E_{\text{final}} < E_{\text{initial}}$ and the energy decreases by the product of the momentum exchange and velocity difference. Equation (7.90) can be obtained directly from the well-known result that the change in energy of a particle is the product of momentum change and velocity, i.e., the equation for kinetic energy

$$E = \frac{m}{2} v^2 \tag{7.91}$$

implies that

$$\delta E = v(m\delta v). \tag{7.92}$$

Let us consider an element of a water column of extending from $z = -h$ to $z = \eta$. Suppose that there is a wind stress applied to its surface and a bottom stress at its base due to friction. Momentum is passing from the surface to the bottom. This will involve momentum passing from regions of higher to lower velocity. This is achieved by eddy viscosity and always involves loss of energy. Consider an element of the water column at z of elemental height dz with a vertical flux of momentum $\langle w'u' \rangle$ which is momentum per unit area per unit time, passing in the positive z direction from the bottom of the element to its top face. Considering positive $\langle w'u' \rangle$, the element at z gains momentum through its lower face, and then loses momentum through its upper face. However, this takes place at different horizontal velocities (because of vertical shear in the horizontal velocity). So for *each unit of momentum* passing through the volume, the net *decrease* of *energy* is $(\partial u/\partial z)dz$, i.e., if the top of the element is moving more quickly than the bottom, the element loses energy. The rate of *loss* of energy per *unit mass* denoted by Gdz is

$$G = -\langle w'u' \rangle \frac{\partial u}{\partial z} \tag{7.93}$$

Rate G is positive since $\langle w'u' \rangle$ and $\partial u/\partial z$ have opposite signs, or equivalently,

$$G = \frac{\tau}{\rho} \frac{\partial u}{\partial z}, \tag{7.94}$$

or using (7.61), (7.94) becomes

$$G(z) = K_z \left(\frac{\partial u}{\partial z} \right)^2. \tag{7.95}$$

The total rate of loss of energy per unit mass throughout the water column is (in a one-dimensional system)

$$\int_{-h}^{\eta} G(z)dz = \frac{F}{\rho} \tag{7.96}$$

where

$$F \equiv \int_{-h}^{\eta} \tau_x \frac{\partial u}{\partial z} dz. \tag{7.97}$$

Notice that (7.96) and (7.97) have the dimensions of energy per unit area. If the component of velocity in the y direction v is non-zero and also a function of z, we have an additional contribution to the loss of energy from viscosity. Equation (7.97) may be integrated by parts to give

$$F = \tau_x u|_{-h}^{\eta} - \int_{-h}^{\eta} u \frac{\partial \tau_x}{\partial z} dz \tag{7.98}$$

and if we use the momentum continuity equation (7.87) in (7.98), F becomes

$$F = \tau_x u|_{-h}^{\eta} - \int_{-h}^{\eta} u \frac{\partial u}{\partial z} dz, \tag{7.99}$$

and expanding the first term,

$$F = (\tau_x u)|_{z=\eta} - (\tau_x u)|_{z=-h} - \frac{\rho}{2} \int_{-h}^{\eta} \frac{\partial u^2}{\partial t} dz. \tag{7.100}$$

In (7.100) τ_x is the internal shear stress conventionally measured on the upper surface, and so in terms of the surface wind stress τ_w and bottom stress τ_b, (7.100) gives with (7.97)

$$F = \tau_w u(\eta) + \tau_b u(-h) - \frac{\rho}{2} \int_{-h}^{\eta} \frac{\partial u^2}{\partial t} dz \tag{7.101}$$

$$\frac{\rho}{2} \int_{-h}^{\eta} \frac{\partial u^2}{\partial t} dz = \tau_w u(\eta) + \tau_b u(-h) - \rho \int_{-h}^{\eta} G(z) dz. \tag{7.102}$$

The left-hand side of (7.101) is the rate of production of kinetic energy of the water column. This is balanced on the right-hand side by the rate of input of energy by surface stresses, i.e., wind and bottom friction and the loss of energy to turbulence through viscosity.

7.8.3 Turbulent dissipation rate in the logarithmic layer

An important parameter of the turbulent spectrum is the rate of production, and subsequent dissipation, of turbulent energy (to heat) *per unit mass* denoted by ε. This impotent parameter of the turbulence spectrum has already been discussed in Section 7.5. It is possible to determine ε if we make the assumption that the rate of generation and dissipation of turbulence is equal to the rate of loss of energy by viscous damping of the flow, i.e., from (7.94)

$$\varepsilon = \frac{\tau}{\rho}\frac{\partial u}{\partial z}.$$ (7.103)

If we use the theory of the log layer and write the stress velocity at the surface as u_*, where

$$u_*^2 = \frac{\tau}{\rho}$$ (7.104)

and recalling that for the log layer

$$\frac{\partial u}{\partial z} = \frac{u_*}{\lambda z'}$$ (7.105)

where λ is the von Kármán constant,

$$\varepsilon = \frac{u_*^3}{\lambda z}.$$ (7.106)

A alternative approach is to take (7.93) and write G as

$$G = -\langle w'u'\rangle\frac{\partial u}{\partial z}$$ (7.107)

which with the mixing length model expressed in (7.78) becomes

$$\varepsilon = \frac{w'^3}{Bl}.$$ (7.108)

7.9 Turbulence closure

In this section we consider more precise methods of determining the coefficients of vertical mixing. The full problem involves solving for the turbulence field and this is presently not possible over the spatial scales of interest in a coastal basin. Instead we use a method of moments which is terminated, or *closed*, at low order. We define certain parameters associated with the turbulence in the basin, and develop a model that interrelates these parameters and the way in which they control mixing. This is called a *turbulent closure scheme*. The model to be explored in detail here is the so-called *level 2½ closure scheme of Mellor and Yamada,*[10] which has found extensive application in a variety of different types of basins.

One of the tasks of the model is to write a balance equation for the turbulent energy. We know the rate of loss of energy from the mean flow as determined in Section 7.8. Note that the difference between mean flow and turbulence depends on the space and time scales over which we average, as discussed in Section 7.2. Section 7.9 provides various forms for the rate of loss of energy from the mean flow by viscosity, and our model will assume that that all of the energy appears as turbulence. We have two major processes to consider for turbulent energy. The first is the turbulent dissipation rate, and here we adopt the Prandtl mixing length idea. Let us return to the form

$$\varepsilon = -\frac{\tau}{\rho} \frac{\partial u}{\partial z} \tag{7.109}$$

and express the shear stress as $\tau/\rho = \langle w'u' \rangle$ and recall that $w' \sim u' \sim l\partial u/\partial x$, so that (7.109) becomes

$$\varepsilon \sim \frac{\langle w'^3 \rangle}{l} \tag{7.110}$$

which should apply to any eddy. We need to define a dissipation rate that involves one averaged parameter of the turbulence spectrum. This is taken as the mean turbulent velocity q defined in terms of the mean turbulent energy k as

$$k = \frac{q^2}{2} \tag{7.111}$$

and hence (7.110) becomes

$$\varepsilon = \frac{q^3}{Bl}, \tag{7.112}$$

B being a constant of proportionality. The second process is the diffusion of turbulence, and we propose this is given by a diffusivity for turbulent energy, denoted by $K_z^{(E)}$. This may be different from that for momentum which we shall now denote now by $K_z^{(m)}$, with both having the form of (7.77) $K_z = lw'$, and with the same generalization as adopted in (7.112) this becomes

$$K_z^{(m)} = lqS_m$$
$$K_z^{(E)} = lqS_E \tag{7.113}$$

---

where S_m and S_E are constants (for unstratified flow). So, the dynamics of the turbulent energy k are given by

$$\frac{\partial k}{\partial t} - \frac{\partial}{\partial z} K_z^{(E)} \frac{\partial k}{\partial z} = G(z) - \varepsilon \qquad (7.114)$$

and the most convenient form for G comes from (7.95) with the use of $K_z^{(m)}$, i.e.,

$$G(z) = K_z^{(m)} \left(\frac{\partial u}{\partial z} \right)^2. \qquad (7.115)$$

The terms on the left-hand side of (7.114) are the rate of increase of turbulent energy with time and the gradient of the vertical diffusion of turbulent energy, i.e., the terms in the vertical continuity equation for turbulent energy. The right-hand side contains the rate of production and dissipation of turbulence. It is evident that in a steady state with limited vertical diffusion of turbulence, the left hand side of (7.114) is zero, and so we would propose the production–dissipation balance

$$K_z^{(m)} \left(\frac{\partial u}{\partial z} \right)^2 - \frac{q^3}{Bl} = 0, \qquad (7.116)$$

and with the boundary condition at $z = 0$,

$$K_z^{(m)} \left. \frac{\partial u}{\partial z} \right|_{z=0} = u_*^2 \qquad (7.117)$$

and with (7.113) and (7.116), (7.117) gives

$$\left. \frac{q}{u_*} \right|_{z=0} = \left(\frac{B}{S_M} \right)^{1/4}. \qquad (7.118)$$

If we substitute (7.63), i.e., $K_z = \lambda u_* z$ into the first of (7.113) with (7.80) we find

$$u_* = q S_M \qquad (7.119)$$

hence

$$\frac{1}{S_M} = \left(\frac{B}{S_M} \right)^{1/4} \qquad (7.120)$$

The accepted values for S_m is 0.39 and so $B = 16.6$, and hence (7.118) becomes

$$\left. \frac{q}{u_*} \right|_{z=0} = 2.6. \qquad (7.121)$$

The constant S_E is taken as 0.2. Equations (7.112) to (7.116) represent our turbulence closure system, and only differs from our earlier method by the incorporation of the quantity k and its dynamical equation (7.114). This method is presently widely used in three-dimensional models of basins but it is not the only type of turbulent closure system.

7.10 Dispersion in coastal basins

7.10.1 Processes of dispersion

Horizontal mixing is usually called *dispersion,* and the coefficient of dispersion is usually denoted by K_x. There are two processes involved in horizontal dispersion. The first is a consequence of randomness in the horizontal currents. We refer to this process as *direct dispersion,* and it can be measured directly by deploying a set of drogues at a fixed depth below the surface and watching their subsequence movement. The second is due to vertical mixing, which is itself a stochastic process. It is measured by a tracer injected into the water column, and is a function of the K_z for the tracer. The process evidently vanishes if there is no vertical mixing and, for example, a very buoyant material which stays on the surface will not experience dispersion due to this process. We refer to the process as *Taylor dispersion.*[11] In a numerical model, some considerable care is required in choosing a suitable value for the diffusivity K_x. A two-dimensional model must include both direct dispersion and Taylor dispersion, because the model does not include any of the vertical movement of water mass needed to create Taylor dispersion. A three-dimensional model may, in principle, only include direct dispersion (it does need to be noted that its ability to correctly predict Taylor dispersion is dependent on vertical mixing). Direct dispersion increases with spatial scale, whereas Taylor dispersion is accentuated by vertical current shear, which may be more marked on smaller scales. The total dispersion involves the joint influence of direct and Taylor dispersion, and so care is required not to conclude that the total dispersion is entirely composed of direct dispersion (as measured for example by drogues or surface floats). We defer a detailed derivation of Taylor dispersion to Chapter 10.

7.10.2 Measuring direct dispersion

The easiest way to measure direct dispersion is to release some type of tracer particles at a point and measure their spread with time as they form a cloud. Basically the spread is controlled by the spatial variance σ in one dimension after time t:

$$\sigma = \sqrt{2K_x t}, \tag{7.122}$$

---

[11] After the pioneering work of British scientist Geoffrey I. (known to his colleagues as G. I.) Taylor (1886–1975). Taylor made contributions to physics over a huge range of topics (including a major contribution to fluids come from his work that foreshadowed the statistical theory of turbulence discussed in Section 7.5).

so the spread in any one dimension increases as the square root of time, quickly at first and then more slowly. In two dimensions, assuming that the dispersion is *isotropic* (the same in all directions), the area increases as

$$A = 4\pi K_x t, \tag{7.123}$$

so the spread in two dimensions is linear in time. The square root type of dependence of spread in (7.122) also appears if (as often happens), the basin has a channel structure and the plume spreads to fill the width of the channel and then disperses in a one-dimensional fashion along the channel.

A traditional method of measuring total dispersion was to place fluorescent dye in the water and then measure the concentration change with time using a fluorometer (an instrument that measures fluorescence). The center of the patch moves with the mean advective current (and gives a way of measuring that current) and the spread of the patch provides a measure of K_x. The rate of mixing of the dye through the water column does effect the measured dispersion. An oil when spilt onto the ocean is buoyant (its density is 800 to 950 kg m^{-3} compared with that of seawater which is 1020 to 1020 kg m^{-3}), and initially spreads under its own buoyancy rather than through the turbulence of the water. Furthermore, the surface layers of water may show significant vertical velocity shear, and so it may be important to mix the tracer particles through these surface layers. If the water column shows some surface buoyant layers, it is possible to demonstrate the vertical shear that often occurs near the surface by floaters of marginally different buoyancy. When trying to measure horizontal dispersion, care is required in regard to depth of the tracer; and understanding at what depth, or over what depths, we are measuring the horizontal dispersion.

A method commonly used to measure direct dispersion is to track the position of a number of drogues released at one point in the ocean. The drogues need to have their sheets at one common depth, so that one measures the dispersion at that depth only. The mean position of the drogues will change with time, and this gives a measure of the mean current. The position of the individual drogues differs from the mean because of the fluctuations in u and v. Hence, returning to Section 7.2,

$$u = \bar{u} + u' \quad v = \bar{v} + v' \tag{7.124}$$

and it is the fluctuations u', v' that produce the dispersion or spreading in the positions of the drogues. As the drogues advect and spread, there is an increase in the averaging volume of space-time illustrated in Figure 7.3.

Let us assume that dispersion is isotropic, which we denote simply by K. To measure K, after a particular time t, we need to measure the variance σ in their position at time t:

$$\sigma^2 \equiv \frac{1/2}{N-1} \sum_{i=1}^{N} \left[(x_i - \bar{x})^2 + (y_i - \bar{y})^2 \right] \tag{7.125}$$

where x_i, y_i are the coordinates of the position of drogue i at time t, and $(\bar{x}, \bar{y})$ is the mean position of all the N drogues. There are two possible definitions of a dispersion coefficient based on (7.122). We can either consider the rate of change of variance with time at time t, i.e.,

$$K = \frac{1}{2}\frac{d\sigma^2}{dt} \tag{7.126}$$

or the average rate of change of variance with time over the time interval from 0 to t, i.e.,

$$\langle K \rangle_t = \frac{\sigma^2}{2t}. \tag{7.127}$$

In practice we find K as follows:

$$K = \frac{1}{2}\frac{d\sigma^2}{dt} \to \frac{\sigma_2^2 - \sigma_1^2}{t_2 - t_1} \tag{7.128}$$

where the suffixes 1 and 2 refer to measures of variance at times t_1 and t_2 $(t_2 > t_1)$. In principle, the dispersion may be slightly different in the x and y directions, and one notices that the spread of the drogues may be anisotropic. This can be allowed for by calculating a dispersion coefficient K_{xx} and another K_{yy} derived from

$$\sigma_{xx}^2 = \frac{1}{N-1}\sum_{i=1}^{N}\left[(x_i - \bar{x})^2\right]$$

$$\sigma_{yy}^2 = \frac{1}{N-1}\sum_{i=1}^{N}\left[(y_i - \bar{y})^2\right] \tag{7.129}$$

and we can also define the dispersion coefficient K_{xy} based on

$$\sigma_{xy}^2 = \frac{1}{N}\sum_{i=1}^{N}[(x_i - \bar{x})(y_i - \bar{y})] \tag{7.130}$$

which is zero if the drogues spread uniformly in the x and y directions. The usual reason for this asymmetry in the dispersion is the existence of a mean motion, or current, in one particular direction. The effect is much more pronounced in Taylor dispersion.

7.10.3 Direct dispersion as a function of spatial scale

It is usually found that K does increase with time t, and we anticipate this is due to an increase of dispersion with the volume of the space-time averaging volume. The

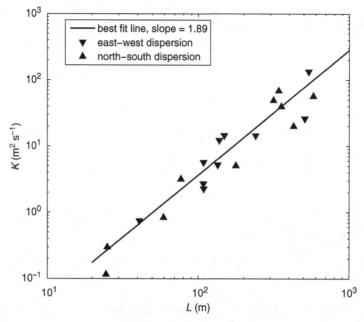

Figure 7.13 Dispersion of a set of drogues in a small coastal basin in south-west Australia. The two types of triangle represent dispersion in two orthogonal directions which for this basin are essentially identical. (Based on reprocessed data from Hearn, 1983.)

increase in K is much simpler if viewed as a function of variance σ. A conventional relation between variance σ and *length scale L* is

$$L \equiv 3\sqrt{2}\sigma. \tag{7.131}$$

The increase of K with L is demonstrated by Figure 7.13 which shows points obtained for a set of drogues in a small bay in south-west Australia.

In Figure 7.13, K varies from about 0.1 to 10^2 m^{2}s^{-1} as L increases from 20 m to 600 m. The reason for the increase in K with L is that the patch of drogues captures increasing turbulence with the increasing spatial scale of the averaging volume, and so the strength of the fluctuations in u and v increases with the distance apart of the drogues. The plot shown in Figure 7.13 is a log against log plot so that the straight line indicates that a relation of the type

$$K \propto L^n \tag{7.132}$$

with a value for the exponent of $n = 1.89$. The various points in the figure are based on K_{xx} and K_{yy} taken with x, y orientated to east and north. Clearly, in this system there is little difference between dispersion in the two directions, i.e., this coastal basin is relatively isotropic (this is not true in all basins). Notice how the spatial scales will expand as we move to larger basins and, for example, in the Great Lakes of North

America spatial scales will reach hundreds of kilometers, so that K would reach some hundreds of square meters per second. There is a suggestion that the exponent n decreases slightly as the scale increases, and values of $n \sim 1$. If K is proportional to L, it can be written as

$$K = v_K L \tag{7.133}$$

where v_K is called the *dispersion velocity* and has a value of order $2 \ 10^{-3} \ \mathrm{m\,s^{-1}}$. So for example, when L is $100 \ \mathrm{km}$ $(10^5 \ \mathrm{m})$, $K = 50 \ \mathrm{m^2\,s^{-1}}$. By contrast when L is small, it appears that n is of order 2, so that

$$K = \frac{L^2}{t_K} \tag{7.134}$$

where the dispersion time is of order $2 \ 10^4 \ \mathrm{s}$ with a crossover value of L of order $50 \ \mathrm{m}$.

In a well-publicized event in the early 1990s,[12] a ship dropped a container full of sports shoes in the North Pacific, and some $10\,000$ shoes were left to drift around on the surface currents. The shoes were boxed as pairs, but the boxes disintegrated and the individual shoes drifted as separate entities. The center of the shoe patch followed the prevailing current, but the dispersion increased with time. Eventually, most of the shoes washed onto beaches and shoe-swaps where set up along the US west coast to find pairs of these valuable shoes. As an application of our model of the variation of dispersion with distance we would predict that if pairs were found up to $100 \ \mathrm{km}$ apart, we would expect that that K would reach about $200 \ \mathrm{m^2\,s^{-1}}$. So, an average K of order $100 \ \mathrm{m^2\,s^{-1}}$ should be used to estimate the time taken:

$$t = \frac{\sigma^2}{2K} = \frac{L^2}{36K} \tag{7.135}$$

i.e., about 1 month.

7.11 A closer look at the logarithmic boundary layer

In this section we consider in more detail the theory of *skin friction*, i.e., friction due to the motion of fluid over a surface.[13] Nearly all descriptions of vertical mixing are based on the theory of the *log layer* also called *law of the wall* (which we discussed in Section 7.7) which is one of the most important observations in all of research on turbulent flow. The law simply states that the gradient of mean velocity varies inversely with distance from a wall. So the closer we are to the wall, the greater is the rate of variation of velocity with distance. The usual representation of the law for a rough wall is

$$\frac{\partial u}{\partial z} \propto \frac{1}{z'} \tag{7.136}$$

---

[12] Ebbesmeyer and Ingraham (1994).   [13] More detail is available in Pope (2000).

where z' is the distance from the surface. If we drive the flow with a pressure gradient parallel to the wall, we can determine the bottom stress in the steady state from the pressure gradient. We shall denote this by τ_0. If the fluid has a free surface at a distance h away from the wall and that surface has zero hear stress, then

$$\tau = \tau_0(1 - z'/h) \tag{7.137}$$

and if we are considering flow over a rough bottom h is equivalent to total depth. As usual we can write

$$u_* \equiv \sqrt{\tau_0/\rho}. \tag{7.138}$$

Suppose we perform a number of experiments in pipes with both smooth and rough surfaces, in which we vary the value of u_* and then plot $(1/u_*)\partial u/\partial z'$ against $1/z'$, we find a straight line in every case, and very remarkably, that the slope is always 2.4. Or put another way, if we plot $(1/u_*)\partial u/\partial z'$ against $1/(\lambda z')$ where $\lambda = 0.41$, the von Kármán constant, we get a straight line of unit slope

$$\frac{\partial u}{\partial z} = \frac{u_*}{\lambda z'}. \tag{7.139}$$

To be more precise, this law is only found at sufficiently high Reynolds numbers and not too close to the surface. For a pipe, (7.137) must be modified by replacing h by the radius R of the pipe. In that case, the argument for τ vanishing at $z' = R$ is simply that flow in the pipe is symmetrical around the central axis, so the radial gradient of velocity and hence, stress, must be zero. If we integrate (7.139),

$$u = \frac{u_*}{\lambda} \int \frac{dz'}{z'} + A \tag{7.140}$$

where A is a constant of integration. We generally prefer now to use a non-dimensional variable $z'/z*$ and so (7.140) becomes

$$u = \frac{u_*}{\lambda} \log\left(\frac{z'}{z_*}\right) \tag{7.141}$$

where $z*$ is the extrapolated value of z' at which u as predicted by (7.141) would vanish. Note that we use the mathematical modeling notation $\log(z)$ to mean the natural logarithm of x, which is strictly written as $\log_e(x)$ and sometimes written as $\ln(x)$. A very important point about the log law is that $z*$ adds a constant to the whole function; it does not affect the *shape* of u with z'. For any experimental data we can plot u against $\log(z)$ which allows us to determine the value of $z*$ and $u*$, i.e., if the straight line is

$$u = C\log(z') + B \tag{7.142}$$

then

$$u* = \lambda C$$
$$z* = \exp(-B/C).$$
(7.143)

For a *smooth surface*, a large number of experiments have established that $z* = \delta/Z$ where δ is the viscous length scale:

$$\delta \equiv \frac{u*}{\nu}$$
(7.144)

in which ν is the molecular viscosity, and many high-quality experiments have shown that

$$Z = 8.4$$
(7.145)

so that

$$u = \frac{u_*}{\lambda} \log\left(\frac{z'}{\delta/Z}\right)$$
(7.146)

or equivalently

$$u = u_* \left[\frac{1}{\lambda}\log\left(\frac{z'}{\delta}\right) + B\right]$$
(7.147)

$$B = \frac{\log Z}{\lambda} \sim 5.2.$$
(7.148)

Inside the viscous sublayer the velocity shear is constant and so has the form

$$\nu\frac{\partial u}{\partial z'} = u_*^2$$
(7.149)

which we solve for current u:

$$u = u * z'/\delta.$$
(7.150)

This form for u is plotted against z'/δ in Figure 7.14 as the full line along with u from the log law for a smooth surface (as a broken line). The intersection of the linear viscous sublayer line (7.150) and the log law (7.146) occurs at $z = \Delta$ which we can see by eye in Figure 7.14 lies near $z'/\delta = 11$.

$$\frac{u* \log(Z\Delta/\delta_\nu)}{\lambda} = \frac{\Delta u_*^2}{\nu},$$
(7.151)

i.e.,

$$\log(Zr) = r\lambda$$
(7.152)

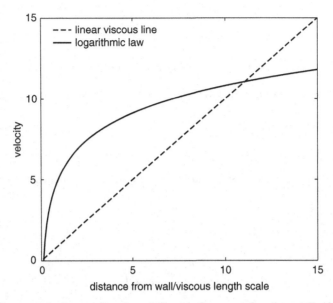

Figure 7.14 Current velocity as a function of distance from a surface (normalized to the viscous length scale), z'/δ. The broken line applies to the linear viscous sublayer and the full line is the log law. Note the intersection at $z'/\delta \sim 11$.

where

$$r \equiv \Delta/\delta \tag{7.153}$$

or

$$\frac{\log(Zr)}{Zr} = \frac{\lambda}{Z}. \tag{7.154}$$

There are two solutions to (7.154), one of which is very close to the origin at $r \sim 1/Z$ (and not of importance here), whilst the other involves

$$\frac{\log(Zr)}{Zr} \sim 93 \tag{7.155}$$

which means that

$$r \cong \frac{93}{Z} \sim 11. \tag{7.156}$$

In other words, the thickness of the viscous sublayer thickness Δ is about 11 times the viscous length scale δ. Different authors provide slightly different values for r ranging from 5 to 15, depending on how the edge of the viscous sublayer is defined. Note that the log law has u vanishing at $z* = \delta/Z$ where $Z \sim 8.4$ and so $z*/\delta = \sim 0.012$.

Rough surfaces have *roughness elements* that protrude above the plane of an idealized smooth surface. In a coastal basin these may be elements of the sediment or vegetation, and are often taken to be the ripples in the sand. On a microscopic level these are the hills and valleys of a terrestrial landscape and produce a *roughness layer*. It is usual to represent the elements by their *mean height s*, although this is clearly not the only important parameter that characterizes the elements. If we define the *Reynolds number* by

$$Re(z') \equiv \frac{u*z'}{\nu} = \frac{z'}{\delta} \tag{7.157}$$

it is clear that the behavior of fluid inside the roughness layer is partially controlled by the value of Reynolds number at the scale of the roughness elements, i.e., by $R(s)$:

$$R(s) \equiv \frac{u*s}{\nu} = \frac{s}{\delta} \tag{7.158}$$

We know from (7.153) and (7.156) that $R = 11$ represents the top of the viscous sublayer and so we could surmise that roughness is only important if $R(s) > 11$. For example, a flow of $0.1 \, \mathrm{m\,s}^{-1}$ over a sandy bottom gives $u* \sim 0.005 \, \mathrm{m\,s}^{-1}$, and so with $\nu = 2*10^{-6} \, \mathrm{m}^2 \, \mathrm{s}^{-1}$, (7.144) gives $\delta = 0.4 \, \mathrm{mm}$ and $\Delta \sim 4.4 \, \mathrm{mm}$.

Johann Nikuradse (1894–1979), in a famous, and pivotal, set of experiments in the 1930s, glued sand grains onto the walls of a flume pipe and measured the properties of the flow. Using s to denote the mean diameter of sand grains, he found that roughness affects the value of B in (7.147), so that B effectively becomes a function of s/δ, so that (7.147) for a rough surface has the form

$$u = u_* \left[\frac{1}{\lambda} \log\left(\frac{z'}{\delta}\right) + B\left(\frac{s}{\delta}\right) \right] \tag{7.159}$$

or equivalently

$$u = u_* \left[\frac{1}{\lambda} \log\left(\frac{z'}{s}\right) + \frac{1}{\lambda} \log\left(\frac{s}{\delta}\right) + B\left(\frac{s}{\delta}\right) \right] \tag{7.160}$$

which we prefer to write as

$$u = u_* \left[\frac{1}{\lambda} \log\left(\frac{z'}{s}\right) + \tilde{B}\left(\frac{s}{\delta}\right) \right] \tag{7.161}$$

where

$$\tilde{B}\left(\frac{s}{\delta}\right) = \frac{1}{\lambda} \log\left(\frac{s}{\delta}\right) + B\left(\frac{s}{\delta}\right) \tag{7.162}$$

or

$$\tilde{B}\left(\frac{s}{\delta}\right) = 5.6 \log_{10}\left(\frac{s}{\delta}\right) + B\left(\frac{s}{\delta}\right). \tag{7.163}$$

When $s/\delta \ll 1$, we have the limit of a smooth surface know and from (7.147) the second term in (7.163) is

$$\lim_{s/\delta \to 0} B \sim 5.2. \tag{7.164}$$

At large s/δ, we would expect that $\tilde{B}$ becomes independent of s/δ, and indeed we find that

$$\lim_{s/\delta \to \infty} \tilde{B} \sim 8.5, \tag{7.165}$$

i.e.,

$$z* = s/32.6 \tag{7.166}$$

with a transition region between (7.164) and (7.166) when s/δ lies between 10 and 100, i.e., $1 < s/\Delta < 10$. For this reason, it usual to say that the surface is *effectively smooth* if $s/\delta < 10$ and the surface is *fully rough* if $R(s) > 100$. Finally, let us write a *general form for the log law* which applies to both rough and smooth surfaces,

$$u = \frac{u*}{\lambda} \log\left(\frac{z'}{z*}\right) \quad z' > z'_{min} \tag{7.167}$$

and express $z*$ in terms of the ratio of the roughness length s (defined more precisely as the mean sand grain diameter) and viscous length scale $\delta \ (= v/u*)$ as

smooth surface: $(s/\delta) \lesssim 10$

$$z* \sim \delta/8.4 \tag{7.168}$$

$$z'_{min} \sim 11\delta$$

rough surface: $(s/\delta) \gtrsim 100$

$$z* \sim s/32.6 \tag{7.169}$$

$$z'_{min} \sim s.$$

Notice that for a smooth surface $z*$ is much less than the thickness of the viscous boundary layer ($\Delta \sim 10\delta$) and for a rough surface $z*$ is much less than the grain size or height of the roughness elements. Importantly z' in (7.167) is the distance from the true *surface* and *not* the top of the roughness elements. For coastal basins, it is more convenient to define a distance from the *top* of the roughness elements called z, so that

$$z' = z + s. \tag{7.168}$$

7.12 Coefficients of skin friction

A very important consequence of the law of the wall is that it shows that the friction created by the wall is quadratic in the velocity field and we have discussed this result in Section 7.7. If we use the depth-averaged current $\bar{u}$ to define a drag coefficient the usual notation is

$$\tau = \rho C_\mathrm{d} \bar{u}^2. \tag{7.169}$$

If we derive the mean current from the log law, we obtain

$$\bar{u} \equiv \frac{u^* z^*}{\lambda h} \int_{p_{\min}}^{h/z^*} \log(p)\,dp \quad p \equiv z'/z^* \quad p_{\min} = z'_{\min}/z^* \tag{7.170}$$

where we are ignoring any contribution below the log layer. The integral in (7.170) is

$$\int \log(p)\,dp = p \log p - p \tag{7.171}$$

and so we obtain

$$\bar{u} = \frac{u^* z^*}{\lambda h} |p \log p - p|_{z'_{\min}/z*}^{h/z*} \tag{7.172}$$

and evaluating the limits

$$\bar{u} = \frac{u^*}{\lambda}\left[\log\frac{h}{z^*} - 1 - \frac{z'_{\min}}{h}\left(\log\frac{z'_{\min}}{z^*} - 1\right)\right] \tag{7.173}$$

and so finally (7.169) and (7.173) give

$$C_d = \lambda^2 \Big/ \left[\log\frac{h}{z^*} - 1 - \frac{z'_{\min}}{h}\left(\log\frac{z'_{\min}}{z^*} - 1\right)\right]^2. \tag{7.174}$$

If we consider a *fully rough* surface, i.e., $R(s) > 100$,

$$C_\mathrm{d} \sim \left(\frac{\lambda}{\log h/s}\right)^2 \tag{7.175}$$

while for a *completely smooth* surface

$$C_\mathrm{d} \sim \left(\frac{\lambda}{\log h/\Delta}\right)^2. \tag{7.176}$$

If the flow is entirely laminar,

$$u_*^2 = v\frac{u}{z} \tag{7.177}$$

and so the depth-averaged speed is

$$\bar{u} = u_*^2 \frac{h}{2\upsilon} \tag{7.178}$$

which means that

$$C_d \equiv \frac{u_*^2}{\bar{u}^2} = \frac{2\upsilon}{\bar{u}h}, \tag{7.179}$$

i.e., the drag coefficient is inversely proportional to the Reynolds number (since in reality the stress is not quadratic in speed for this case).

If we use the velocity along the center line of the pipe, U_0, to define a drag coefficient the usual notation is

$$\tau = \frac{1}{2}\rho c_f U_0^2. \tag{7.180}$$

Clearly the log layer does not extend to the center line (since the gradient of u is zero by symmetry on the center line) and a possible model is

$$\frac{\partial u}{\partial z'} = \frac{u_0^* \sqrt{1 - z'/h}}{\lambda z'} \tag{7.181}$$

so that

$$du = -\frac{2u_0^* p^2}{\lambda(1 - p^2)} dp \tag{7.182}$$

where

$$p^2 \equiv 1 - \frac{z'}{h}, \tag{7.183}$$

and integrating

$$u = \frac{u_0^*}{\lambda}\left[2(p - p^*) + \log\left(\frac{1 + p^*}{1 + p}\frac{1 - p}{1 - p*}\right)\right]. \tag{7.184}$$

When $z' \ll h$, (7.184) reduces to

$$u \sim \frac{u_0^*}{\lambda}\log\frac{z'}{z*}. \tag{7.185}$$

Since $z* \ll h$, it follows that $p*$ is close to unity or

$$p^* \sim 1 - \frac{z*}{2h}. \tag{7.186}$$

Hence, (7.184) becomes, for all z,

$$u \sim \frac{u_0^*}{\lambda} \left[2(p-1) + \log\frac{h}{z^*} + \log\left(\frac{1-p}{1+p}\right) \right] \tag{7.187}$$

so that when z' is close to h, i.e., $p \ll 1$,

$$u \sim \frac{u_0^*}{\lambda} \left[2(p-1) + \log\frac{h}{z^*} + \log\left(\frac{1-p}{1+p}\right) \right] \tag{7.188}$$

which we can expand as a power series in p and (7.188) gives

$$u \sim \frac{u_0^*}{\lambda} \left[2(p-1) + \log\frac{h}{z^*} + (-p - p^2/2 - p^3/3) - (p - p^2/2 + p^3/3) \right] \tag{7.189}$$

and collecting terms,

$$u \sim \frac{u_0^*}{\lambda} \left[-2 + \log\left|\frac{h}{z^*}\right| - 2p^3/3 \right]. \tag{7.190}$$

Note that we have already ensured that the velocity shear is zero at $p=0$ through (7.181), i.e.,

$$\frac{du}{dp} \sim \frac{2hu_0^* p^2}{\lambda} \qquad z' \sim h \tag{7.191}$$

which agrees with (7.189).

For a rough surface C_d also increases with increasing s, as we would expect. Note that C_d is not otherwise a constant and increases with decreasing h. For a smooth surface, Δ decreases with increasing frictional stress, and so C_d also decreases with stress. We can add a term to represent the volume flow inside the bottom boundary layer, but it is of second order importance. The logarithmic terms in the denominators show that the total current increases as h increases and s (or Δ) decreases, but this change is very slow. For a typical coastal basin with $h=5$ m and s, or Δ, $=0.1$ to 10 mm, C_d from (7.175) or (7.176) gives $C_d = 0.001$ to 0.004, which are the commonly quoted drag coefficients for *smooth* or *sandy* bottoms. These friction coefficients are greatly changed by the presence of waves in coastal basins. This does not occur so markedly on the continental shelf, simply because typical wind and swell waves do not penetrate to the bottom of water of depth 50 to 100 m.

There are several other different ways of representing bottom friction which have come from studies of flow in pipes and channels. These alternative formulations do provide different (perhaps superior) ways of incorporating the depth dependence of the drag coefficient. The *Chezy* formula derived by Antoine Léonard de Chézy (1773–1832) relates the mean flow velocity in a pipe to the radius of the pipe and slope of the surface. It has the form

$$\bar{u}^2 = C^2 R \frac{\partial \eta}{\partial x} \tag{7.192}$$

where C is the *Chezy coefficient* which has units of $\mathrm{m}^{1/2}\,\mathrm{s}^{-1}$ and R is the *hydraulic radius* of a channel defined as the cross-sectional area divided by the wetted parameter and an alternative description of R is *effective depth*. For a wide coastal basin R is effectively the depth h, and this is also true of the tidal channels which find in coastal basins, since depth is always much less than width. If we replace (7.192) by a balance between longitudinal channel pressure gradient using our conventional bottom drag coefficient

$$C_d \bar{u}^2 = gh\frac{\partial \eta}{\partial x} \tag{7.193}$$

we see that

$$C = \sqrt{\frac{g}{C_d}}. \tag{7.194}$$

In a formula used for pipe flow, it is common to express the drag coefficient C_d as

$$C_d = \frac{f}{8} \tag{7.195}$$

where f is called the *friction factor*. The relation for wall stress, expressed in terms of f rather than C_d, is called the *Darcy–Weisbach formula*, i.e., from (7.169)

$$\tau = \rho f \frac{\bar{u}^2}{8}. \tag{7.196}$$

The equation is a variant on an equation developed by Gaspard de Prony (1755–1839) and by Henry Darcy (1803–58), and further refined by Julius Weisbach (1806–71) into the form used today. The use of a drag coefficient is clearly not appropriate in any but the simplest models and a much superior model is based on the roughness length and mixing length with bottom stress calculated at the boundary of the roughness layer. In other words, we should not consider the definition of a drag coefficient as anything more than an artifact for use in depth-averaged models. Importantly, C_d is dependent on h and generally increases as h decreases. The reason for this dependence is simply that the velocity increases away from the surface, and so that the ratio of bottom stress velocity to depth-averaged decreases.

7.13 Further reading

See further reading from earlier chapters. Additional texts for this chapter are:
Batchelor, G. K. (2000). *An Introduction to Fluid Dynamics*. Cambridge, UK: Cambridge University Press.
Baumert, H. Z., J. H. Simpson, and J. Sündermann (2005). *Marine Turbulence: Theories, Observations, and Models*. Cambridge, UK: Cambridge University Press.

Burchard, H. (2002). *Applied Turbulence Modeling in Marine Waters*. New York: Springer.

Davidson, P. A. (2004). *Turbulence: An Introduction for Scientists and Engineers*. Oxford, UK: Oxford University Press.

Frisch, U. (1995). *Turbulence: The Legacy of A. N. Kolmogorov*. Cambridge, UK: Cambridge University Press.

Kantha, L. H. and C. A. Clayson (2000). *Small Scale Processes in Geophysical Fluid Flows*. New York: Academic Press.

Lewis, R. (1998). *Dispersion in Estuaries and Coastal Waters*. New York: John Wiley.

Rutherford, J. C. (1994). *River Mixing*. New York: John Wiley.

Schlichting, H. and K. Gersten (2004). *Boundary-Layer Theory*. New York: Springer.

Tennekes, H. and J. L. Lumley (1972). *A First Course in Turbulence*. Cambridge, MA: MIT Press.

8

Advection of momentum

8.1 Introduction

This chapter discusses some of the basic ideas behind our modeling of advection in coastal basins. We have mentioned advection in previous chapters, but not yet considered any of the unique properties of advection. The word *advection* simply means the transport of some property of the water column or material from one point in a coastal basin to another. As such, all advection requires a current (and we speak of currents as *advecting* material). In the flux of any quantity in a coastal basin there are two terms, one due to advection, i.e., current, and the other due to diffusion or dispersion. So, when we have found the current, or *advective velocity* (u, v, w), we have solved the advective problem to the best of our ability (provided that we do not neglect certain terms in the momentum equation that arise from the advection of momentum). The advection of quantities such as salt is naturally taken into account by the continuity equation for salt by merely expressing the salt flux in the way which we done in previous chapters. So, this chapter is concerned with the advection of momentum and the momentum continuity equation.

Let us consider a one-dimensional system, so that the mathematics look simpler (although there is really no difference for a three-dimensional system). If the component of current is u, then (ignoring diffusion) the flux density of any quantity described by concentration c is uc. If the basin has constant depth and zero elevation, the continuity equation for the quantity described by c is

$$\frac{\partial c}{\partial t} = -\frac{\partial uc}{\partial x}. \tag{8.1}$$

If we differentiate the right-hand side of (8.1) as a product we obtain

$$\frac{\partial c}{\partial t} + u\frac{\partial c}{\partial x} = -c\frac{\partial u}{\partial x}. \tag{8.2}$$

The linearized continuity equation for mass or volume is, in general, for a basin of constant depth,

$$\frac{\partial \eta}{\partial t} = -h\frac{\partial u}{\partial x} \tag{8.3}$$

and so in the steady state

$$\frac{\partial u}{\partial x} = 0 \tag{8.4}$$

which is called a *solenoidal* vector field, i.e., it has zero divergence. If we substitute (8.4) in (8.2) we have

$$\frac{\partial c}{\partial t} + u\frac{\partial c}{\partial x} = 0 \tag{8.5}$$

and the second operator on the left, i.e., $u\partial/\partial x$, is usually referred to as the advective operator. Let us now consider the quantity represented by c to be momentum per unit mass in the x direction. In this case, c simply reduces to u and so (8.5) becomes

$$\frac{\partial u}{\partial t} + u\frac{\partial u}{\partial x} = 0 \tag{8.6}$$

where $u\,\partial u/\partial x$ describes the advection of momentum. It is this term that we have omitted from all of our simple model codes and the major focus of this chapter.

8.2 Coordinates for many-particle models

8.2.1 Lagrangian coordinates

As a particle moves in a coastal basin, the values of x, y, and z associated with that particle also change with time t. So if we tag a particular particle and call its position X, Y, Z, these quantities will be functions of time t. In the eighteenth century, Daniel Bernoulli investigated the forces that control a moving fluid. He considered the major problem of how the motion of a fluid differs from the motion of a finite number of individual particles like the planets. When Newton considered the motion of the planets, he applied his laws of motion to a set of discrete bodies. So he would just consider the motion of one body with coordinates X, Y, Z. In a sense, a fluid is like a collection of such bodies but in a fluid, one has to follow an *infinite number* of such particles. Furthermore, one is often not interested in particular particles, but in the speed and direction of the fluid at a particular point. There are two views of the motion of a fluid. The first, which follows individual particles, is called the *Lagrangian* view introduced by Joseph Louis Lagrange (1736–1813) who was one of the outstanding mathematicians of the eighteenth century, and made major contributions to celestial mechanics. For this reason our particle variables X, Y, Z are called *Lagrangian coordinates*.

8.2.2 Drogues

One of the commonly used field methods for tracking the position of a water mass with time, i.e., its Lagrangian motion, is to deploy a *drogue*. This is illustrated in Figure 8.1.

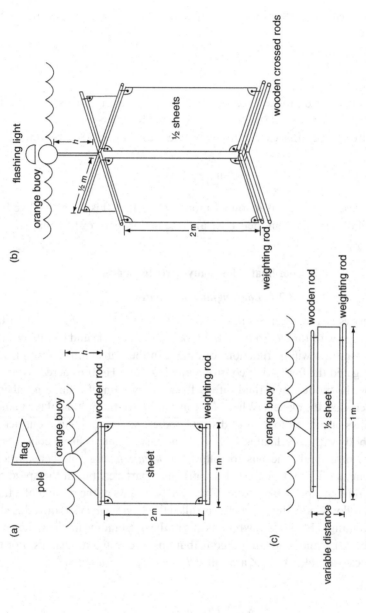

Figure 8.1 Illustration of three traditional types of drogue used to follow the Lagrangian motion of water in a coastal basin: (a) is a *blind drogue* in which the window-blind billows outward under the influence of the current, (b) is a *cross-sheet drogue* which maintains an approximately constant cross-sectional area to the current, (c) is a *shallow drogue* designed to follow motion close to the surface. All of these drogues can have *GPS position fixing* devices attached. The colored buoys, flags, and lights aid recovery and navigational safety.

The basic idea of a drogue is that it remains within a particular water mass as the water mass moves around the coastal basin. Drogues that are used in deep water oceanography have the form of a large parachute but in shallow coastal basins drogues usually consist of a large sheet of cloth attached to a buoy at the surface and weighted below. Drogues were originally used as *sea anchors* by sailing ships, and designed to anchor the boat within the local water mass, dragging it along with the current eliminating wind drift (the origin of the word is similar to *drag*). The efficacy of drogues in maintaining zero drift relative to the local water mass is dependent on a number of factors, including wind drag on the buoy. Drogues can be suspended at any point in the water column and can, in principle, measure the Lagrangian motion of water anywhere in the column. The size of the sheet determines the cross-sectional area of the current, which controls the drift of the drogue drifts. This generally means that the drogue is not measuring the mean current as usually determined by numerical models which work at minimum spatial scales of 10 to 20 m (whereas drogue sheets used in coastal basins are rarely greater than 1 to 2 m across). If we use a large number of drogues, their mean motion does provide some measure of the mean current. Additionally, the variance in the motion of individual drogues is a measure of turbulence or *dispersion* in the basin. This was considered in some detail in Chapter 7. Nevertheless, these caveats aside, drogues do (at least in principle) measure the Lagrangian motion of the water within a basin. They are certainly more effective under strong current conditions in which the mean current greatly exceeds the effects of local dispersion.

8.2.3 Eulerian coordinates

The other view of the motion of a fluid involves calculating the speed and direction of the fluid at every point (x, y, z). This is called the *Euler* approach, after the brilliant mathematician Leonhard Euler (1707–83) who is justifiably described as one of the most brilliant, and certainly prolific, mathematicians of his time, and usually credited with invention of the concept of a mathematical function. If we measure the current at a particular point in a fluid over some time interval, we are measuring the *Eulerian current*, i.e., the current associated with all the particles that move through that point in that time interval. The major difference between the Lagrangian and Euler systems is encountered when we attempt to find the *acceleration* of a particle.

8.2.4 Transforming coordinates

Fortunately, there is a simple relation between the two coordinate systems. It is vital to understand the difference between Euler and Lagrangian coordinates, because we can only apply Newton's laws of motion to individual particles (in the Lagrangian sense).

In the Lagrangian system, suppose we follow a particle moving with speed U along the x axis; our speed changes with time, i.e., U is a function of t. The acceleration of the

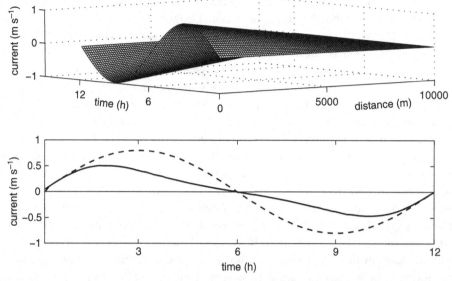

Figure 8.2 Comparison of Lagrangian and Eulerian systems for describing the motion of a fluid in a one-dimensional basin 10 km long closed at one end and subject to an oscillatory tidal amplitude of period 12 h at the other open end. Upper panel shows a surface corresponding to the Eulerian speed $u(x, t)$ as a function of time t and position x along the basin. The full curve in the lower panel shows the Lagrangian speed $U(t)$ of a particle near the open mouth of the basin, which has position $x_0 = 2000$ m at $t = 0$. The broken curve shows the Eulerian speed at the particles' initial position: $u(x_0, t)$.

particle is dU/dt in the x direction. In a small interval of time dt the particle moves a distance $dx = U dt$, and so we can also write the acceleration as $du/(dx/U) = U dU/dx$. In a one-dimensional model, the velocity U_1 and position X_1 of a particle labeled I are functions of one independent variable t. For a second particle, U_2 and position X_2, are also functions of t. To describe the fluid totally we would need to describe $X_n(t)$ for $n = 1$ to N where N is the total number of particles in the fluid.

An observer in the Eulerian system watches all the particles as they pass each point (x, y, z), and in a one-dimensional system, describes the current at point x, and time t, as $u(x, t)$. Notice that we are using lower-case u as distinct from U. So u is a function of two independent variables, x and t. We can, for example, make a three-dimensional plot with current u as the vertical axis on a plane with axes x and t. This is shown in the top panel of Figure 8.2 for a tidal variation in a one-dimensional narrow basin of length 10 000 m closed at one end, and subject to a simple sinusoidal variation at the open end. To find the time variation of the Eulerian current u at any value of x, simply consider the plane normal to the x axis at that value of x. Notice that this variation of u with time t decreases as we move towards the closed end of the basin. To find the variation of u with x at a particular time t, take a plane normal to the t axis at that

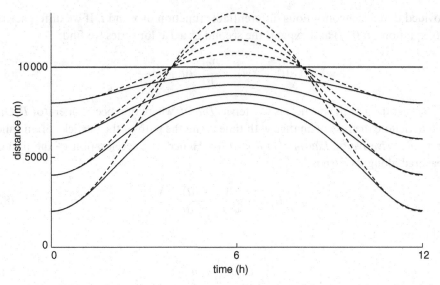

Figure 8.3 Full lines show the Lagrangian coordinates X of four particles released from locations along the basin in Figure 8.2 at time $t=0$. The broken lines show their coordinates based on their Eulerian speeds at their points of release as a function of time (which may leave the physical boundaries of the basin).

value of t. This shows that u simply decreases linearly with distance to become zero at the head (the closed end).

The full curve in the lower panel shows the Lagrangian speed $U(t)$ of a particle near the open mouth of the basin which has position $x_0 = 2000$ m at $t = 0$. The broken curve shows $u(x_0, t)$, the Eulerian speed $u(t)$ of particles at a distance of 2000 m from the mouth and note that this is a simple sinusoidal variation and just corresponds to the variation in a plane normal to the x axis in the upper panel. $U(t)$ drops noticeably below $u(x_0, t)$ because as the particle advects to higher values of x, the local speed u at its position decreases. All currents in the basin drop to zero at a time of 6 h and then U changes sign and the particle returns to its starting point at a time of 12 h. Figure 8.3 shows the trajectories of four particles released from $x_0 = 2000$, 4000, 6000, and 8000. The full lines represent the Lagrangian positions $X(t)$ of the four particles, while the broken lines represent their coordinates based on their Eulerian speeds $u(x_0, t)$ for the four values of x_0. Notice that the Eulerian track would leave the closed ends of the basin which is of course incorrect and a consequence of the model not using the correct local speed for the particles.

The Lagrangian trajectory can be found from the Eulerian velocity field using

$$X(t) = u(x_0, t)$$
$$X(t + dt) = u(x_0 + udt, t + dt)$$

(8.7)

provided that u is a continuous differentiable function of x and t. If we difference the two equations in (8.7) and expand the second as a Taylor series we find

$$X(t + dt) - X(t) = \left.\frac{\partial u}{\partial t}\right|_{x_0} dt + \left.udt\frac{\partial u}{\partial x}\right|_{x_0}.$$
(8.8)

The acceleration of the particle is written as Du/Dt, where the special operator D/Dt is used to denote the rate of change with time along the path of the particle; often called the *total derivative* or *Lagrangian derivative*. Hence, the acceleration of the particle measured along its path is

$$\frac{D}{Dt}U = \lim_{dt \to 0} \frac{X(t + dt) - X(t)}{dt}$$
(8.9)

and using (8.8), (8.9) gives

$$\frac{D}{Dt}U = \frac{\partial u}{\partial t} + u\frac{\partial u}{\partial x}.$$
(8.10)

The first term on the right of (8.10) is called the Eulerian or *local acceleration* and the second term referred to as the *advection* or preferably as the advective contribution to the acceleration, or advection of momentum. If u changes much more rapidly with time t than it does with distance x, the local acceleration is by far the most important term. However, if the spatial variation of u is greater than the local time variation, the advective term becomes more important and the local acceleration can be zero, while the advective term is non-zero; a good example is a set of particles traveling around in closed orbits, such as the inertial oscillations that we met in Section 4.6. For this reason, the advective term is sometimes referred to as the *inertial term*. The velocities do not change with time at a particular x, y but do change with position. The acceleration is not zero, and a force is required to hold the particles in these orbits. If the system is in any steady state, the local acceleration is zero.

It is valuable to note that the second term in (8.10) can be written as the derivative of u^2, by a purely mathematical transformation (derivative of a product),

$$\frac{D}{Dt}U = \frac{\partial u}{\partial t} + \frac{1}{2}\frac{\partial u^2}{\partial x}$$
(8.11)

so that the second (advective) term is the spatial rate of change of energy per unit mass. This is, of course, an application of the simple principle that the rate of change of energy is the product of the force (or acceleration) and distance moved. As an example, the flow of water across a coral reef is in a steady state (does not change with time). Suppose that the speed of the current is very small on the fore-reef but increases to a maximum of $0.3\ \mathrm{m\ s^{-1}}$ on the reef flat which is 300 m from the fore-reef. The average acceleration of a particle over this distance is simply the spatial rate of kinetic energy, i.e., $0.5\ (0.3)^2/300 = 1.5\ 10^{-4}\ \mathrm{m\ s^{-2}}$. This is a very large horizontal acceleration, and is physically due to the difference in the height of water between the fore-reef and reef flat. An example of a much weaker acceleration (but possibly also due to a similar change in the height of water but over much larger distances) is a regional current in which the speed might change by an amount of order 0.1 m s^{-1} over a distance of 100 km.

8.2.5 Lagrangian methods in coastal basins

The Lagrangian approach has great appeal for some aspects of modeling. It allows us to follow individual particles, and therefore consider the fate of such particles as they undergo all of the processes which operate inside a basin. This approach is mechanistic and has some advantages over the Euler method, insofar as it allows us to visualize the processes in a coastal basin. The most immediate of these processes is advection itself, and especially insofar as it relates to the flushing of coastal basin by the ocean, i.e., the exchange processes between the basin and the ocean.

Numerical models determine the Eulerian currents, i.e., (u, v, w) as functions of (x, y, z, t), usually on a spatial grid of cells that cover the coastal basin. Such models can include the advection of momentum. From the Eulerian velocities it is possible to determine the Lagrangian track of a particle via

$$X(t) = x_0 + \int_0^t u(X, t)dt \tag{8.12}$$

which represents the average track of an a particle which starts at x_0 at $t = 0$. Models that create Lagrangian trajectories are called *particle tracking models*. These models solve (8.12) by a time-stepping technique so that

$$\Delta X = u(X, t)\Delta t. \tag{8.13}$$

In order to simulate dispersion, the model has to recognize that $u(x, t)$ only represents the mean motion of the fluid and ignores the fluctuations that we discussed in Chapter 7. This aspect of the motion is usually included numerically by adding a stochastic term to (8.13):

$$\Delta X = (u(X, t) + r)\Delta t \tag{8.14}$$

where r is a random number in some small range $(-\varepsilon, \varepsilon)$. The magnitude of ε controls the extent of dispersion. If one starts with a group of particles in one cell of the model and applies (8.14) to each particle, the stochastic term will ensure that the group spreads out into a cloud with time. This means that the particles sample a region of the basin, and so they experience differential advection which adds to the dispersion.

8.2.6 Drogue models

Determining water circulation by the tracking of drogues involves the reverse problem of converting Lagrangian trajectories to an Euler current field.[1] Let us suppose that the basin has a dominant type of water circulation, for example, currents are primarily due to a daily sea breeze. It is necessary to assume that the wind-driven circulation is reasonably consistent between days, so that we only need to plan the drogue sampling over a finite number of days. The deployment of drogues needs to be widespread if most of the basin is to be included in the circulation map. It is then necessary to assimilate the drogue tracks into a *synthesis model* by overlaying a *universal spatial grid* on the basin with cells of finite size. Each drogue that passes through a particular cell contributes to the sample of local velocities in that cell, and allows us to build up a statistical distribution of velocities in that cell in exactly the same way that we envisaged in Section 7.2. That is, we have a mean value of the components of velocity and stochastic fluctuations which are related to the natural degrees of freedom of the fluid:

$$u = \bar{u} + u'$$
$$v = \bar{v} + v' \tag{8.15}$$

and the mean of each distribution provides the Euler circulation map (Figure 8.4). Some cells will have insufficient passes of the drogues, and so it is necessary to perform statistical tests within each cell to determine whether the mean is significant; otherwise we have no information for that cell. Some caution needs to be exercised in drawing conclusions about circulation based on a too limited number of drogues' tracks. In situations where the forcing from winds and tides (plus river flow) is more varied, it is further necessary to determine statistical distributions for sub-intervals of the forcing functions. This clearly increases the amount of drogue tracking that is necessary, but fortunately that task has been eased by self-recording position fixing onboard the drogues.

8.3 Role of advection in coastal basins

8.3.1 Inertial force

In order to understand the role of the momentum advection terms in coastal basins, we need to formalize the type of argument discussed in Section 8.2. Let us denote the space scale as L, the timescale by T, and the velocity scale by V. The ratio of advection to local acceleration is then

$$\text{ratio} = \frac{VT}{L}. \tag{8.16}$$

---

[1] Hearn and Prince (1983).

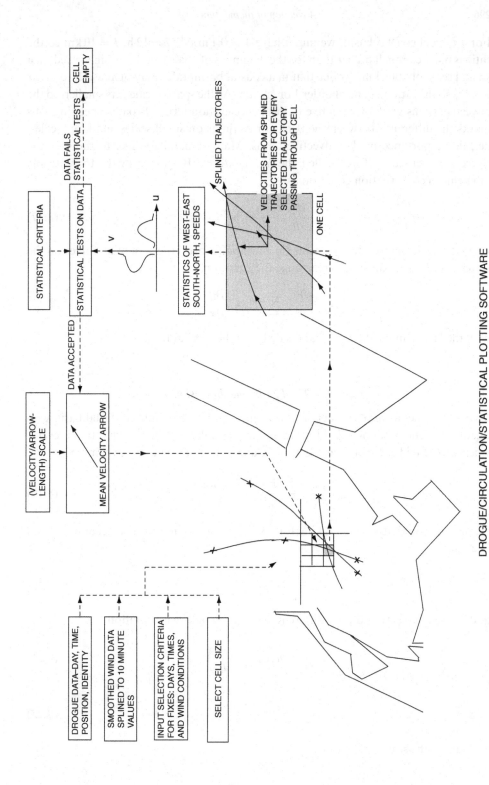

Figure 8.4 Illustration of drogue model of circulation in a coastal basin. The drogue tracks are interpolated onto a universal spatial grid, statistically analyzed, and classified according to external driving forces such as wind and tide.

For a typical coastal basin, we might have $V = 0.1\,\mathrm{m\,s^{-1}}$, $T = 12\,\mathrm{h}$, $L = 10\,\mathrm{km}$ so the ratio is 0.4, i.e., much higher than for the dynamics of a large ocean basin viewed on a spatial scale of 1000 km. Notice that in a coastal basin, tides may increase V to order $1\,\mathrm{m\,s^{-1}}$ so the ratio may be of order 1 or greater. As the spatial scale gets smaller, so the advective terms get larger relative to local acceleration terms. Broadly speaking, this marks the difference between the behavior of fluids on small scales and large scales, i.e., the importance of the advective terms. Many dynamicists like to think of the advective terms as a force called the *inertial force*. If we use (8.10) to write the momentum conservation equation as

$$\frac{\partial u}{\partial t} + u \frac{\partial u}{\partial x} = \sum F_i \tag{8.17}$$

where the right-hand side represents (in a symbolic form) the sum of all forces on the water column, we can simply rearrange the equation as

$$\frac{\partial u}{\partial t} = \sum F_i - u \frac{\partial u}{\partial x} \tag{8.18}$$

with the last term on the right-hand side being the inertial force.

8.3.2 Tidal jet in one dimension

Let us construct a simple model of flow from a tidal channel into a coastal basin with quadratic friction, but no other forcing and in particular, the surface elevation is everywhere zero. Fluid enters the basin with $u = 0$ and is then subject to the dynamic equation:

$$\frac{\partial u}{\partial t} + \frac{1}{2} \frac{\partial u^2}{\partial x} = -C_d \frac{|u|u}{h + \eta}. \tag{8.19}$$

Let us suppose that we are looking for a steady state solution for the channel; (8.19) becomes

$$\frac{du^2}{dx} = -\frac{2C_d}{h} |u|u \tag{8.20}$$

and we now have an ordinary differential equation in u^2 with solution

$$u^2 = u_0^2 \exp\left(-\frac{2x}{L_i}\right) \tag{8.21}$$

$$L_i \equiv \frac{h}{C_d}, \tag{8.22}$$

and taking the square root of both sides of (8.21),

$$u = u_0 \exp\left(-\frac{x}{L_i}\right). \tag{8.23}$$

Equation (8.23) shows that u decreases with distance x into the basin to a factor of $1/e$ (about 30%) in the *inertial distance* L_i. If $h = 10$ m and C_d has the typical value for sand of 0.0025 the inertial distance is 4 km. For $h = 1$ m this reduces to 400 m. This effect is related to the *inertia* of the particles, in the sense that without friction they would continue to travel within the basin at the speed of u_0 (in the absence of any forcing). Particles entering a tidal channel with zero velocity would similarly take a finite distance to be accelerated up to some steady state speed, at which they experience a balance between pressure and friction. For example, consider the solution to a balance between pressure gradient and friction,

$$\frac{1}{2}\frac{du^2}{dx} = -g\frac{d\eta}{dx} - C_d\frac{|u|u}{h}, \tag{8.24}$$

which is

$$u = u_0\left(1 - e^{x/L_i}\right) \quad u_0^2 \equiv -\frac{gh}{C_d}\frac{d\eta}{dx}, \tag{8.25}$$

i.e., it takes a distance L_i for u to reach the steady speed u_0 where $d\eta/dx$ is negative, i.e., the surface slopes downward in the positive x direction. This phenomenon is known as a *tidal jet*, and can sometimes be seen in the coastal ocean as the tide ebbs through a tidal channel from an inland lagoon. The jet is simply a consequence of the unidirectional currents within the channel persisting into the coastal ocean. This is illustrated in Figure 8.5 using reprocessed data collected by the author for Koombana Bay in south-west Australia. The Koombana Bay jet occurs during summer nights due to spring-tidal ebb flow into the bay from Leschenault Inlet which is a coastal lagoon connected to the bay through a dredged tidal channel. The jet is caused by very strong tidal currents in the channel which are a consequence of its limited width and walled sides. Because of the strong currents, the bed of the channel is mobile as are many dredged channels. Inside the channel, the basic force balance is between the along-channel pressure gradient and quadratic friction. As was discussed in Section 4.5.6, such a balance produces a *square root singularity* that causes the tidal current to change direction suddenly from flood to ebb at almost constant speed. This behavior greatly facilitates the formation of the jet on the ebb tide. Interestingly, its discovery in the early 1980s was the result of a prediction by numerical model developed by the Center for Water Research of the University of Western Australia followed by the tracking of drogues placed in the channel.[2] Although the jet is dominated by the advection of momentum, created in the narrow tidal channel, it is slightly buoyant (due to summer heating and remnants of winter river flow into the inlet) and therefore

[2] Hearn (1983).

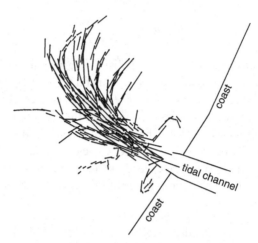

Figure 8.5 A tidal jet from a coastal lagoon in south-west Australia. .The arrows are derived from drogue tracks processed by the drogue synthesis model shown in Figure 8.4 with square cells of side 100 m but without statistical tests. (Data from Hearn *et al.*, 1985.)

does eventually lift off from the bottom providing a buoyancy signal across the bay.[3] The flood tide into the channel has minimum momentum advection in the ocean and consequently is fairly isotropic; inside Leschenault Inlet itself the water is very shallow and so the dynamics are dominated by friction.[4]

8.3.3 Tidal jet in two dimensions without rotation

For a jet to be observable, the currents in the channel must be considerably greater than the alongshore currents in the coastal ocean. The usual situation is that there is an alongshore pressure gradient which induces an alongshore component of current. Let us suppose this has a steady value of v_0 when those forces come into balance with friction. Hence, the value of u and v are governed by the equations (in the absence of rotation effects, i.e., the Coriolis force)

$$\frac{1}{2}\frac{\partial u^2}{\partial x} = -g\frac{\partial \eta}{\partial x} - C_{\mathrm{d}}\frac{\sqrt{|u^2 + v^2|}u}{h} \tag{8.26}$$

$$\frac{1}{2}\frac{\partial v^2}{\partial y} = -g\frac{\partial \eta}{\partial y} - C_{\mathrm{d}}\frac{\sqrt{|u^2 + v^2|}v}{h}. \tag{8.27}$$

The simplest solution is to assume that the alongshore pressure gradient is constant and that there is no offshore gradient:

---

[3] Luketina and Imberger (1987).   [4] Hearn (1983).

$$\frac{1}{2}\frac{\partial u^2}{\partial x} = -C_d \frac{\sqrt{|u^2 + v^2|}u}{h} \qquad (8.28)$$

$$\frac{1}{2}\frac{\partial v^2}{\partial y} = C_d \frac{v_0^2 - \sqrt{|u^2 + v^2|}v}{h}, \qquad (8.29)$$

where we have chosen the sign of the alongshore gradient to be negative, so that the equilibrium offshore speed is positive and given by

$$v_0^2 \equiv -g\frac{\partial \eta}{\partial y}. \qquad (8.30)$$

We will scale all distances to the inertia distance L_i and all speeds to u_0 (the speed inside the tidal channel), so that $v_0 = \alpha$ (a typical value would be 0.1). At $x = 0$, $y = 0$, $u = 1$ and $v = 0$. Since u and v are not functions of time, they essentially represent the components of Lagrangian velocity U, V of a particle at point (X, Y):

$$\frac{1}{2}\frac{\partial U^2}{\partial X} = -C_d \frac{\sqrt{|U^2 + V^2|}U}{h} \qquad (8.31)$$

$$\frac{1}{2}\frac{\partial V^2}{\partial Y} = C_d \frac{v_0^2 - \sqrt{|U^2 + V^2|}V}{h}. \qquad (8.32)$$

We can code the model in Matlab. The simplest approach is to follow a particle in time, and for this purpose we use a dimensionless time defined in terms of L_i/u_0 (Figure 8.6):

$$U = \frac{dX}{dt} \qquad V = \frac{dY}{dt} \qquad (8.33)$$

$$\frac{\partial U}{\partial t} = -\sqrt{|U^2 + V^2|}U \qquad (8.34)$$

$$\frac{\partial V}{\partial t} = v_0^2 - \sqrt{|U^2 + V^2|}V. \qquad (8.35)$$

8.3.4 Tidal jet with rotation

The model can easily be adapted to include Coriolis terms by replacing the time loop in Table 8.1 as shown in Table 8.2 where f' is the Coriolis parameter scaled to L_i/u_0 which has a typical value for $h = 10$, $C_d = 0.0025$, of $f' = 0.29$ at latitude $30°$ N.

This is shown in Figure 8.7. Note that the jet firstly turns to the right due to the Coriolis force, and completes inertial oscillations around the jet trajectory before

Table 8.1. *Matlab code for a coastal jet*

```
alpha = 0.1; u = 1; v = 0; x = 0; y = 0;
dt = 0.01; t = 0; N = 10000
for n = 1 : N;
    t = t + dt; dx = u*dt; dy = v*dt;
    speed = sqrt(u^2 + v^2);
    u = u - speed*u*dt; v = v + (alpha^2-speed*v)*dt;
    x = x + dx; y = y + dy;
end;
```

Table 8.2. *Matlab code for a coastal jet with rotation*

```
alpha = 0.1; u = 1; v = 0; x = 0; y = 0;
dt = 0.01; t = 0; N = 10000
    for n = 1 : N;
    t = t + dt; dx = u*dt; dy = v*dt;
    speed = sqrt(u^2 + v^2);
    u = u - (speed*u-v*fprime)*dt; v = v + (alpha^2-speed*v-u*fprime)*dt;
    x = x + dx; y = y + dy;
end;
```

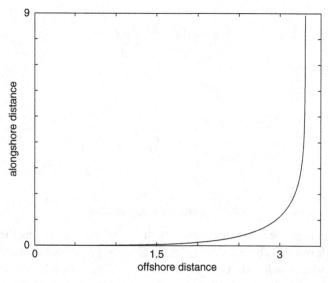

Figure 8.6 A model trajectory of a tidal jet on a non-rotating Earth when the alongshore current has a speed equal to 10% of that of the tidal channel. Distances are scaled by the inertial length h/C_d.

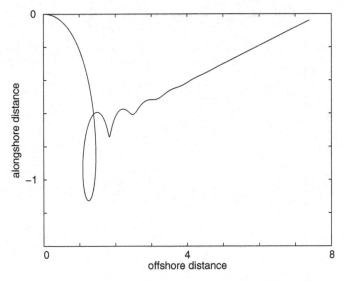

Figure 8.7 Repeat of Figure 8.6 with the inclusion of rotation in the northern hemisphere and the reduced Coriolis parameter f' set at 0.3.

coming to a steady path under the influence of the Coriolis force and friction. The reader can experiment with the model and, for example, run it with no alongshore pressure gradient as shown in Figure 8.8.

The ebb flow from the tidal jet produces a coastal jet whereas the flood tidal inflow to the channel is more symmetrical (depending again on the alongshore pressure gradient). This means that the water ejected on the ebb tide tends to travel beyond the local region, which supplies the water for the next flood tide. This evidently improves the flushing of the basin, in the sense that water ejected on the ebb does not simply return on the flood. The tidal jet is akin to the jet of air we can produce from our lips to blow out the candle on a cake. This is in stark contrast to the slower, more symmetrical inflow of air when we breathe in through our mouths. The tidal jet is often enhanced by training walls. At the lagoonal end of the channel the water is usually shallower, and so the inertial distance is shorter.

The tidal jet is influenced by processes other than bottom friction. An important process is the mixing of momentum in the horizontal (usually called dispersion), or in the case of jets we usually talk of the entrainment of water into the jet. This slows and widens the jet. Usually, the coastal ocean deepens offshore, and in this case the jet may *lift off* from the seabed and exist only in the surface water. Such a jet then tends to mix momentum downward or equivalently *entrain* water from below its lower surface. As shown by Figure 8.6, a tidal jet usually attaches to the coast by turning to its right or left after leaving the tidal channel. This may be a consequence of the Coriolis force as in Figure 8.7 or the alongshore pressure gradient as in Figure 8.6. A typical circulation around a tidal jet consists of a trapped eddy or vortex between the coast and the

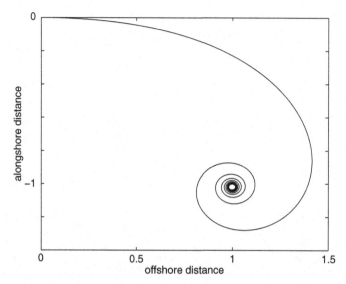

Figure 8.8 Repeat of Figure 8.7 on a frictional rotating Earth but without an alongshore pressure gradient so that particles in the jet spiral to rest.

attached jet. Modeling such an eddy can be achieved in two dimensions using one of the codes in Chapter 6 and such features are subject to exactly the same dynamical equations as all other features in a coastal basin.

8.3.5 Frictional Reynolds number

The ratio of the inertial terms to the friction terms forms a Reynolds number (see Chapter 7) which is here based on bottom friction (rather than viscosity):

$$R_{C_d} = \frac{h}{2C_d u^2} \frac{du^2}{dx} \tag{8.36}$$

which has the equivalent form

$$R_{C_d} = \frac{h}{C_d} \frac{d\log(u)}{dx} \tag{8.37}$$

(and note that as always in this book "log" means the natural logarithm). If we assume that the logarithm of u varies on a space scale of d, (8.38) becomes

$$R_{C_d} = \frac{h}{dC_d} \tag{8.38}$$

where R_{C_d} is called the bottom friction Reynolds number. We see from (8.38) and (8.22) that the Reynolds number is unity when $d = L_i$, and so flow problems with

$d \ll L_i$ are dominated by the inertia force with the generation of vortices and eddies. For $d \gg L_i$, inertia tends to be small and friction removes eddies of such size.

8.3.6 Bernoulli model

The inertial terms are responsible for changes for changes in pressure along a streamline in a fluid. If we consider that the force F on the right of (8.17) is due to pressure,

$$\frac{Du}{Dt} = \frac{\partial u}{\partial t} + u \cdot \nabla u = -\frac{1}{\rho} \nabla p - g\hat{z} \tag{8.39}$$

where p denotes pressure, ρ is density, u is the vector velocity, and $\hat{z}$ is the unit vector in the positive z direction. In the steady state (8.39) becomes

$$u \cdot \nabla u + \frac{\nabla p}{\rho} + g\hat{z} = 0 \tag{8.40}$$

and if integrated along a streamline

$$\frac{u^2}{2} + \frac{p}{\rho} + gz = \text{constant} \tag{8.41}$$

which is clearly a statement of the conservation of total energy. This is called the *Bernoulli equation* or *principle* and plays a vital role in fluid dynamics. It is named after the Dutch mathematician Daniel Bernoulli (1700–82), who came from a family of talented mathematicians. Bernoulli worked with Leonhard Euler, spending much of his life in Switzerland. The equation is of very fundamental significance in fluid flow, and examples are evident in everyday life. An often-quoted example is the jet that springs from a hole in the side of a bucket filled with water. The pressure in the jet and at the open top of the bucket are equal to atmospheric pressure, and so the very much greater speed of water in the jet compared to the rate of fall of the surface in the bucket is due to the difference in vertical height z. Physically, this is due to the conversion of gravitational potential energy to kinetic energy.

Bernoulli's principle has relevance to stagnation points in fluid flow. An example of a stagnation point occurs on the leading edges of the blades of a ceiling fan. This is more easily visualized if we consider a jet of air striking a stationary blade at right angles. There is one streamline that divides the flow in half: above this streamline, the air flows over the blade and below it the flow goes under the blade. On the streamline, the flow comes to a halt as it meets the blade, i.e., there is a stagnation point. Clearly, the stream of air brings with it suspended dust particles. Whether a dust particle can escape from the stagnation point depends on its size. Fine particles have much greater area per unit volume than larger particles, and so have less inertia relative to friction and tend to stick more easily. We shall see this effect when we consider cohesive sediments in Chapter 12. This sticking problem is exacerbated by Bernoulli's principle (8.41), which tells us that minimum u involves maximum pressure p.

It essentially states that if we follow a streamline, regions in the fluid with high-speed flow will have lower pressure than regions with lower speeds. The reason is simply that Newton's law requires a force in order to achieve a change of velocity. Often this velocity change is demanded by the geometry of the flow, such as the flow of air over an airfoil: the higher speed on one side of the wing of an aircraft lowers pressure, and hence, provides lift. A similar effect is also used in producing wind force on a sailing boat. Similar forces are responsible for the motion of a spinning ball used in many sports, in which the motion of the ball slows a layer of air on one side of the ball and accelerates it on the other side.

If we use the hydrostatic approximation in a coastal basin:

$$p = p_0 - g\rho(z - \eta) \tag{8.42}$$

and ignore w, (8.41) becomes

$$\frac{u^2}{2} + g\eta = \text{constant.} \tag{8.43}$$

If we multiply (8.43) through by the total height of the water column H, we see that (8.43) is simply a statement of conservation of energy, i.e., the sum of the kinetic and potential energy of a particle is a constant (potential energy here being measured relative to the surface datum). In the presence of friction the changes of kinetic energy and potential energy

$$\frac{u^2}{2} + g\eta + \frac{C_d}{h + \eta} \int u^2 dx = \text{constant} \tag{8.44}$$

so that the loss of kinetic energy and potential energy is due to work done against friction. Without friction, the loss of potential energy naturally leads to an increase in kinetic energy. The usual situation in coastal basins is that the kinetic energy is very small in relation to changes in potential energy, and this is a general result for most of the world's oceans. It only fails in basins with very strong tidal currents.

Equation (8.43) can, of course, be derived directly by integrating

$$u\frac{du}{dx} = -g\frac{d\eta}{dx} \tag{8.45}$$

with respect to x. There are alternative ways of expressing (8.43) for a one-dimensional steady state flow, which are very useful in helping our understanding of conservation properties. These use the conservation of volume flux Q in such flows which is simply

$$T = uH \tag{8.46}$$

where H is total depth and

$$H = h + \eta. \tag{8.47}$$

The advection terms describe processes that are fundamental to much of the dynamics of fluids. However, these processes are usually (but not always) very small in a coastal basin or, in other words, the non-linear terms are small. To understand why it is necessary to use the scaling that we adopted in the last chapter, in which all velocities are scaled to c the speed of long waves and all horizontal dimensions to the length of the basin. In that form, (8.6) involves the variation of u on spatial and temporal scales of some fraction of unity. These space and time scales are directly related by the speed of gravity waves, and so the importance of the inertial term relative to the local acceleration depends on the ratio of u relative to c, the speed of gravitation waves. In typical basins, c is of order a few meters per second, and so u is rarely sufficient for inertial terms to be of major importance, except in special situations. However, it is important to note that the inertia terms represent a distinct physical process.

8.4 Hydraulic jumps

8.4.1 Froude number

Hydraulic jumps are very common in fluids that move with a speed in excess of the propagation speed of long waves which we denote by c. They can be easily observed in rivers and streams, and even in water flowing down a pathway. They appear as sudden upward jumps in the height of the fluid with a corresponding drop in speed. This change conserves the volume flux. A very simple way to view a hydraulic jump is to run the water from a tap onto the center of a flat dinner plate held horizontally beneath it. The water will flow outward from the center of the plate as a thin layer with gradually diminishing speed (due to conservation of volume). Then, at a certain critical radius the water will form a much thicker flow, and one sees a circular ridge. On the high-velocity side of the jump, the current u exceeds c, i.e., $u > c$ and on the other side of the jump, $u < c$. This type of instability is very common in physics, in the sense that c plays the role of the maximum speed of propagation of information. The instability is always initiated by a perturbation of some form, and this is usually a change in topography, such as a roughness element or obstacle in the path of the flow. The flow in the region before the jump is called *supercritical* (because the speed exceeds the wave speed) and after the jump the flow is *subcritical* (as the flow speed is less than that of a wave). This ratio of current to wave speed is called the *Froude number*, as we shall see in more detail later.

Hydraulic jumps are fairly uncommon on the surface of the ocean because current speeds rarely exceed $1 \, \mathrm{m\,s^{-1}}$ and the speed of propagation of waves, c, is typically much higher except in very shallow water; c only becomes greater than $1 \, \mathrm{m\,s^{-1}}$ when the height of the water column is about 0.1 m. Nevertheless, hydraulic surface jumps can and do occur in coastal basins and may have a dominant effect in limiting flow through tidal channels.[5] However, a much more likely place to find hydraulic jumps is on the pycnocline between the bottom denser water and the lighter surface water. Because the density difference between the two water layers is very much less (about a factor of 100 to 1000) than that which occurs across the interface between water and air at the surface, the speed of waves on the interface (called internal waves) is much lower than the speed c of surface waves. The speed of internal waves is determined by a similar relation to that for surface waves, but the acceleration due to gravity g is replaced by $g' = g\Delta\rho/\rho$, where $\Delta\rho$ is the density difference between the layers. So c' is typically a few percent of c, so that *internal hydraulic jumps* are relatively common.

The following sections discuss a model of hydraulic surface jumps in a coastal basin, and we have deliberately chosen a very particular problem in order to produce a definitive model. It is relatively easy to find the relationship between the change of height and speed in a hydraulic jump, but more detailed modeling is needed to determine the exact point in the system at which the jump will occur. We have treated this model in some detail, because it provides an excellent introduction to the ideas of momentum, energy conservation, and the role of advection, and also to the value of mathematical analysis in model construction.

8.4.2 Reynolds number and bottom slope

The dynamics of small coastal basins are controlled by spatial scales of hundreds of meters (or less). Over such distances, bottom slopes are greatly in excess of the mean slopes experienced over the larger spatial scales encountered on the continental shelf, or in the deep ocean. Consequently, coastal basins often have currents which flow down bottom slopes whose magnitude exceeds the drag coefficient C_d, for quadratic bottom friction, i.e., $\partial h/\partial x > C_d$ where h is the depth and x is distance measured in the direction of the current flow. The typical value of C_d (for sand) is 0.0025 (which is commonly used in shallow water hydrodynamic models) while bottom slopes typically have values of at least 0.01 in coastal regions. The situation in which the slope $|\partial h/\partial x|$ is less than 0.001 (1 m change of depth in 1 km) is comparatively rare. Some coastal systems experience much higher friction coefficients: as an example, some coral reefs have C_d as high as 0.1, and interestingly, coral reef lagoons show small-scale bathymetric features with bottom slopes of a similar magnitude. This can produce an important instability in the relationship between current flow and the surface elevation specified at the boundary of a model. This instability requires $\partial h/\partial x \geq C_d$, and results in the flow making a hydraulic jump.

For simplicity, consider a one-dimensional basin with horizontal and vertical coordinates x and z, respectively (Figure 8.9). The steady state, depth-averaged, dynamical equation assuming constant density and quadratic friction, is

---

[5] The science of flow through pipes and along open channels is usually described as *hydraulics*. Coastal basins are a meeting area of many of the sciences of water flow and in Chapter 1, we commented on some of the names in use.

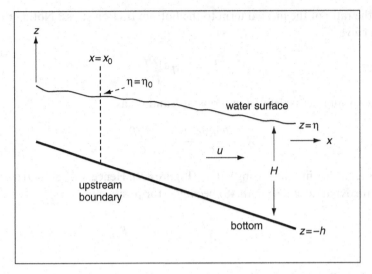

Figure 8.9 Schematic of variation of depth and surface elevation in the one-dimensional model of this section.

$$-g\frac{d\eta}{dx} = u\frac{du}{dx} + C_d\frac{u|u|}{H} \tag{8.48}$$

where H is the total height of the water column

$$H = h + \eta \tag{8.49}$$

with η the elevation of the surface, and h the depth of the basin; both η and h are relative to some height datum, which for the present model will be defined later. In (8.48) the basin is forced by the pressure gradient developed by surface slope, and this is balanced by bottom friction and the advective term.

Continuity implies that the volume flux (per unit width of the basin) Q,

$$Q \equiv uH, \tag{8.50}$$

is independent of position x. The Froude number is denoted by F:

$$F \equiv \sqrt{\frac{u^2}{gH}} \tag{8.51}$$

and in view of (8.50), F^2 varies as H^{-3}:

$$F^2 = \frac{Q^2}{gH^3}. \tag{8.52}$$

The present instability is a consequence of the inertial terms exceeding the frictional terms, and therefore can be expressed by the Reynolds number which, in a shallow basin, is the ratio of the inertial term to the bottom friction stress. Noting that (8.50) and (8.51) give

$$u\frac{\partial u}{\partial x} = -gF^2\frac{\partial H}{\partial x} \tag{8.53}$$

the Reynolds number is defined in terms of h and C_d as

$$R \equiv \frac{h|u\partial u/\partial x|}{C_d u^2} = \frac{\partial h/\partial x}{C_d} \tag{8.54}$$

and our interest lies in h increasing with x (Figure 8.9). Hence, using (8.49) to (8.52) the steady state dynamic equation (8.48) becomes, for positive u,

$$(F^2 - 1)\frac{\mathrm{d}\eta}{\mathrm{d}x} = C_d F^2(1 - R). \tag{8.55}$$

When $R < 1$,

$$R < 1 \quad \begin{cases} F < 1 & \partial\eta/\partial x < 0 \quad \text{(subcritical)} \\ \\ F > 1 & \partial\eta/\partial x > 0 \quad \text{(supercritical).} \end{cases} \tag{8.56}$$

The situation when $F < 1$ corresponds to our usual perception of flow in a shallow basin with a current driven by a surface elevation that *decreases with x* and is mainly balanced by friction. For the other (supercritical) situation, $F > 1$, the surface elevation *increases with x*, producing a pressure force that acts in the same direction as friction to oppose the advective terms. When $R > 1$, and the current is again positive, (8.55) shows that

$$R > 1 \quad \begin{cases} F < 1 & \partial\eta/\partial x > 0 \quad \text{(subcritical)} \\ \\ F > 1 & \partial\eta/\partial x < 0 \quad \text{(supercritical)} \end{cases} \tag{8.57}$$

so that the sign of the surface slope $\partial\eta/\partial x$ is opposite to that of the $R < 1$ case, and for subcritical flow, $\partial\eta/\partial x > 0$ with the pressure force acting in the same direction as friction. This reversal of the surface slope is clearly due to the inertial force, i.e., the advective terms, changing relative to friction as the bottom slope changes. For this reason we say when $R > 1$, the bottom slope is *supercritical* and otherwise *subcritical*. Hence, any attempt to drive a current down a supercritical slope with a surface elevation that decreases in the direction of the current will produce a supercritical flow; to obtain subcritical flow, the elevation must increase with x.

8.4.3 Model of flow down a slope

The theoretical development will consider the simplest possible coastal basin in which the bottom slope (and hence R) is independent of x (see Figure 8.9). The variation of H

with x can be found by using (8.49) to express the left-hand side of (8.55) in terms of $\partial H/\partial x$. The subsequent mathematical development depends on the definition of a characteristic water column height H_T,

$$H_Q^3 \equiv \frac{Q^2}{gR},$$

(8.58)

where the subscript Q emphasizes that H_Q is dependent on the total volume transport. We now use H_Q as the basis of all dimensionless lengths which are denoted by the addition of a prime to the previously defined variable, i.e.,

$$H' \equiv H/Q_T \quad \eta' \equiv \eta/Q_T \quad h' \equiv h/Q_T.$$

(8.59)

It is noted that the cube of the dimensionless total height H' is the ratio of R to the square of the Froude number F^2, i.e.,

$$H'^3 = \frac{R}{F^2}$$

(8.60)

and hence (8.55) becomes

$$H_Q\left(H'^3 - R\right)\frac{dH'}{dx} = RC_d\left(H'^3 - 1\right).$$

(8.61)

The height datum is now set so that the downstream boundary condition involves vanishing elevation η, i.e., the model fixes the height of the water surface at the downstream boundary, which is then chosen to be the height datum. It is simplest to start the analysis by considering a semi-infinite basin, and in this case the downstream boundary condition is

$$x \to \infty, \quad \eta \to 0 \text{ (semi-infinite-basin)}.$$

(8.62)

A finite length basin will be considered later. A convenient origin for x is shown schematically in Figure 8.9, and with this origin the depth h has the simple form,

$$h = RC_d x,$$

(8.63)

i.e., the origin is a point, outside the basin, at which the depth extrapolates to zero. The upstream edge of the basin is set at $x = x_0 > 0$ with an elevation boundary condition (see Figure 8.9):

$$x = x_0, \quad \eta = \eta_0.$$

(8.64)

It is convenient to define a dimensionless horizontal distance by

$$x' \equiv RC_D x/H_Q$$

(8.65)

so that by (8.59) and (8.63), x' is identical to h' (for the present special case of a basin of constant bottom slope); hence using (8.59), (8.63) with (8.49) gives

$$\eta' = H' - x'. \tag{8.66}$$

Using (8.65), (8.61) becomes

$$\frac{dx'}{dH'} = \frac{H'^3 - R}{H'^3 - 1}. \tag{8.67}$$

Equation (8.67) has two solutions. The first solution is singular at H', and is obtained by direct integration of (8.67). The second solution is simply a flow of constant total depth H', as can be easily checked by looking at the earlier form of (8.67), which was (8.61), and noting that the derivative of dH'/dx' is zero for this second solution. Integrating (8.67) gives the first solution for H' as

$$x' = H' + (\alpha - 1)f(H') \quad H' \neq 1 \tag{8.68}$$

where f is the function

$$f(H') \equiv \frac{1}{6}\log\left(\frac{1 + H' + H'^2}{(1 - H')^2}\right) + \frac{1}{\sqrt{3}}\left[\tan^{-1}\left(\frac{2H' + 1}{\sqrt{3}}\right) - \frac{\pi}{2}\right]. \tag{8.69}$$

For this first solution, (8.66) and (8.68) give

$$\eta' = -(R - 1)f(H'). \tag{8.70}$$

Note that $f(H')$ vanishes at large $H' = 1$, so (8.70) shows that this first solution satisfies the downstream boundary condition (8.62) for the semi-infinite basin.

For the second solution, in which, as discussed above, $H' = 1$, (8.66) gives

$$H' = 1; \quad \eta' = 1 - x'. \tag{8.71}$$

Equation (8.60) shows that this second solution has a Froude number $F = R$ (for all x) so that the solution is supercritical whenever the bottom slope is supercritical $R > 1$, and otherwise subcritical. Note that this solution represents the only flow in which the inertial term $u\,\partial u/\partial x$ vanishes (since the velocity u is constant), and the momentum flux is spatially constant. This means that the negative pressure gradient must exactly balance the bottom frictional stress:

$$-\frac{\partial\eta}{\partial x} \to \frac{\partial h}{\partial x} = RC_d = C_d\frac{u^2}{gH}. \tag{8.72}$$

However, this constant depth solution can only extend a finite distance from the upstream boundary since (8.71) shows that η' diverges as $x' \to \infty$ and cannot satisfy the downstream boundary condition.

R=0.1

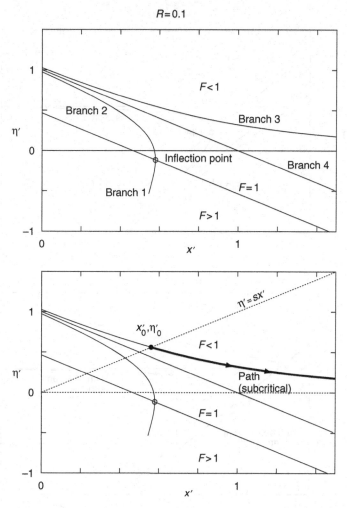

Figure 8.10 Both panels are plots of normalized surface elevation η' against distance x' for $R=0.5$. The three branches of the solution are numbered 1, 2, and 3, and the constant-depth solution, $H' = 1$, is labeled as branch 4 and appears as a straight line (with slope −1).

For the first solution represented by (8.68), (8.69), x' is a single-valued function of H'. When H' is plotted as a function of x', or equivalently, when η' is plotted against x', the logarithmic divergence of f at $H' = 1$ has the effect of producing three branches. Figure 8.10 contains a plot of η' against x' for $R=0.5$. The three branches are numbered 1, 2, and 3, respectively in the upper panel. These curves of $\eta'(x')$ correspond to the variation with x' of the ocean surface. The second, constant-depth solution, $H' = 1$, is labeled as branch 4 in the upper panel of Figure 8.10, and appears as a straight line (with slope −1). Another straight line in Figure 8.10, also with slope −1, is labeled as $F=1$ in the upper panel and corresponds to

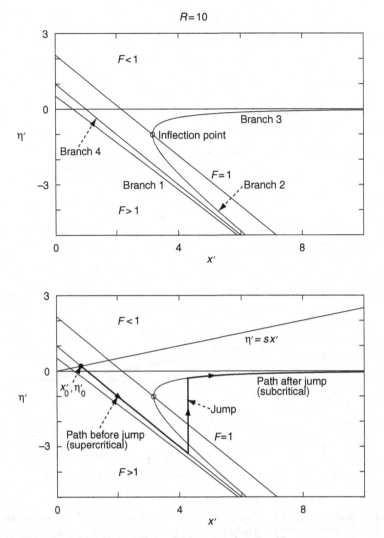

Figure 8.11 This plot is identical to Figure 8.10 except that $R = 10$.

$$\eta' = R^{1/3} - x' \tag{8.73}$$

which according to (8.60) and (8.66), represents a Froude number of $F = 1$, so that this line divides the plot into supercritical and subcritical regions as labeled in both panels of Figure 8.10. Importantly, the figure shows that when $R < 1$, branches 2 and 3 tend asymptotically to branch 4 as $x' \to \pm\infty$; we note here that although only the region $x > x_0$ is physically significant (as shown by Figure 8.9), the behavior of $\eta'(x')$ overall x helps an understanding of the properties of the various branches.

Figure 8.11 is identical to Figure 8.10 except that the value of R has been increased to 10, so that Figure 8.11 represents a supercritical bottom slope (while Figure 8.10

corresponds to a subcritical bottom slope). For Figure 8.11 where $R > 1$, note that it is now branches 1 and 2 that tend asymptotically to branch 4 as $x' \to \pm\infty$.

The boundary condition at the upstream edge of the basin is given by (8.64), and this can now be written in the dimensionless forms introduced by (8.59) and (8.65) as

$$x' = x'_0, \quad \eta' = \eta'_0 \tag{8.74}$$

$$x'_0 \equiv x_0/H_Q \quad \eta'_0 \equiv \eta_0/H_Q \tag{8.75}$$

and recall that x' is identical to h'. In this dimensionless form, (x'_0, η'_0) can be visualized as a point somewhere in the $\eta'(x')$ plot. This point must lie on one of the four branches, and this is always possible because the value of H_Q has not yet been fixed; the value of H_Q, i.e., the total volume transport through the basin, is controlled by this upstream boundary condition. Since the dimensional values x_0 and η_0 are already fixed, the position of (x'_0, η'_0) varies as the value of H_Q is changed with the effect that this point must always lie on a line through the origin, i.e.,

$$\eta' = sx' \quad s \equiv \eta_0/x_0 \tag{8.76}$$

where the slope s of the line has a value determined by the fixed values of x_0 and η_0. The intersection of this line $\eta' = sx'$ with one of the branches in Figure 8.10 or Figure 8.11 then gives the actual value of (x'_0, η'_0), and thereby determines the value of H_Q and hence, the volume transport through the basin. The line $\eta' = sx'$ is shown schematically in the lower panels of both Figure 8.10 and Figure 8.11, and note that the slope of the line has been drawn quite arbitrarily; its actual slope s depends on the values of x_0 and η_0.

Because the solution $\eta'(x')$ has four branches, the line $\eta' = sx'$ can have more than one intersection. All of these intersections represent possible starting points for the upstream flow (x'_0, η'_0) and each intersection effectively gives a different value for H_T via

$$H_Q = \eta_0/\eta'_0, \tag{8.77}$$

and note that this also determines the current u_0 at the upstream boundary via (8.49), (8.50), and (8.58). The flow through the basin can be visualized as the movement of a point (x', η') in the lower panel of Figure 8.10, starting at the upstream boundary (x'_0, η'_0) (given by one of the intersections) and proceeding to the downstream boundary $x' \to \infty$. Each intersection gives a solution for the flow which simply follows the curve $\eta'(x')$ for that particular branch, and this is illustrated in the lower panel of Figure 8.10. However, the chosen intersection (x'_0, η'_0) must lead to a solution $\eta'(x')$ which satisfies the downstream boundary condition (8.62), i.e., η' must tend to zero as $x' \to \infty$. Only branch 3 in Figure 8.10 satisfies this downstream boundary condition, and so a continuous solution must lie on that branch.

The bottom slope in Figure 8.10 is subcritical, i.e., $R < 1$, and branch 2 is seen to join branch 1 at a point of inflection which lies at $F = 1$, so that branches 1 and 2 have a maximum value of x' at this inflection point. When the bottom slope is supercritical ($R > 1$) as in Figure 8.11, the point of inflection occurs at the junction of branches 2 and 3, so that both branches have a *minimum* value of x'. Hence, if the upstream boundary corresponds to x_0' less than the value at the point of inflection, the flow must adopt either branch 1 or branch 4. However, Figure 8.11 shows that neither of these branches satisfies the downstream boundary condition (8.62), i.e., that η' must tend to zero as $x' \to \infty$. The only branch which does satisfy that downstream condition is again branch 3, and it is noted that it does so with η' negative (rather than positive as in Figure 8.10). This means that the solution must jump from branch 1 or 2 to branch 3 at a value of x', which is beyond the inflection point. Notice that branches 1 and 4 in Figure 8.11 are supercritical, while branch 3 is subcritical, so that a hydraulic jump occurs from supercritical to subcritical flow.

The jump in the lower panel of Figure 8.11 is shown as a vertical line (representing a discontinuous change in the surface elevation η') at the point of inflection (which is the minimum value of x' for the jump). In principle, the upstream path can follow either branch 1 or branch 4 prior to the jump. However, the Froude number F of branch 1 is higher than that of branch 4, and so the expected path starts on branch 4 (which is the constant depth solution, $H' = 1$), which means that the solution for H_Q is simply the value of the height of the water column at the upstream boundary, i.e.,

$$H_Q = h_0 + \eta_0 \tag{8.78}$$

so (8.58) gives the volume flux as

$$T = \sqrt{gR(h_0 + \eta_0)^3} \tag{8.79}$$

which means that above the threshold of $R = 1$, the flux increases as $\sqrt{R}$. This is to be compared with the subcritical region, $R < 1$, where as $\eta_0 \to 0$, $Q \to 0$. This sudden jump in volume flux is evident on changing either the bottom slope or the friction coefficient.

As the value of the slope s in (8.76) changes, so the intersection point (x_0', η_0') moves over the various branches. When $R < 1$, there is always an intersection lying on branch 3, provided that $s > 0$, and this involves $\eta_0 > 0$, so that the upstream elevation is positive. There are also intersections with the branches 2 and 4 in Figure 8.10 but, presumably, there would need to be some other constraint to force the flow to follow one of these alternative paths, since it would evidently need to jump to branch 3 to satisfy the downstream boundary condition. By contrast, when $R < 1$ as in Figure 8.11, the line $\eta' = sx'$ only has an intersection with branch 3 when $s < 0$, i.e., the upstream elevation is *negative* (and provided that $-s$ does not exceed some critical value). However, the steady state associated with positive current u and negative η_0 is probably not realizable, and a negative surface elevation at x_0 would lead to a negative current. Hence, when $R > 1$, as η_0 is increased from small negative to small positive values, the volume flux Q increases discontinuously from infinitesimally small negative values to $Q = \sqrt{gRh_0^3}$.

If the basin is of finite length, L, the downstream boundary condition becomes

$$x = L \quad \eta = 0 \text{ (finite length basin)} \tag{8.80}$$

and in this case (8.70) must be generalized to

$$\eta' = \Gamma - (R - 1)f(H') \tag{8.81}$$

where Γ is an arbitrary constant (arising from the integration of [8.67]), the value of which can be chosen so that η' vanishes at any prescribed value of x'_L. The effect of positive Γ is simply to move the plots of $\eta'(x')$ upward so that branch 3 cuts the $\eta' = 0$ axis at finite x'. This is illustrated for a supercritical bottom slope, $R = 1.5$, in Figure 8.12 with $x'_L = 1.5$ and $x'_L = 0.5$. These figures again show the intersection of the line $\eta' = sx'$ with one of the branches, which determines the value of H_Q. The figures show that the supercritical behavior previously encountered when $\eta_0 > 0$ also occurs in finite length basins, but may involve branches other than 3 and 4 (as occurred for the semi-infinite basin). For example, when $x'_L = 0.5$, the path must follow branch 1.

8.4.4 A numerical study of supercritical bottom slopes

The model of the last section is illustrated in this section by a two-dimensional numerical study of a coastal basin.[6] One of the elevation boundary conditions is adjacent to a region of slope 0.0034 downward from the boundary, which is supercritical when the bottom friction coefficient $C_d = 0.0025$. Results are presented in Figure 8.13 and Figure 8.14. The domain is a channel with elevation specified at each open end. The inertial terms are included using *upstream differencing*. The effect of Coriolis acceleration is excluded from the model run shown in the lower panel of Figure 8.13, and included in some of the runs shown in Figure 8.14; this produces only a small change in the results. The horizontal grid size is 100 m square, and the model dimensions are 3000 m (30 grid cells) along the channel and 300 m (three grid cells) across the channel. The depth is shown in the upper panel of Figure 8.13, and varies linearly from 20 m at the shallow end to 30 m at the upstream end, yielding a constant bottom slope $\partial h/\partial x$ of 0.0034. The surface elevation is prescribed as 0.1 m at the upstream boundary and zero at the downstream boundary. Different runs cover a

---

[6] The model runs described here were performed by Dr. John R. Hunter of the University of Tasmania and are previously unpublished.

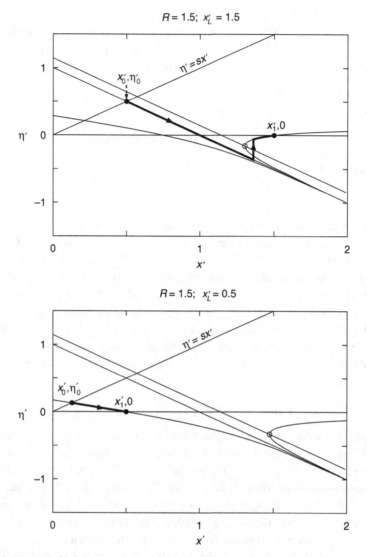

Figure 8.12 Plots of $\eta'(x')$ for a supercritical bottom slope, $R = 1.5$, with $x'_L = 1.5$.

range of bottom friction coefficients from 0.0015 to 0.005. The model is run to steady state, which occurs in under $2 \ 10^5$ s.

The lower panel of Figure 8.13 shows the elevation η for a friction coefficient of $C_d = 0.002$ giving $R = 1.67$. Based on (8.78) and (8.79) the analytical model gives $H_Q = 20.1$ m and $Q = 365$ m^2 s^{-1}, in good agreement with the numerical model. This value of H_Q allows the dimensionless length of the basin to be determined as $x'_L = 30/20.1 \approx 1.5$ (recall that x_L is identical to the depth at the downstream end of the basin). For this value of R, the inflection point (compare Figure 8.12) lies at $x' \sim 1.3$

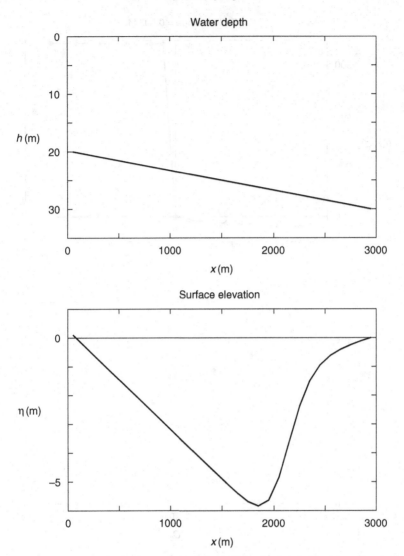

Figure 8.13 The upper panel shows the variation of depth h with distance x for the channel used in the numerical experiment with bottom slope $dh/dx = 0.00345$ and $C_d = 0.0023$, giving $R = 1.50$. The lower panel shows the variation of elevation η which follows a solution with constant total depth H until it exhibits a hydraulic jump at about $x = 1800$ m. The upstream boundary has an elevation 0.1 m yielding a water column height of 20.1 m. (The model runs were performed by Dr. John R. Hunter of the University of Tasmania and are previously unpublished.)

(which is lower than the value of 1.52 for a semi-finite basin), so that using (8.63) and (8.65) the hydraulic jump would occur at a depth greater than 26 m. The numerical model shows a jump at a depth of about 27 m, which suggests that the path of the flow may remain on branch 4 after passing the inflection point before jumping to branch 3.

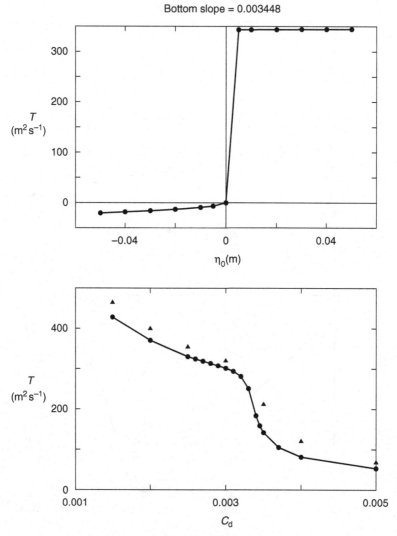

Figure 8.14 The variation of total volume transport T with changing η_0 (upper panel) and C_d (lower panel) in the numerical experiment shown in Figure 8.13, illustrating the sudden change in T when $\eta_0 = 0$ or $C_d = dh/dx$. The circles represent an irrotational basin and the triangles show the results with a Coriolis parameter $f = 0.0001\ \text{s}^{-1}$.

In many situations, at low values of the current u, the friction law is linear in u and only becomes quadratic in current at much higher values of u, i.e., the frictional stress is

$$
\tau_b = \begin{cases} -C_d u_b u & u \leq u_b \\[2mm] -C_d u^2 & u > u_b \end{cases}
\tag{8.82}
$$

where u_b is an effective background speed due to other motions which dominate the generation of turbulence. This has the effect of modifying (8.55) to read

$$(F-1)\frac{\partial\eta}{\partial x} = C_d F\left(\frac{u_b}{u} - R\right) \tag{8.83}$$

so that when $R \leq 1$, the bottom slope is always frictional but for $R > 1$ the bottom slope does not become inertial until the current exceeds u_b/R. The reason for this change in behavior is that a linear frictional stress increases more rapidly with current than the inertial term.

Hydraulic jumps due to the phenomena of the critical slope are observed in models of flow over coral reefs as discussed in Chapter 11.[7]

8.5 Further reading

See further reading from earlier chapters. Additional texts for this chapter are:
Chanson, H. (2004). *The Hydraulics of Open Channel Flow: An Introduction*. Oxford, UK: Butterworth-Heinemann.
Chow, V. T. (1959). *Open-Channel Hydraulics*. New York: McGraw-Hill.
Pedlosky, J. (1987). *Geophysical Fluid Dynamics*. New York: Springer.

[7] Hearn and Schulz (in press).

9

Aspects of stratification

9.1 Solar heating

9.1.1 Solar radiation

Much of the modeling of coastal basins is a study of currents, turbulence, and mixing, some part of which is due to temperature gradients produced by heating from the Sun. The Sun is a driving mechanism for most of the dynamics of the oceans and atmosphere and we feel its effect in coastal basins not only through solar heating but through winds, barometric pressure, regional currents, rainfall, and river flow. The Sun is not the only source of energy on our planet: we also have energy released by geological processes, fueled by the heat inside the Earth's mantle, and the effect of terrestrial processes some of which are induced by humans.

Solar energy, i.e., heat from the Sun, strikes the upper atmosphere as radiation of short wavelength (visible light, ultraviolet light, and even higher-energy radiation and particles) at an average rate near $1367\,\mathrm{W\ m}^{-2}$ and is returned back into space as infrared radiation (or as reflected short-wavelength radiation). In a steady state the inward and outward heat fluxes are equal. The intensity of the incoming radiation from the Sun is fairly constant at the top of the atmosphere although there is some variation over the year due to the Earth's orbit, and variations on longer timescales which have a major influence on the long-term variability of climate. The *average* temperature of the Earth (the oceans, atmosphere, and land) is nearly constant, and so the Earth's heat budget is in an approximate steady state. If we reduce that rate of loss of heat by the Earth, we would find that the Earth starts to heat up and if we increase that rate of loss, we get a drop in temperature on Earth.

Of all the energy arriving from the Sun, 47% of the energy penetrates to the surface of the land or ocean, i.e., 53% is stopped on its way down through the atmosphere: 23% is absorbed by clouds, oxygen moisture, and dust (only 4% by clouds) and 30% gets reflected by molecules in the air, clouds, and surface of the Earth. So the original $1367\,\mathrm{W\ m}^{-2}$ is reduced to $640\,\mathrm{W\ m}^{-2}$. This 47% must now go back into the atmosphere and space: 16% is radiated from the surface of the Earth (lands and oceans) and of this 10% makes it out into space; the other 6% is absorbed by clouds, oxygen, and carbon dioxide (the "greenhouse" effect) and warms the atmosphere. Another 7%

goes to heating the atmosphere by direct contact with the surface of the land and sea (called conduction or *sensible* heat loss). But most of that 47% is lost by the evaporation of water, i.e., 24%. Finally the 7% from conduction and 24% from evaporation, which heat the atmosphere, are lost back into space by radiation from the gases that form the atmosphere.

Evaporation is a process by which water molecules escape from water to become vapor. As they do so, they take energy from the liquid and this heat loss is called *latent heat loss*. The heat is simply transferred to the atmosphere. Note that evaporation accounts for about 50% of all heat arriving at the Earth's surface. So it is a very important process. As water evaporates, it creates vapor in the atmosphere and hence clouds and rain; this is part of the Earth's *hydrological cycle*.

Because the Earth is approximately spherical, there is an uneven distribution of heat over the globe. If the Earth were a sphere with its axis at right angles to the plane containing the Earth and the Sun, there would be no heat arriving at the poles. The average heat would still be 1367 $\mathrm{W\,m^{-2}}$, but the heat at the equator would be much higher and drop to zero at the poles. The Earth's axis is not at right angles to the Earth–Sun plane, but has a tilt of 23.5°. That tilt is responsible for our seasons and variation of day length. The tilt gives the southern hemisphere more heat for half the year and the northern hemisphere more heat for the other half year, due to the annual motion of the Earth around the Sun.

9.1.2 Model of temperature lag

We can construct a model that illustrates the seasonal variation of the temperature of a region (or more specifically a coastal basin) just by considering the seasonal variation of input of heat, and a rate of loss of heat that is a function of the temperature of the basin. A simple modeler's tool is to define the so-called *equilibrium temperature* T_e of the basin (this was discussed in Section 3.6). It is the temperature that would be achieved by an isolated basin if it were in equilibrium with all the incoming radiation. That equilibrium invokes all of the possible heat loss processes by the basin, and is itself a function of meteorological parameters including wind speed, air temperature, humidity, and cloud cover. The incoming radiation varies with time of day and day of the year. The actual temperature of a coastal basin will in general differ from the equilibrium temperature, both as a result of failure to reach a steady state and because the basin is not isolated. The failure to reach a steady state is a combined effect of the rate of change of T_e with time t, and the rate of response of the temperature of the basin to changes of incoming radiation. The departure of the steady state temperature from T_e is a result of other processes such as heat exchange with the open ocean.

The speed of response of a basin is related to its total depth, or more exactly to the depth to which heat can penetrate through attenuation of incoming radiation and downward diffusion of heat. We denote this time to reach equilibrium by t_e. The total outgoing energy (heat) flux per unit surface area of the basin is denoted by q. In the present model we assume that q is a function of a characteristic temperature T which is, broadly speaking, the surface temperature of the basin and is also a function of meteorological variables. The incoming heat flux at the surface of the basin is denoted by Q, and our model has the form

$$h\rho C_p \frac{\partial T}{\partial t} = Q - q(T) \tag{9.1}$$

where we have excluded any interaction of the basin, or region, with any other zone or agency. Equation (9.1) only considers heat gain and loss through the surface. In (9.1) the quantity $h\rho C_p$ is the heat capacity of the basin per unit area of the surface, i.e., the heat per unit area required to raise T by 1 °C. This is derived from the specific heat per unit volume ρC_p and the effective depth h of the basin, or the depth to which heat penetrates. The equilibrium temperature T_e is defined as the temperature at which heat loss and heat gain are equal, i.e.,

$$q(T_e) = Q. \tag{9.2}$$

For T near to T_e, we assume that q can be expanded in a Taylor series, i.e., q is a simple continuous function of T. In general, we would not expect the q to reach a maximum or minimum at T_e. So the first derivative of q with respect to T will not vanish, hence defining

$$\frac{1}{t_e} \equiv \frac{1}{h\rho C_p} \frac{\partial q}{\partial T}\bigg|_{T=T_e}, \tag{9.3}$$

and noting that q increases with T, we obtain

$$q(T) = Q - h\rho C_p \frac{T - T_e}{t_e} \tag{9.4}$$

so that

$$\frac{\partial T}{\partial t} = -\frac{T - T_e}{t_e} \tag{9.5}$$

where we recall that t_e is a function of T and meteorological conditions. Note the sign in (9.5) so that when $T > T_e$, T decreases with time toward T_e.

We now construct a model in which t_e is assumed constant over the whole range of T that is encountered. This provides a first appraisal of the annual variation of T, and is also physically acceptable as a broad first attempt to look at an isolated coastal basin. One of the basic precepts of all modeling is to start from the simplest assumptions and then include more sophistication, after the simplest models and basic processes have been explored. We can interpret T_e as the temperature that would be achieved in a very shallow basin, in the sense that h in (9.3) is small, and consequently t_e is very small, or simply that the *thermal inertia* of the basin tends to zero. Let us firstly look at what happens if we have a periodic variation in Q and which results in T_e, having the form,

$$\frac{T_e}{T_0} = 1 + \alpha \sin(2\pi t / T_Q) \tag{9.6}$$

where T_0 is the long-term average of T_e and αT_0 the variation over time period T_Q. If (9.6) is used in the model (9.5), we find a transient solution which depends on our starting condition for T and this transient decays away with a characteristic time equal to t_e. The remaining long-term solution is oscillatory with a period of T_Q, and determined as follows. Firstly define

$$\theta \equiv 2\pi t_e / T_Q \qquad t' \equiv t / t_e, \tag{9.7}$$

assume that the periodic solution has the form

$$\frac{T - T_0}{T_0} = A \sin \theta t' + B \cos \theta t' \tag{9.8}$$

and substitute into (9.5) and (9.6):

$$A = \frac{\alpha}{1 + \theta^2} \qquad B = -\frac{\alpha \theta}{1 + \theta^2} \tag{9.9}$$

$$T = T_0 \left\{ 1 + \frac{\alpha}{\sqrt{1 + \theta^2}} \sin\left(\theta \frac{t}{t_e} - \phi\right) \right\} \qquad \phi \equiv \tan^{-1} \theta. \tag{9.10}$$

For large t_e / T_Q, i.e., a thermal equalization time which is much longer than the characteristic period of variation of heat input, the temperature lags the cyclic input of heat by 90°. The reason is that the rate of temperature *increase* is maximal when the heat *flux* is maximal, so that temperature itself lags by a quarter of a cycle. If T_Q corresponds to 12 months, i.e., the annual cycle of heating from the Sun, the rate of heating Q is about maximum at the summer solstice (near June 21–22, or December 21–22, according to the hemisphere) and so the following two, or perhaps three, months have the *highest temperatures* of the year and constitute summer.

Most coastal basins respond to changes in T_e over a time period of days, and therefore develop a surface temperature which is an average of T_e over that period with superimposed diurnal fluctuations. The surrounding land has a very much faster response, since heat has comparatively limited penetration. As a result, the land shows much stronger diurnal temperature variations than coastal basins.

9.1.3 Vertical mixing and potential energy

The present section extends the ideas on mixing discussed in Chapter 7 to the case of a stratified coastal basin and it assumes some familiarity with ideas presented in Chapter 7.

Kinetic energy is simply the energy contained in a system as a consequence of its mechanical motion. Potential energy is the energy available to a system as a result of its weight. Total energy is the sum of these two forms of energy. We must always specify the origin used for potential energy, and that energy can be negative. One of the great laws of physics is that systems tend to move towards a state of lower potential energy if such a state is *available*. As an example, a water column denser at the top than bottom is unstable since it can turn over, in what we call it *convective overturning*, so as to reach a state of lower potential energy. The lost of potential energy is taken up as kinetic energy so that the total energy remains constant. The best way to observe vertical overturning is to watch water in a lake being cooled at night in winter. Water at the surface is cooled by the cold air. If the cooled water stayed on the surface, it would get to air temperature and stop cooling. Instead it overturns to be replaced by warm water from below and the lake may appear to be *steaming* in early morning. Overturning is a *stirring process* and may completely mix the water column. This is often observed in coastal basins that are cooled at night, and the resulting vertical profile of temperature may show no measurable variation in temperature over many tens of meters. The opposite of surface cooling is surface heating by the Sun which creates a surface heated layer that is quite stable. The reason is that the surface water is less dense than the underlying unheated water.

Consider a water column which is heated at the surface. Suppose we try to mix it. This involves taking two particles of water and interchanging them vertically. Let us take their volumes to be the same so that the upper one will have greater mass than the bottom one. So interchange requires a net lifting of mass in the vertical and must increase the potential energy of the water column. If we continue this process of interchange, the potential energy will increase until the water column becomes vertically mixed; at that stage interchange of particles can produce no further increase in potential energy. The mixed state therefore has maximum potential energy of all available *stable states*. The potential energies of states which have higher density at the top than bottom are higher than the mixed state but these are not stable and indeed are likely to decay to the mixed state. Stratified states have lower potential energy than the mixed state but the mixed state does not decay back to a stratified state because mixing is *irreversible*, and systems only tend to move to *available* states of lower potential energy. The stratified state is just not available to the mixed state just as mixing milk into a cup of coffee is an essentially irreversible process.

Heating (usually from the surface) changes the temperature profile $T(z)$ of the water column and for any temperature profile we can calculate the potential energy of the water column. In doing this calculation we assume that the mass of the water column is constant; so the water expands as it is heated. Let us consider that the water column has height h, and is divided into an upper layer of thickness h_1 and a lower layer of thickness h_2,

$$h = h_1 + h_2, \tag{9.11}$$

with densities ρ_1 and ρ_2 with $\rho_1 < \rho_2$ so that the water column is *stably stratified*. The interface between the layers forms a *pycnocline*. If the difference in density is due to temperature the interface also forms a *thermocline*. The mean density

$$\bar{\rho} = \frac{\rho_1 h_1 + \rho_2 h_2}{h}, \tag{9.12}$$

lies between ρ_1 and ρ_2. Complete vertical mixing of this stratified state would produce a mixed state with density $\bar{\rho}$ everywhere, because mixing conserves mass (since it only involves interchange of particles). If the density difference between the two layers is due to either a temperature or a salinity difference, the mixed state will have a temperature and a salinity equal to the mean temperature and salinity of the stratified state. If density varies linearly with temperature and salinity, this means that the volume and hence, height, of the water column are unchanged by mixing. We need to assign one state of the water column as an arbitrary origin for potential energy, and we choose to use the most clearly defined state which is the vertically mixed state. Since this has the highest potential energy of all stable states, this choice means that all other stable states will have negative potential energy.

The difference between the energy of the stratified and mixed state is

$$E = -g\frac{\rho_2 - \rho_1}{2}h_1 h_2 \tag{9.13}$$

and note that E is energy per unit area. The energy in (9.13), measured relative to the mixed state, is sometimes referred as the *potential energy anomaly* and for a general water column with density $\rho(z)$ has the form

$$E = g\int_{-h}^{\eta} [\rho(z) - \bar{\rho}]z\,\mathrm{d}z, \tag{9.14}$$

the mean density being

$$\bar{\rho} \equiv \frac{g}{h+\eta}\int_{-h}^{\eta} \rho(z)\,\mathrm{d}z. \tag{9.15}$$

Surface heating in a stratified water column would be expected to decrease the density of the upper layer, and so differentiating (9.13) with respect to time, and assuming that ρ_2 is constant,

$$\left.\frac{\partial E}{\partial t}\right|_{\text{heating}} = g\frac{\partial \rho_1}{\partial t}\frac{h_1 h_2}{2}, \tag{9.16}$$

and note that the left-hand side of (9.16) is negative. It is easy to calculate the rate at which the density of the upper layer decreases with time if we know the net rate of heating. Using a coefficient of thermal expansion

$$\alpha \equiv -\frac{1}{\rho}\frac{\partial \rho}{\partial T}$$ (9.17)

(note that α has the units of $°C^{-1}$), (9.5) gives

$$\frac{1}{\rho_1}\frac{\partial \rho_1}{\partial t} = \alpha\frac{T_1 - T_e}{t_e}$$ (9.18)

and so finally

$$\left.\frac{\partial E}{\partial t}\right|_{\text{heating}} = \frac{g\alpha\rho_1 h_1 h_2}{2}\frac{T_1 - T_e}{t_e}.$$ (9.19)

Mixing between the top and bottom layers involves the interchange of particles, with particles from the bottom layer moving upward and particles from the top layer moving downward. This represents a transfer of heat downward from the warm surface layer to the colder bottom layer and a net input of potential energy to the water column from mixing. Hence, vertical mixing requires that the turbulence field has sufficient energy to supply the requisite increase in potential energy. Let us suppose that the mixing supplies energy at a rate denoted by $\partial E/\partial t|_{\text{mixing}}$ and the rate of change of energy due to surface heating is $\partial E/\partial t|_{\text{heating}}$ so that

$$\frac{\partial E}{\partial t} = \left.\frac{\partial E}{\partial t}\right|_{\text{heating}} + \left.\frac{\partial E}{\partial t}\right|_{\text{mixing}}.$$ (9.20)

Remember that the rate of change of energy due to heating is negative, and that due to mixing is positive. So that whilst heating tends to reduce E, increasing the strength of mixing reduces the net rate of decrease of E and if mixing is strong enough it is possible that

$$\left.\frac{\partial E}{\partial t}\right|_{\text{mixing}} = -\left.\frac{\partial E}{\partial t}\right|_{\text{heating}}$$ (9.21)

and this may mean that the water column remains almost vertically mixed, i.e., that the rate of heating is insufficient to cause stratification.

We can also view the effect of mixing in terms of the temperature of the upper layer. Let us suppose that the coefficient of mixing between the two layers is K_z. The rate of mixing of heat is dependent on the temperature gradient and gives

$$\left.\frac{\partial T_1}{\partial t}\right|_{1\to 2} = -\frac{T_1 - T_2}{t_{1\to 2}}$$ (9.22)

since the distance between the centers of the layers is $h/2$

$$\frac{1}{t_{1\to 2}} \equiv \frac{2K_z}{hh_1},$$ (9.23)

and note that the heat flux is negative, i.e., downward since $T_1 > T_2$. Hence the rate of change of temperature of the upper layer becomes

$$\left.\frac{\partial T_1}{\partial t}\right|_{1\rightarrow 2} = -\frac{T_1 - T_e}{t_e} - \frac{T_1 - T_2}{t_{1\rightarrow 2}} \qquad (9.24)$$

and this reaches zero, i.e., the surface temperature reaches a steady state when

$$T_1 = \frac{T_e + rT_2}{1 + r} \qquad (9.25)$$

where

$$r \equiv \frac{t_e}{t_{1\rightarrow 2}}. \qquad (9.26)$$

Note that in the limits of small and large r, T tends to T_2 and T_1, respectively. The mixing coefficient K_z is sensitive to the density difference between the upper and lower layers which means that K_z decreases as $T_1 - T_2$ increases. We see later in this chapter that this can be approximated by a inverse power dependence on $T_1 - T_2$ so that

$$\frac{r}{r_0} = \left[1 + \left(\frac{T_1 - T_2}{T_c}\right)^n\right]^{-1} \qquad (9.27)$$

where T_c is a critical temperature difference, r_0 is the limit of r at very small temperature differences, and the exponent $n \geq 1$. Equation (9.27) implies that the warm upper layer is self-stabilizing. The reason that the mixing is reduced by the temperature difference is that the mixing has to provide greater potential energy. Herein lies the basic reason that coastal basins can be thermally stratified. If we assume that $n = 1$ and define

$$\delta_1 \equiv \frac{T_1 - T_2}{T_c} \qquad (9.28)$$

$$\delta_e \equiv \frac{T_e - T_2}{T_c} \qquad (9.29)$$

we find

$$\delta_1 = \frac{\delta_e(1 + \delta_1)}{\delta_1 + (1 + r_0)} \qquad (9.30)$$

and if T_c in (9.27) is small in relation to the temperature differences $T_1 - T_2$ and $T_e - T_2$, we can see that the effect of stratification is to severely reduce mixing, and (9.30) gives

$$\delta_1 \sim \delta_e \qquad (9.31)$$

$$T_1 \sim T_e \qquad\qquad (9.32)$$

so that effectively, heat conduction down through the thermocline between the two layers almost ceases and the heated surface layer comes into equilibrium with incoming heat, i.e., heat loss through the surface exactly matches heat gain. The first order term in $T_c/(T_e - T_2)$ is

$$\frac{T_1 - T_2}{T_e - T_2} \sim 1 - \frac{r_0 T_c}{T_e - T_2}, \qquad\qquad (9.33)$$

i.e.,

$$T_1 - T_e \sim -r_0 T_c. \qquad\qquad (9.34)$$

9.1.4 Seasonal thermal fronts in macroscopic tidal basins

In temperate waters, the continental shelf is usually vertically mixed in winter, but may be stratified in summer. This is due to summer heating combined with usually higher winds in winter (which produce stirring). In tropical waters we often find permanent stratification, and in polar waters there is insufficient solar heating for any thermal stratification. In macro-tidal temperate continental seas (parts of the continental shelf partly enclosed by land) we may find that in summer some parts of the water column are stratified, and in others vertically mixed. The demarcation line between the two regions is called a *thermal front* because there is a sudden change in surface temperature. Thermal stratification due to summer heating is very important in shallow coastal basins, but has much shorter timescales than on the continental shelf. So, we are less likely to see seasonal thermal fronts, although a variety of other short-term frontal phenomena are often evident. But the phenomenon of seasonal thermal fronts is sufficiently important to warrant a study here, and we will derive the u^3/h rule[1] which controls the position of fronts, where u is the tidal streaming velocity,[2] and h is the water depth. This rule is a direct result of our study of energy in Chapter 7 which showed that energy input by bottom friction varies as u^3 whilst the energy of the stratified state (9.13) varies as h (per unit mass of the water column). The regions on either side of a thermal front experience the same meteorological conditions, and hence, have the same equilibrium temperature T_e. The equalization time t_e, however, is shorter for the stratified state (if we assume that only the surface layer is heated). Returning to (9.3), it is evident that on the continental shelf t_e may reach one or more weeks or months. This means that the stratified surface water will become much warmer than the vertically mixed water.

---

[1] In common parlance this is called the *h on u-cubed* rule although the present author prefers the term *u-cubed on h* since it is dominated by the strength of the tidal current. The rule is also called the *Simpson and Hunter* rule after the pioneering work by John H. Simpson, a British physical oceanographer and John R. Hunter, an innovative modeler in coastal oceanography (Simpson and Hunter, 1974).

[2] In the original derivative this is the streaming velocity at spring tides although we naturally associate the criterion with neap tides following Hearn (1985).

The basic reason is that in the stratified water column, the incoming heat is mixed over a smaller part of the water column. In places on the continental shelf of north-western Europe, which has optimal conditions for these fronts, it is noticeable that surface temperatures at the coast do vary sharply according to the position relative to such fronts, and this does affect coastal basins.

Thermal fronts are visible on an infrared image of surface temperature, and also as well-defined lines to an observer on board ship, sometimes, with some distinct changes in surface color and wave state on either side of the front. These effects are due to the buoyancy of the upper layer on the stratified side of the front, and are similar to the effects of a buoyant plume from a river discharge. The buoyant layer affects water circulation, which creates a region of surface convergence near the thermal front. Such convergence has the effect of sweeping buoyant surface material towards the front, where it remains as the water *downwells*. This action includes surface-dwelling marine life and surface de-tritus, and is attractive to fish and birdlife. This is true of all types of fronts on the shelf and in coastal basins. The congregation of birds is often an indication of some type of surface convergence associated with a front. These include buoyant river discharge and tidal streaming fronts associated with the intrusion of marine water into estuaries.

Let us now derive the u^3/h rule in more detail. Consider the rate of energy dissipa-tion (see Chapter 7) in the water column per unit mass due to bottom friction for a current u (which was derived in Chapter 7):

$$\varepsilon = C_\mathrm{d} \frac{u^3}{h} \tag{9.35}$$

where ε is in W kg^{-1}, u in m s^{-1}, and h in m; C_d is the coefficient of bottom friction. So if $u = 1\,\mathrm{m\,s}^{-1}$ and $h = 100\,\mathrm{m}$, $\varepsilon = 2.5 * 10^{-5}\,\mathrm{W\,kg}^{-1}$. We can compare this with a typical value of ε for a wind speed of 15 m s^{-1} which is $10^{-6}\,\mathrm{W\,kg}^{-1}$. The cubic dependence on u in (9.35) is a very important part of the physics of energy dissipation by bottom friction, and plays an important role in the sharpness of tidally controlled thermal fronts. For example if we halve the value of u, the dissipation is ε is reduced by a factor of eight (nearly an order of magnitude lower). The reason for this cubic dependence is twofold. To calculate ε, we find the bottom friction stress τ_b. This is the frictional force per unit area of the bottom. To find the rate at which this force does work, we multiply by the speed u, i.e., the rate of working of frictional force is $u\tau_\mathrm{b}$. Since the bottom stress has a quadratic dependence on u, we find a cubic dependence of the rate of working on current speed.

Revisiting the model of (9.3), we recall that the quantity of heat required to raise the temperature of a layer of water of thickness h_1 by ΔT is $\rho C_p h_1 \Delta T$ per square meter of surface. If the *net* rate of solar heating is Q_surface (with units W m^{-2})

$$\frac{\partial T_1}{\partial t} = \frac{Q_\text{surface}}{\rho_1 C_p h_1} \tag{9.36}$$

and the rate of change of potential energy is

$$\left.\frac{\partial E}{\partial t}\right|_{\text{heating}} = -\frac{\alpha g h_2 Q_{\text{surface}}}{2C_p}. \tag{9.37}$$

Let us assume that we are starting with the water column initially vertically mixed and determine the rate of change of E when the surface layer thickness is very small, i.e., $h_2 \sim h$,

$$\left.\frac{\partial E}{\partial t}\right|_{\text{heating}} = -\frac{\alpha g h Q_{\text{surface}}}{2C_p}. \tag{9.38}$$

At the thermal front, the input of potential energy by mixing exactly matches the rate of decrease of potential energy by surface heating; we need to stop that stratification from forming and so,

$$\beta C_d \frac{u^3}{h} \rho h = \frac{g h}{2} \frac{\alpha Q_{\text{surface}}}{C_p} \tag{9.39}$$

where ρh is the mass (per unit area of water column). We need to introduce a parameter β which is the co-called the mixing efficiency, and identified as the fraction of the turbulent dissipation energy that can be converted to potential energy in mixing. This is essentially a phenomenological parameter which can be calibrated against observational data although more detailed models of mixing efficiency have been proposed.[3] Equation (9.39) provides an explicit for u^3/h:

$$\frac{u^3}{h} = \frac{g}{2\beta\rho C_d} \frac{\alpha Q_{\text{surface}}}{C_p}. \tag{9.40}$$

Taking $g = 9.81\,\text{m}^2\,\text{s}^{-1}$, $\alpha = 1.6\ 10^{-4}\,\text{C}^{-1}$, $C_p = 4.2\ 10^3\,\text{J kg}^{-1}$, $\rho = 1026\,\text{kg m}^{-3}$, and $Q_{\text{surface}} = 170\,\text{Wm}^{-2}$, (9.40) becomes

$$\frac{u^3}{h} = \frac{10^{-5}}{\beta} \tag{9.41}$$

which is the *u-cubed on h rule*. The actual tidal current varies during the tide (and also from spring to neap tides), and so we really need to define more carefully how we choose u. The most realistic model is to assume that vertical mixing requires that the tidal currents are able to overcome the effect of vertical mixing throughout the spring–neap cycle so that the current in (9.40) should be chosen at neap tides. If tidal mixing at neaps is unable to overcome the tendency to stratify, then the higher currents during the remainder of the cycle will also be ineffective, since

---

[3] Hearn (1985).

stratification impedes the mixing process as we saw in Section 9.1.3. However, a contrary argument is that once spring tides are able to break up the stratification, it may take longer than the period of the spring–neap cycle for the stratification to return. What is clear, and the point of most physical significance, is that the stratified state, once fully formed, is self-stabilizing and does not collapse until major wind mixing from winter storms. Equation (9.41) is called the Simpson–Hunter criterion, and evaluated for spring tides on the European shelf has an empirical value of $\beta \sim 10^{-3}$. It should be emphasized that the major variation in the u^3/h parameter comes from u^3, and as we mentioned earlier, this is very sensitive to changing tides on a regional spatial scale.

There are competitive models of tidal frontal formation. The first is called the u/h model which differs from the u^3/h model that we discussed above in relating the vertical mixing from tidal currents to the thickness of the bottom Ekman layer that we met in Chapter 7. The thickness of the bottom layer relative to water depth is

$$\frac{u}{h} = \frac{\kappa f}{\sqrt{C_\mathrm{d}}} \tag{9.42}$$

which does give good agreement with observations. The model is not dependent on the rate of surface heating, but simply says that stratification can occur during summer heating if the bottom stress from tidal currents is not able to mix up through the entire water column. Another major model of tidal fronts claims that the phenomenon is simply the same as *shelf break fronts*. This type of front is often observed, as the name implies, at the break of flat shelves and forms a demarcation between the less saline, and often cooler, shelf waters and the surface waters of the deep ocean. This can create a situation of *density compensation* between the effects of salinity and temperature, which we will meet later in this chapter. Density compensation tends to reduce the horizontal transport that might otherwise occur through buoyancy effects, and so we may find two regions of differing physical properties coexisting side by side. This is almost certainly not the situation for the original fronts studied in relation to the u^3/h and u/h rules, but may be relevant in other shelf waters. The basic processes of fronts reaffirm the basic lessons of previous chapters, concerning our scaling of processes in the vertical and horizontal, and especially the need for positive mechanisms to produce mixing.

9.2 Effect of stratification on vertical mixing

The formulae for the coefficient of vertical mixing, K_z, which we developed in Chapter 7, apply to unstratified water columns and do not allow for the reduction in mixing produced by stratification, or by any change of density with height. This reduction has a form that we considered briefly in Section 9.1.

9.2.1 Stabilization of the pycnocline

In Chapter 7 we considered the simplest picture of vertical mixing in which two particles interchange positions vertically due to turbulence, i.e., the fluctuations w' in the vertical speed w. It is important now to emphasize that this interchange does not complete the mixing process, because the two particles must first come into equilibrium with their surroundings, i.e., each particle must integrate with its new local fluid environment. If that equilibrium is not reached, the particles can, for example, simply move back to their original positions or to other positions within the water column. The tendency not to remain in their new environments is especially true if the water column has a vertical gradient of density, since the interchange necessarily involves an increase of potential energy. This process of particles returning to their original positions is nicely illustrated by considering the motion of a sphere in a water column whose density decreases with height. Suppose the sphere has a density the same as the water column at a height $z = z_0$. Suppose that we drop this sphere from the surface into the water. After a long time the sphere will come to rest at the height $z = z_0$ where it is in *stable equilibrium*. But initially, it oscillates in the vertical around that position. The frequency of this oscillation is controlled by the way in which the density varies with height. It is called the *buoyancy (angular) frequency* (also known as the *Brunt–Väisälä* frequency) and controlled by the way in which the density varies with height. Consider the force acting on a parcel of water of volume dV at a height z in the water column. The mass of the particle is $\rho(z_0) dV$ where z_0 is the point in the water column where the particle has *neutral buoyancy*. At any point z in the water column, the particle experiences a net buoyancy force given by

$$g[\rho(z) - \rho(z_0)]dV \sim g\frac{\partial \rho}{\partial z}(z - z_0)dV \qquad (9.43)$$

and so the equation of motion (recognizing that the vertical derivative of density is negative in a stable water column) becomes

$$\rho(z)\frac{d^2z}{dt^2}dV = -g\left|\frac{\partial \rho}{\partial z}\right|(z - z_0)dV, \qquad (9.44)$$

and using the theory of the simple harmonic oscillator which we developed in Section 4.7, the particle oscillates with angular frequency ω_b:

$$\omega_b^2 = \frac{g}{\rho}\left|\frac{\partial \rho}{\partial z}\right| \qquad (9.45)$$

and in the literature this angular frequency is often denoted by the symbol N. To find the order of magnitude of ω in a coastal basin, we would take a change in density of order a few kilograms per cubic meter over a depth of (say) 1 m, which yields $\omega_b \sim 0.2\,\text{s}^{-1}$ or a period of $2\pi/\omega_b \sim 30$ s, although the frequency could be higher in a sharper pycnocline and if we had almost fresh water overflowing marine water $\omega_b \sim 1 \text{ s}^{-1}$ or a period of $2\pi/\omega_b \sim 6$ s.

When we interchange two particles to achieve mixing, the particles may just continue to oscillate back and forth at exactly the buoyancy frequency. It is easy to see these oscillations as waves in the interface between two fluids of different density. They are called *internal waves*. For a sharp interface between two layers of fluid, the theory of the formation of internal waves is essentially the same as that which we developed in Chapter 4 for surface gravity waves, provided that we make the replacement of the acceleration due to gravity, g, with the reduced gravity g' defined by

$$g \rightarrow g' \equiv g \frac{\rho_2 - \rho_1}{\frac{1}{2}(\rho_1 + \rho_2)} \tag{9.46}$$

where ρ_1 and ρ_2 are the densities of the upper and lower layers, respectively. The justification for this replacement is simply that the pressure gradient induced by an elevation of the interface is

$$\Delta p \rightarrow g(\rho_2 - \rho_1)\eta. \tag{9.47}$$

Internal waves are observed on the continental shelf, often excited by astronomical tides to create *internal tides*. Particle motion for internal waves follows the same elliptical paths that we met for surface waves.

Internal waves usually move at much more sedate speeds than those associated with surface waves, since by (9.46) g' is typically a factor of 1 to 5 10^{-3} lower than g providing speeds of some meters per second. The waves take a time of order 12 h (about the period of the M2 tide) to cross a continental shelf of width 100 km. Internal tides are perhaps the most prevalent (and largest amplitude) internal gravity waves in the ocean. They can reach very large amplitude comparable with water depth, and have a surface manifestation so that they are visible by remote sensing of surface temperatures. At that stage they are well outside of the simple linear assumptions made in deriving (9.46) and the study of *non-linear internal wave* phenomena is an area of active research.

If we study the waves on the interface between two fluids, we find that the waves become unstable if there is too much velocity shear across the interface. This is called the *Kelvin–Helmholtz instability*.[4] The Kelvin–Helmholtz instability produces a billowing motion. It is widely observed in the Earth's atmosphere, and also in the atmosphere of other planets such as Jupiter. It is also visible in strong ocean currents, and these structures can appear when there is a sudden widening of a river bed. The *instabilities* are basically caused by the existence of different horizontal velocities on

---

[4] It is named after two distinguished physicists. William Thomson (1824–1907) was an Irish–Scottish mathematical physicist, engineer, and outstanding leader in the physical sciences of the nineteenth century. He did important work in the mathematical analysis of electricity and thermodynamics, and did much to unify the emerging discipline of physics in its modern form; he later became Baron Kelvin. Hermann Ludwig Ferdinand von Helmholtz (1821–94) was a German physicist and one of the foremost scientists of the nineteenth century.

either side of the interface. This happens in many types of physical systems (not just an interface in a fluid). A well-known example is the *flapping of a flag*. At slow wind speeds, flags just have waves running through them. But at high speed, the flag develops a very uneven sharp flapping motion. You can get the same effect shaking the dust out of a mat. Generally the instability is called the *two-stream instability*. It is claimed to be responsible for the meandering in the course taken by rivers.

A stable internal wave on a pycnocline means that a fluid particle will remain on one side of the interface. Waves need to break to produce mixing, and this can be caused by an instability, which essentially means that particles mix rather than oscillate in the vertical. So, by studying the conditions for these instabilities, we may learn more about mixing in the presence of stratification.

9.2.2 Richardson number

If we place a stick in a stratified water column through which an internal wave is passing, it will be seen to have a very unstable tumbling motion unless it is horizontal. The characteristic angular frequency for this motion is of order the velocity shear $\partial u/\partial z$. The restoring force at the pycnocline which creates internal waves is due to buoyancy, and quantified as the buoyancy frequency defined in (9.45), and this also stops instabilities on the pycnocline from growing without limit. The instability itself is produced by the velocity shear across the interface, and hence, the ratio of the buoyancy frequency to the velocity shear should be a measure of the stability of the interface. It is convenient to work with the square of that fraction, known as the *Richardson number*,

$$\mathrm{Ri} = \omega_b^2 \left/ \left(\frac{\partial u}{\partial z}\right)^2 \right. \tag{9.48}$$

and using (9.45), (9.48), this becomes

$$\mathrm{Ri} = -\frac{g}{\rho}\frac{\partial \rho}{\partial z} \left/ \left(\frac{\partial u}{\partial z}\right)^2 \right. \tag{9.49}$$

The Richardson number is an accepted measure of how successful the buoyancy force is in damping out instabilities and is named after Lewis Fry Richardson, whose work we met in Chapter 7. It is one of the important fundamental non-dimensional numbers of stratified flow. Buoyancy acts to suppress instability and force the system toward internal waves. In other words, buoyancy tends to prevent interchanged particles from remaining in their interchange positions whilst instabilities allow that interchange and we achieve mixing. At large Richardson number we find no instabilities (either because the shear is small or because the buoyancy forcing is large) and hence no mixing. For flow at small Richardson numbers the effect of the shear is dominant, and any disturbance breaks into instabilities and provides mixing. It is possible to express the Richardson number as the inverse square of an *internal Froude number* (we met the Froude number in Chapter 8). This involves using (9.46) to express the buoyancy frequency in terms of a reduced gravity g' as

$$\text{Ri} \rightarrow \frac{g'h}{u^2} \tag{9.50}$$

where we have expressed the velocity shear as u/h, and g' is $g\Delta\rho/\rho$.

9.2.3 Density gradient models

The rate of energy dissipation per unit mass is obtained from Chapter 7 as

$$\varepsilon = \tau_x \frac{\partial u}{\partial z}. \tag{9.51}$$

If we assume a coefficient of eddy viscosity K_z, (9.51) becomes

$$\varepsilon = K_z \left(\frac{\partial u}{\partial z}\right)^2 \tag{9.52}$$

and we note the presence of a term of the same form as the denominator of (9.49) and related to the transfer of energy from the mean motion into turbulence. The numerator of (9.49) is connected with the change in potential energy of the water column. The rate of increase of potential energy per unit volume is mass flux density times the acceleration due to gravity g. This mass flux can be obtained from the mixing coefficient, and so the increase of energy per unit mass is

$$\left.\frac{\partial E}{\partial t}\right|_{\text{mixing}} = -\frac{gK_z}{\rho}\frac{\partial \rho}{\partial z}. \tag{9.53}$$

So the Richardson number is the fraction of the available turbulent energy that is needed to supply the potential energy required by mixing. When Ri is small, the turbulence can supply sufficient energy relative to that needed to provide the potential for mixing. When Ri is larger there is a shortage of energy. Hence, the stratification is *stable at high Richardson number*, i.e., at low internal Froude number. The Kelvin–Helmholtz instability is found for $\text{Ri} < 0.25$, i.e., a Froude number greater than 2. So, we would expect that weakly stratified basins tend to become vertically mixed, and that there is some critical Richardson number above which stratification becomes stable against the effects of mixing. If is now immediately evident in a mechanistic sense how that stability occurs, we can just consider (9.52) and (9.53) with a fixed value of K_z obtained. For example, by using the methods discussed in Chapter 7. The crux of the problem with the loss of mixing at higher Ri is that there is insufficient energy to supply potential energy to the water column, and therefore if

(9.53) is to continue to be true, the mixing coefficient K_z must decrease with increasing Richardson number. This is generally known as the Munk–Anderson model.[5] Munk and Anderson proposed that we model

$$K_z = \frac{K_z^0}{f(\text{Ri})} \tag{9.54}$$

where K_z^0 is the eddy viscosity at very low Richardson number and $f(\text{Ri})$ is the stability function

$$f(\text{Ri}) = (1 + \sigma \text{Ri})^p \tag{9.55}$$

where the value of the exponent p and coefficient σ are dependent on the quantity being mixed, i.e., they are different for momentum (viscosity) and heat. Munk and Anderson used $p = \frac{1}{2}$ for momentum and $p = 3/2$ for heat. Other authors have suggested other values of p. The specific form used by Munk and Anderson (Figure 9.1) is

$$f(\text{Ri}) = (1 + 3.33\text{Ri})^{1.5}. \tag{9.56}$$

Clearly the value of $1/\sigma$ determines the value of the Richardson number at which the stability function starts to increase substantially above unity. For the Munk–Anderson value of $\sigma = 3.33$, this is around $\text{Ri} = 0.3$, which we recall is the value of Richardson number that corresponds to the damping out of Kelvin–Helmholtz instabilities, i.e., to the demise of mixing.

9.2.4 Munk–Anderson model

If we take a specific form for the velocity shear in

$$\text{Ri} = \alpha g \frac{\partial T}{\partial z} \Big/ \left(\frac{\partial u}{\partial z}\right)^2 \tag{9.57}$$

and write

$$\text{Ri} = \frac{1}{s(z)} \frac{\partial T}{\partial z} \tag{9.58}$$

where

$$s \equiv \left(\frac{\partial u}{\partial z}\right)^2 \Big/ \alpha g \tag{9.59}$$

---

[5] After a seminal paper by Walter Munk (b. 1917), a major contributor to the field of physical oceanography and geophysics, and his student Ernest R. Anderson. See Munk and Anderson (1948).

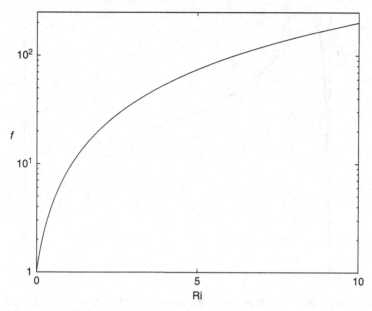

Figure 9.1 Plot of the stability function f for heat conductance based on the form used by Munk and Anderson (1948).

we can solve the vertical heat conduction equation

$$\frac{\partial T}{\partial t} = -\frac{\partial q}{\partial z} \tag{9.60}$$

$$q(z) = -K_z(\mathrm{Ri})\frac{\partial T}{\partial z}, \tag{9.61}$$

or using (9.58)

$$q(z) = -s(z)K_z(\mathrm{Ri})\mathrm{Ri} \tag{9.62}$$

as uncoupled from the momentum equation. Equations (9.58) to (9.61) give

$$\frac{\partial T}{\partial t} = \frac{\partial}{\partial z}s(z)G(\mathrm{Ri}) \tag{9.63}$$

where we define the function $G(Ri)$ by

$$G(\mathrm{Ri}) \equiv K_z(\mathrm{Ri})\mathrm{Ri}, \tag{9.64}$$

i.e.,

$$\frac{-q(z)}{s(z)} = G(\mathrm{Ri}). \tag{9.65}$$

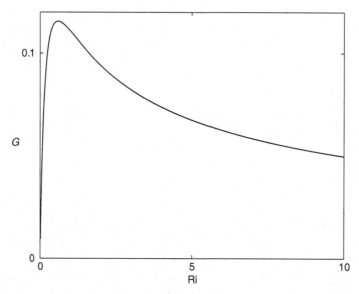

Figure 9.2 The heat flux G as a function of Richardson number, from the form used by Munk and Anderson (1948).

Figure 9.2 shows the function G plotted against Ri.

The critical aspect of this form of G is that it passes through a maximum at $\text{Ri} = \text{Ri}_{\max}$. We can show that the solutions corresponding to $\text{Ri} > \text{Ri}_{\max}$ are unstable in that a small change in Ri will create a runaway perturbation in T. If we have a steady state solution for $T(z)$ and then perturb T to

$$T \to T + \Delta T \tag{9.66}$$

and write

$$\Delta T \sim e^{i(\omega t - kz)} \tag{9.67}$$

it follows from (9.58)

$$\Delta\text{Ri} \sim -ike^{i(\omega t - kz)} \tag{9.68}$$

where we are assuming that the s varies more slowly with z than does ΔRi. Rewriting (9.63) to include the perturbation in Ri,

$$\frac{\partial T}{\partial t} = \frac{\partial}{\partial z} s(z)G(\text{Ri} + \Delta\text{Ri}) \tag{9.69}$$

and ignoring the spatial variation of s,

$$i\omega \sim -k^2 s(z)G'(\text{Ri}) \tag{9.70}$$

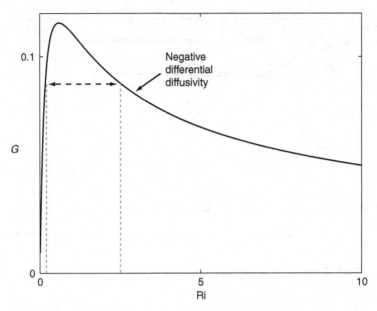

Figure 9.3 Schematic of a sudden jump in the Richardson number from the stable to unstable part of the G curve (Figure 9.2) and immediately back again. Two such jumps are shown in Figure 9.4. The jumps allow essentially discontinuous changes in temperature and have a sharpness which grows with time due to the negative differential diffusivity associated with the negative slope region.

where the prime indicates differentiation with respect to the argument of the function. Equation (9.70) shows that if $G' < 0$,

$$\omega \sim -iak^2 \tag{9.71}$$

where $a > 0$, if $\mathrm{Ri} > \mathrm{Ri}_{max}$ and so that the perturb must grow with time. The larger values of k grow most quickly and so the perturbation becomes increasingly sharper. We contrast this situation with that for positive G' when the perturbation decays with time, and the components with higher k decay most rapidly so that the perturbation spreads spatially as one would expect from a diffusive process. The slope G' is called the *differential diffusivity* and the case of $G' < 1$ corresponds to *negative differential diffusivity*. This situation is found in many branches of non-linear physics, i.e., problems in which rate processes for some quantity (heat in the present case) are controlled by coefficients, which are themselves dependent on the magnitude of the quantity (or its spatial gradient in our case). In most of these cases, modelers have to deal with an instability that leads to a distinctive physical phenomenon, and in our case this is the development of a thermocline.

The unstable nature of the negative slope branch of the G curve means that solution of the vertical temperature profile must lie on that part of the function with positive slope. It is possible for a single point in the profile to lie on the negative slope region as illustrated in Figure 9.3.

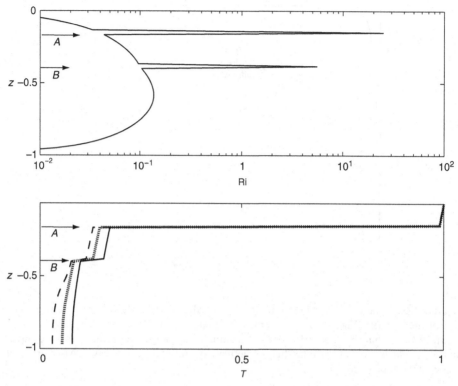

Figure 9.4 The lower panel shows the temperature profile in a water column with normalized temperature fixed at $T = 1$ at the surface $z = 0$. All physical variables are dimensionless. The water column shows thermoclines at depths A and B. The upper panel shows the corresponding profile of the Richardson number. The full lines are profiles at dimensionless time $t = 10$. In the lower panel the broken curve shows the profile at $t = 4.5$ and the dotted curve at $t = 7$.

In that figure the horizontal line shows a jump at a particular value of z to the negative slope region, followed by an immediate return to the positive slope portion of the curve. The other arrows on the curve indicates that as z varies from the top of the water column, the value of Ri moves from zero to some point at which it jumps to a higher value and then back again with Ri decreasing back to zero. During the jump, the value of q remains unchanged to ensure its continuity, and hence by (9.60) the time dependence of T is finite. The usual situation in a water column is that dT/dz is positive; the larger its value, the more rapid is the change of T with z, and as dT/dz tends to infinity the profile becomes a line parallel to the T axis. If we look at the profile in Figure 9.4 we see two examples (A and B) of such a jump. Physically this is a thermocline: with sudden change in T from a high surface value to lower value for the bottom water. The question is: why does a model based on a G curve that exhibits a maximum show such jumps, i.e., thermoclines? The jump clearly allows the temperature to drop suddenly to a lower value. It therefore produces a discontinuity in temperature, so that the temperature of the upper water column can increase relative to that of the bottom of the water column. This means that the temperature jump across the thermocline can change with time. This clearly occurs for both thermoclines in Figure 9.4. Let us suppose that we start from a condition in the water column in which the heat flux q is everywhere zero, and then impose a heat condition at the surface. That condition may be either a downward heat flux or an imposed value of surface temperature. Both cases are viable simulations of conditions in a coastal basin. In either case, q starts to increase with time t and we can see from (9.62) that the Richardson number Ri must then increase away from its initial value of zero, i.e., away from the origin in Figure 9.2 or Figure 9.3. We can see from (9.62) that the value of Ri at any point depends not only on q, but also on the value of s for the particular z. As s decreases as z moves away from the surface, so Ri must also increase (for given q). The function s(z) will start to increase as we approach the bottom of the water column, and so Ri will then start to decrease back towards the origin of the G curve. However, that simple continuous behavior of Ri through the water column requires that the maximum value of Ri does not reach Ri_{max}, i.e., the top of the G curve, which means that it is only possible for sufficiently small value of flux q. Once q exceeds some critical value, Ri reaches Ri_{max} provided the Ri is a continuous function of z. Further increase of Ri means that $-q$ must decrease as z moves away from the surface. The profile is unstable for $\text{Ri} > \text{Ri}_{max}$, and has negative diffusion, so that eventually that part of the profile in the negative slope region contracts to a single point. Continuity of q means that the residual point must be connected to the region of positive differential diffusivity by a jump parallel to the Ri axis as shown in Figure 9.3 The effect of the jump is a discontinuity in T. It is possible for a number of such jumps to occur as seen in Figure 9.4. The Matlab code that has led to Figure 9.4 is shown in Table 9.1. This uses a constant temperature boundary condition at the surface of $T = 1$ at $z = 0$. The model has been run for a dimensionless time of $t = 10$. Figure 9.5 shows the profiles of T, Ri, and q just after the boundary condition $T = 1$ is applied at $z = 0$. Notice the distinct points marked N and M, which correspond to the beginning and end of the excursion of the water column into the negative slope part of the G curve. As time progresses the region NM will contract to a sharp transition at a single value of z.

Notice that at high Ri, the vertical transfer of momentum, or internal friction, becomes low due to the momentum stability function. This allows layers to slip over one another easily since stress, which a vertical flux of momentum, has reduced downward propagation through the thermocline. However, it is important to note that the momentum stability function is weaker than that for heat conduction. Importantly, the G function for momentum does not exhibit a maximum.

9.3 Wind-driven currents in stratified basins

In a stratified basin the downward penetration of the wind stress is limited by the pycnocline. The pressure gradient due to the surface elevation is effective at all depths,

Table 9.1. *Matlab code for the Munk–Anderson model*

```
N = 200; r = (1:N); dz = 1/N; KzT(r) = 1; dt = 0.1* dz^2;
T(r) = 0; time = 0; Cp = 1; g = 9.81; z = -1 + dz*r;
s = sqrt(1/(z*(1 - z)));
while time < 10;
        time = time + dt;
        for n = 2:N;
            Tgrad(n) = (T(n) - T(n - 1))/dz;
            q(n) = - Cp*KzT(n)*Tgrad(n);
        end;
        Tgrad(1) = 0;
        q(1) = 0;
        for n = 1:N;
            T(n) = T(n) - (q(n + 1) - q(n))*dt/dz;
        end;
        T(N) = 1;
        for n = 1:N;
            Ri(n) = 1;
            Ri(n) = 1e3*g*alpha*abs(Tgrad(n)/s(n)^2);
            KzT(n) = Kz/(1 + 3.33*Ri(n))^(3/2);
        end;
end
```

but can be changed and nullified below the pycnocline by a slope of the pycnocline itself. The pressure gradient produced by the combined slopes of the surface and the pycnocline is easily determined by writing the hydrostatic pressure at depth, at a point z in the water column below the pycnocline which is at height z_p

$$p(z) = g\rho_1(\eta - z) + g(\rho_2 - \rho_1)(z_p - z) \quad z < z_p \tag{9.72}$$

where ρ_1, ρ_2 are the densities of the upper and lower layers. So that differentiating (9.72) with respect to distance x along the basin,

$$\frac{\partial p}{\partial x} = g\rho_1 \frac{\partial \eta}{\partial x} + g(\rho_2 - \rho_1) \frac{\partial z_p}{\partial x} \quad z < z_p. \tag{9.73}$$

If we assume that there is no wind induced motion in the lower layer, i.e., that the wind stress cannot penetrate the pycnocline, the pressure gradient will also be zero in that layer. Equation (9.73) shows that the pressure gradient in the lower layer becomes zero if

$$\frac{\partial z_p}{\partial x} = -\frac{\rho_1}{\rho_2 - \rho_1} \frac{\partial \eta}{\partial x} \tag{9.74}$$

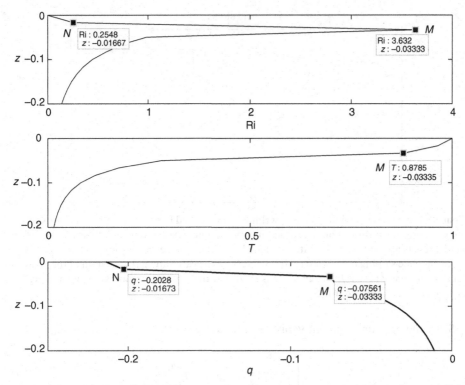

Figure 9.5 The profiles of Richardson number Ri, temperature T, and heat flux, q (all in dimensionless form) for the Munk–Anderson model (Munk and Anderson 1948) shown in Figure 9.4 at a very early stage of thermocline formation (dimensionless time $t = 0.15$). The point marked N shows the water column entering the negative (unstable) slope region of the G curve which requires a reduction in $-q$. This continues up to point M which marks the genesis of the thermocline.

in which the pycnocline slopes downward in the downwind direction; as we move downwind the pressure at a fixed height in the lower layer is increased due to the greater thickness of the upper layer, but reduced (by the same amount) as a result of the lesser thickness of the lower layer. Inside the upper layer there will be a circulation cell; downwind at the surface and upwind just above the pycnocline. This is illustrated in Figure 9.6.

If the pycnocline slope in (9.74) becomes comparable with the depth of the basin per unit length of the basin, it is possible for the pycnocline to break through the surface at the downwind end of the basin. This means that the lower layer occupies the whole of the water column at that end of the basin. The line, normal to the wind direction, along which the pycnocline meets the surface, is a density front. In a rotating basin, there can be geostrophic currents that balance the slope.

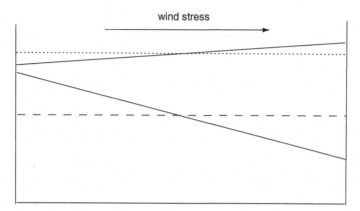

Figure 9.6 Side view of wind-driven upwelling in a stratified basin. The dotted and broken lines show the original positions of the surface and pycnocline (respectively) in a closed basin. After a wind stress is applied a steady state is reached with the surface sloping upward downwind and the pycnocline sloping downwind (two full lines). In the case shown the strength of the wind and density difference across the pycnocline is such that the pycnocline almost breaks through the surface to form a surface front.

Since the pressure gradient vanishes in the lower layer,

$$\int_{-h}^{\eta} \frac{\partial p}{\partial x} dz = g\rho_1 \left(\eta - z_p\right) \frac{\partial \eta}{\partial x} \tag{9.75}$$

and writing the wind stress as $\rho_1 u_*^2$

$$\frac{\partial \eta}{\partial x} = \frac{u_*^2}{gh_1} \tag{9.76}$$

and from (9.74)

$$\frac{\partial z_p}{\partial x} = -\frac{u_*^2}{g'h_1} \tag{9.77}$$

which gives a change in height of the pycnocline over the length L of the basin

$$\Delta z_p = \frac{u_*^2 L}{g'h_1}. \tag{9.78}$$

Limnologists call the quantity $h_1/\Delta z_p$ the *Wedderburn number* after E. R. Wedderburn, who made early measurements of the stratification and seiching of Loch Ness[6] in Scotland in the early twentieth century. The Wedderburn number is denoted by We:

[6] Loch Ness is a steep-sided fresh water lake of great depth, with preferred longitudinal wind stress of the type discussed in Section 5.4 and a large ratio of length to width. The hydrodynamics of Loch Ness was intensely studied by Stephen A. Thorpe in the late twentieth century and a very readable account is given in his book (Thorpe, 2005).

$$\text{We} = \frac{g' h_1^2}{u_*^2 L}. \tag{9.79}$$

Upwelling is very likely if $\text{We} < 1$. The Wedderburn number also measures the strength of stratification.

The Wedderburn number is closely related to the Richardson number Ri defined in (9.50) so that

$$\text{We} = \frac{h}{L} \text{Ri}. \tag{9.80}$$

Stratification is highly stable when the $\text{Ri} > \frac{1}{4}$ which means that we require $\text{We} > h/4L$.

9.4 Classification based on vertical stratification

Coastal basins tend to have variations in salinity in both the horizontal and vertical.[7] These are due to a variety of causes including river flow, evaporation, and variations in the salinity of the neighboring oceans. There are also temperature variations in the vertical and horizontal due to surface heating and temperature variations in the surrounding ocean. The extent of these variations depends on many factors including the magnitude of river flow, tidal and wind regimes, and bathymetry or topography of the basin. In terms of processes within the basin, a very important factor is stratification of the water column. This forms the basis of a classification system widely used by modelers. Figure 9.7 shows a schematic of the four classes of coastal basins based on stratification. Most coastal basins tend to contain sub-basins that exhibit different types of stratification regime, and so we prefer to think of the stratification classification as more appropriate to sub-basins. Furthermore, the stratification tends to be seasonal in most coastal basins and so a particular basin will change classes with season. This is simply a consequence of the rapid response of coastal basins to meteorological conditions and rainfall.

9.4.1 Vertically mixed basins

Basins that are classified as *vertically mixed* have minimal variation in density from top to bottom, or more precisely a variation that is judged to be *small* for the purposes of a particular study or model; see panel (a) of Figure 9.7. The vertically mixed basin is the focus of the simplest hydrodynamic models of the early chapters of this book and is realized in basins with limited input of buoyancy relative to vertical mixing. Coastal basins are usually only judged to be vertically mixed over short periods. Basins that apparently vertically mixed may properly fall into the next category which we consider here, i.e., slightly stratified or partially mixed. Basins that are apparently vertically mixed usually have horizontal variations in salinity and temperature, and as we shall

---

[7] A good reference to many of the processes in this section is Officer (1976).

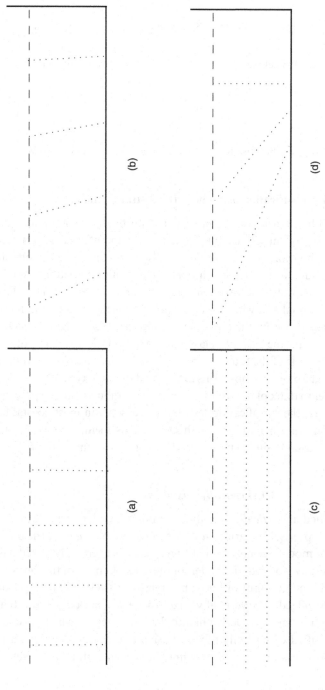

Figure 9.7 The dotted lines are isopycnals inside a coastal basin, or sub-basin, with the surface shown by the broken horizontal line and the thicker lines representing the bottom and closed head of the basin. The mouth of basin is to the left of each panel. Panel (a) illustrates the hypothetical vertical isopycnals inside a vertically mixed basin. Panel (b) shows a partially mixed basin with the isopycnals sloping upwards towards the mouth (classical basin). Panel (c) is a stratified basin with the hypothetical isopycnals horizontal and panel (d) shows a density front in the form of a salt wedge entering from the ocean.

see, these necessarily produce small changes in the vertical. A typical basin will be fresher towards river inlets and more saline close to the mouth. This horizontal density gradient usually produces a convective cell, which induces a small (*second order* in modeling parlance) difference in density between top and bottom.

9.4.2 Slightly stratified or partially mixed basins

In these coastal basins the contours of constant density (*isopycnals*) are not vertical, but slope away from the vertical, so that in moving downwards from the surface we cross isopycnals of higher density; see panel (b) of Figure 9.7. The word *stratification* is not strictly appropriate to such basins because there are no well-defined *strata* or *layers*. However, the word has come to mean loosely any type of vertical density gradient. The difference between this class and the previous *vertically mixed* class is a change in the relative strength of the stratifying influence, i.e., buoyancy input, compared with vertical mixing. Basins is this class would usually be supposed to lie in the range of small Richardson number, i.e., Ri < 0.25 so that vertical mixing coefficients are not significantly reduced by density stratification and the slight stratification is not being stabilized by a reduction in vertical mixing. The slope of the isopycnals is usually due to salinity variations, and salinity is taken as a tracer of water movement with the isopycnals sloping away from the freshwater source at the surface and away from the mouth at the bottom. Apart from the change of class with season many basins change their mixing status with the astronomical tides. In particular many partially mixed basins exhibit a phenomenon known as *tidal straining*[8] which is due to velocity shear in tidal currents so that top and bottom water has different tidal excursions. For example if the surface tidal current is 0.2 m s^{-1} and the bottom tidal current is 0.1 m s^{-1}, during a semidiurnal ebb tide, the surface water will be advected from a point 1.08 km further up the basin than the bottom water. If the basin is vertically mixed at slack tide, the ebb tide, in this example will produce stratification with a vertical density difference equivalent to the original horizontal difference over a distance of 1 km. Vertical mixing will tend to remove this difference or (if straining has put denser water at the surface) convective overturn will be active. This process is considered in more detail in Chapter 10. When mixing is incomplete, the effects of tidal straining can be evident as stratification of the water column which varies through the tidal cycle.

9.4.3 Highly stratified basin

In these basins, there are clearly defined layers, or strata of water of different densities; see panel (c) of Figure 9.7. Richardson numbers are high enough to greatly reduce vertical mixing and so the stratification is stable. In coastal basins, buoyancy forcing

---

[8] See Simpson, Brown, Matthews, and Allen (1990).

can change very rapidly and so the stratification may only exist over limited periods. Vertical mixing can also change with weather conditions and through the spring–neap tidal cycle and this may make the stratified state intermittent. We have already seen a situation in this chapter with thermal fronts in Section 9.1 where stratification changes seasonally due to solar heating (plus winter storms) and varies spatially due to the spatial variation of tidal mixing.

9.4.4 Basins with density fronts

Many basins show fronts between vertically mixed and stratified waters and a particular example are the so-called *intrusion fronts* of ocean water which we find in many shallow basins. These fronts are usually formed by tidal advection. In some cases the fronts are not vertical and the interface between ocean and basin water slopes downward from the surface (at the ocean side of the front) to the bottom (at the basin side of the front). This is often called a *salt wedge*. It is a familiar phenomenon at the entrance to large rivers and is usually assumed to represent a balance between river advection and vertical mixing; as salt is mixed into the upper water column from underlying ocean water, it is conveyed back to the ocean by the advection in the upper water column associated with the river flow. The interfacial friction between the two layers of water is also important so that the outflowing river water balances the pressure gradient forcing salt water inward and this *arrests* the salt wedge.

9.5 Further reading

See further reading from earlier chapters. Additional texts for this chapter are:

Gill, A. E. (1982). *Atmosphere–Ocean Dynamics*. New York: Academic Press.

Curry, J. A. and P. J. Webster (1999). *Thermodynamics of Atmospheres and Oceans.*
New York: Academic Press.

Drazin, P. G. and W. H. Reid (2004). *Hydrodynamic Stability*. Cambridge, UK:
Cambridge University Press.

Fedorov, K. N. (1986). *The Physical Nature and Structure of Oceanic Fronts.*
New York: Springer.

Nash, J. M. (2002). *El Niño: Unlocking the Secrets of the Master Weather-Maker.*
New York: Warner Books.

Simpson, J. E. (1997). *Gravity Currents in the Environment and the Laboratory.*
Cambridge, UK: Cambridge University Press.

10

Dynamics of partially mixed basins

10.1 Transport of heat and salt

This section looks at the dominant process of salt and heat transport in partially mixed basins, which is an application of the mechanism proposed by G. I. Taylor which we mentioned in Chapter 7. Taylor pointed out that vertical mixing, in the presence of vertical shear, and a horizontal concentration gradient, must lead to horizontal diffusion simply because particles mixed in the vertical will experience a range of horizontal advective velocities. The vertical shear in velocity in most partially mixed basins is due to tidal and wind driven flow. In the next section, we shall determine the basic model of Taylor dispersion.[1]

10.2 Taylor shear dispersion

One of the most important causes of horizontal dispersion in coastal basins is vertical mixing through vertically sheared horizontal currents. This mechanism is called *Taylor dispersion*, after G. I. Taylor who first analyzed the process which is illustrated in Figure 10.1.

In the presence of strong vertical mixing, and horizontal advection, the coastal basin is, to a first approximation, in a steady state balance between advection and vertical mixing of a tracer, with concentration c, given by

$$K\frac{\partial^2 c'}{\partial z^2} = u(z)\frac{\partial \bar{c}}{\partial x} \tag{10.1}$$

where $\bar{c}$ is the depth-averaged concentration of a tracer and c' is the departure of c from $\bar{c}$ at any point in the water column, i.e.,

$$c = \bar{c} + c'. \tag{10.2}$$

---

[1] Our discussion is adapted from Taylor (1953).

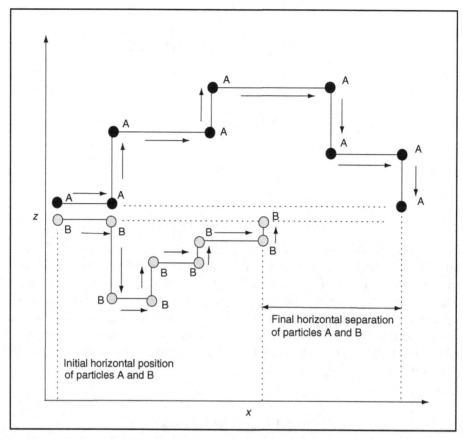

Figure 10.1 Illustration of Taylor shear dispersion for two particles labeled A and B in the diagram. Particle A is stochastically mixed into the upper water column and experiences greater advection in the x direction than does particle B, which mixes into the lower water column where the currents are lower. Particle A subsequently mixes back into the lower water column and the figure shows that it has moved further in the x direction than particle B.

We should note that this steady state may be involve a time average over periods of many days to smooth out variations in tidal currents and wind forcing. The right-hand side of (10.1) is the advection of tracer into an element of the water column at x, z (it is really a first order approximation since it ignores the small departure of c from $\bar{c}$). The left-hand side of (10.1) is the gradient of the vertical (upward) diffusive flux of c, which is given by $-K\partial c/\partial z$ and the fully time-dependent form of (10.1) is

$$\frac{\partial c}{\partial t} = u\frac{\partial \bar{c}}{\partial x} - K_z\frac{\partial^2 c'}{\partial z^2} \tag{10.3}$$

Equation (10.1) states that the variation of c in the vertical is the steady state response to a source term that arises from advection. Note that this source term is a prescribed function of z since we know the function $u(z)$; the unknown quality is c'.

Our aim is to determine the diffusive flux of tracer in the horizontal direction, i.e., $\int c'u\,dz$ (note that $\int \bar{c}u\,dz$ is the advective flux). The proportionality of the left-hand side of (10.1) to the horizontal gradient of the depth-averaged concentration of c means that the diffusive flux of tracer will also be proportion to that gradient, and so can define a coefficient of diffusivity for the tracer. Equation (10.1) is a very fundamental relation in coastal basins that are nearly vertically mixed, and there are many examples of its application all of which lead to *effective* longitudinal diffusivities. The use of such effective diffusivities in two-dimensional models is one of the most powerful methods of treating basins that are almost vertically mixed. The reason is that three-dimensional numerical models may lack the vertical resolution to distinguish between advective processes at neighboring levels.

Equation (10.1) is easily integrated. As a simple example, we will assume that u is a linear function of z, i.e., the velocity shear $\partial u/\partial z$ is constant. Any other dependence of u on z is equally easily treated. There is no loss of generality in assigning the mean current as zero. Let us take a water column of height h, and let there be a velocity difference of Δu between the top and bottom of the water column. For analytical convenience, let us take the height datum ($z = 0$) not at the top of the water column (as is our usual practice) but at the center of the water column, i.e., $-h/2 < z < h/2$ and write $u(z)$ as

$$u = \Delta u \frac{z}{h} \tag{10.4}$$

so that u varies between $-\Delta u/2$ and $\Delta u/2$. The simplest non-dimensional height z' is defined as

$$z' \equiv z/h \tag{10.5}$$

and so (10.1), with (10.4), becomes

$$\Gamma z' = \frac{\partial^2 c'}{\partial z'^2} \tag{10.6}$$

where

$$\Gamma \equiv \frac{h^2 \Delta u}{K_z} \frac{\partial \bar{c}}{\partial x}. \tag{10.7}$$

The solution of (10.6) is

$$c' = \Gamma z'^3/6 + Az' + B \tag{10.8}$$

where A and B are integration constants. We need to ensure that the average of c' is zero, i.e.,

$$\int_{-h'/2}^{h'/2} c'\,dz' = 0 \tag{10.9}$$

and since the first and second terms in (10.8) are odd functions of z', it follows that B must vanish. The vertical flux of c must be zero at both the bottom and the surface of the water column, so the derivative with respect to z must be zero at $z = h/2$, and $z = -h/2$ (since there can be no diffusive flux through either of those surfaces) and this condition fixes the value of A, and hence

$$c' = \frac{\Gamma}{6} z' \left(z'^2 - \frac{3}{4} \right).$$

(10.10)

Finally we obtain the flux of c as

$$\int_{-1/2}^{1/2} u c' \, dz' = u_0 \frac{\Gamma}{6} \int_{-1/2}^{1/2} \left(z'^2 - \frac{3}{4} \right) z'^2 \, dz' = -\frac{h^2 \Delta u^2}{120 K_z} \frac{\partial \bar{c}}{\partial x}$$

(10.11)

and so we have the Taylor shear diffusion coefficient as

$$K_x = \frac{h^2 \Delta u^2}{120 K_z}$$

(10.12)

where the factor of 120 is specific to our choice of the function $u(z)$, i.e., the specification that we have used for the velocity profile within the water column. Other specifications would alter this number but only by a multiplicative factor of order one so that (10.12) is always a good first approximation in determining the order of magnitude of Taylor dispersion.

If K_z is large, in a partially mixed basin, c' is small, i.e., the basin becomes nearly vertically mixed so that c is close to $\bar{c}$ everywhere, and Taylor dispersion becomes very small. For example, consider a coastal basin 10 m deep with tidal currents having a difference $\Delta u \sim 0.1 \, \text{m s}^{-1}$ between top and bottom, and $K_z = 10^{-3} \, \text{m}^2 \, \text{s}^{-1}$, (10.12) gives K_x as $8 \, \text{m}^2 \, \text{s}^{-1}$. It is not necessary for the mixing to occur throughout the water column; if there is considerable vertical shear, the dispersion on the surface may be affected by mixing through a thin surface layer.

Many students, on first meeting Taylor dispersion, consider that the inverse proportionality of the dispersion coefficient on the coefficient of vertical mixing is in some sense counter-intuitive. Their reasoning is that the vertical mixing is essentially a dispersive processes arising from stochastic fluctuations and so should aid, and not inhibit, horizontal dispersion. Vertical mixing is due to turbulence and so turbulence is reducing Taylor dispersion. The way that this occurs is through its forcing of the basin to remain vertically mixed and hence negating the effects of velocity shear. This does not remove the effects of advection, i.e., the tracer continues to be advected and this provides the transport of the tracer which is the fundamental process that enters the tracer dynamics. As we stressed in Chapter 2, the distinction between advection and diffusion is dependent on spatial scale and is not in itself a fundamental property of the fluid.

The velocity shear used in (10.12) can come from either wind-driven or tidal currents. There is a direct link between Taylor dispersion and variation of the tracer concentration in the vertical. One of the obvious candidates for producing Taylor dispersion is *tidal straining* which is a process by which vertical shear in tidal current creates variations in tracer concentration in the vertical.[2] For example if the surface tidal current is 0.2 m s^{-1} and the bottom tidal current is 0.1 m s^{-1}, during a semidiurnal ebb tide, the surface water will be advected from a point 1.08 km further up the basin than the bottom water. If the basin is vertically mixed at slack tide, the ebb tide, in this example will produce stratification with a vertical density difference equivalent to the original horizontal difference over a distance of 1 km. Vertical mixing will tend to remove this difference or (if straining has put denser water at the surface) convective overturn will be active. If the tracer is salt the effects of tidal straining will include density forcing of the circulation, i.e., the density affects associated with the vertical profile of salinity.

Taylor originally formulated (10.12) to describe dispersion along the axis of a channel (*longitudinal dispersion*) due to lateral velocity shear (across the width of the river). We can use the equation in the same way for a basin that is long and narrow, i.e., has the shape of a channel. If we orientate the x axis along the channel and the y axis laterally across the channel, then (10.12) gives for the Taylor shear diffusion coefficient

$$K_x = \frac{b^2 \delta u^2}{120 K_y} \tag{10.13}$$

where b is the half width of the channel, i.e., the distance from the center line to one of the sides, and δu is the difference in current between the center and sides of the channel. In this case, we rely on lateral dispersion and lateral shear to enhance longitudinal dispersion. For example, if $K_y = 50 \text{ m}^2 \text{ s}^{-1}$, $\delta u = 0.05 \text{ m s}^{-1}$, $b = 5 \text{ km}$, (10.13) gives K_x as $10 \text{ m}^2 \text{ s}^{-1}$.

It is possible to use these derived dispersion coefficients to determine flushing times of coastal basins and we pursue this topic in more detail later in this chapter. For the moment we note that in time t a tracer will spread a distance $2\sqrt{K_x t}$ so that we would *estimate* the time t for an acceptably near uniform concentration of a tracer to occur over a basin of L as $L = 2\sqrt{K_x t}$. As an example, if $L = 20 \text{ km}$, and $K = 10 \text{ m}^2 \text{ s}^{-1}$ we find $t = 115$ days. This a typical flushing time for coastal basins which are almost vertically mixed and have a tidal excursion (the distance through which the tide advects in half a tidal cycle) that is small relative to the length of the basin.

10.3 Convection

A longitudinal variation in density within a coastal basin which is initially very nearly vertically mixed naturally tends to produce gravitational circulation, i.e., circulation

---

[2] See Simpson, Brown, Matthews, and Allen (1990).

due to variations of density, and this circulation results in some degree of stratification. If we were to turn off all vertical mixing (a modeler's prerogative) the basin would become totally stratified. In the real world, the stratification depends on the strength of vertical mixing and in the basins under consideration here always remains small. Stratification lowers the potential energy of the basin and so the system is tending towards a state of greater stability with denser water at the bottom and less dense at the surface. The flow itself is a consequence of a horizontal pressure gradient which is response to the greater density of water at one end of the basin. This tends to produce a *convection cell* with denser water moving in the lower water column (from the region of higher density towards the region of lower density), and lighter water moving in the upper part of the water column towards the region of higher density. This convection cell is analogous to the motion of air in a room which has a heated wall on one side of the room and a cooled wall on the other side: cold air moves along the floor and warm air across the ceiling. Convection cells of this type are common through the world's oceans and are responsible for some of the major currents that transport heat and salt.

This type of stratification is often observed in basins that experience diurnal mixing due, for example, to a sea breeze. If the basin is shallow, and has experienced recent river flow, it will be almost vertically mixed during strong mixing from a sea breeze. However, the basin will retain a longitudinal density gradient with lowered salinity near the point of entry of the river, and almost marine salinity at the mouth of the basin. After the sea breeze dies off in the evening, convection produces a stratified state which intensifies during the night and is then dismantled by vertical mixing from a sea breeze the next day. An very good example is the Harvey Estuary in south-west Australia prior to dredging work in the 1990s.[3] The nighttime convection produces a transport of salt from the mouth up the basin. Although there is no net transport of volume, i.e., the depth-averaged current is zero, there is a transport of salt since the salinity varies through the water column. Notice the way in which vertical mixing impedes horizontal transport; despite the horizontal gradient of density there will be very little transport while vertical mixing is strong. The reason for this result is, of course, that vertical mixing reduces velocity shear. This example also illustrates the general phenomenon in coastal basins that greater transport tends to occur under stratified conditions. The extent of stratification is a balance between convection and vertical mixing. Horizontal transport due to convection will be proportional to the product of the vertical shear in current and also be proportional to the vertical gradient of density, because vertical shear in a water column without a vertical density gradient will clearly not produce density transport. The vertical density gradient and velocity shear are both proportional to the powers of the longitudinal density gradient, so that we would guess that the convective density flux has the form

---

[3] Hearn and Robson (2001).

$$Q \propto \left(\frac{1}{K_z}\frac{\partial \rho}{\partial x}\right)^n \tag{10.14}$$

and we shall find that $n = 3$.

The longitudinal density gradient forces the current by producing a longitudinal pressure gradient. The net total force on an element of the water column lying between z and $z + dz$ contains two terms. The first is due to the longitudinal pressure gradient $-\partial p/\partial x$, and the other is due to the viscous drag,

$$F = \frac{\partial}{\partial z}\rho K_z \frac{\partial u}{\partial z}, \tag{10.15}$$

and so the dynamical equation is

$$\rho\frac{\partial u}{\partial t} = -\frac{\partial p}{\partial x} + \frac{\partial}{\partial z}\rho K_z \frac{\partial u}{\partial z} \tag{10.16}$$

where K_z is the coefficient of vertical diffusivity. Adopting the hydrostatic approximation

$$p = g\rho(\eta - z) \tag{10.17}$$

where the density ρ is assumed independent of z to first order. Hence

$$\frac{\partial p}{\partial x} = g\rho\frac{\partial \eta}{\partial x} + g(\eta - z)\frac{\partial \rho}{\partial x} \tag{10.18}$$

Notice the difference between (10.18) and (5.1.3) for the pressure gradient in a basin for which density is independent of x. In such a basin only the first term on the right of (10.18) is present. The longitudinal gradient of density is responsible for the second term. There is an essential difference in the form of the two terms. The first term is independent of z, i.e., it is the same at all points in the water column. By contrast, the second term increases in magnitude as z moves down through the water column, i.e., $-z$ increases. It is this z dependence of the second term in the pressure gradient that produces convection.

The convective motion in the basin in the basin is concerned with the variation of the horizontal current with height z, or in other words with the vertical shear in the current u. We can therefore remove the depth-averaged current from our model. The actual value of the depth-averaged current depends on whether the basin is closed or open and on all forcings that exist in the basin (in addition to the longitudinal density gradient). For simplicity we shall assume that the depth-averaged current is zero. In the same sense we take the depth average of the gradient of pressure to be zero.[4] Using $p = g\rho(\eta - z)$ we obtain, assuming $\eta \ll h$,

---

[4] There is no loss of generality here and readers can replace the present analysis using u with a quantity u' representing the departure of the current from the depth-averaged value.

$$g\bar{\rho}\frac{\partial\eta}{\partial x} = -g\frac{h}{2}\frac{\partial\bar{\rho}}{\partial x} \tag{10.19}$$

which is simply the set-up of the water surface against the density gradient. If one runs a numerical model of a basin with longitudinal density gradient, the surface will be seen to set up in this way. Equation (10.19) represents the actual surface slope that we would obtain if we solved this problem for a closed basin since the depth-averaged current is zero in that.

The coefficient of vertical mixing K_z is initially assumed independent of z (variable K_z is treated later) and K_z is also assumed independent of any vertical density gradient, since the Richardson number is assumed to be small. Let us define the characteristic velocity $\tilde{u}$:

$$\tilde{u} \equiv g\frac{(h/2)^3}{\rho K_z}\frac{\partial\rho}{\partial x} \tag{10.20}$$

and continuing to assume that η is small, we define the relative height coordinate z' as

$$z' \equiv 1 + 2\frac{z}{h} \tag{10.21}$$

so that z' has the range -1 (the bottom) to 1 (surface). We impose the condition that the shear stress is zero at the surface (no wind stress), i.e.,

$$\left.\frac{\partial u}{\partial z'}\right|_{z'=1} = 0 \tag{10.22}$$

and so (10.15) to (10.22) give

$$\frac{1}{\tilde{u}}\frac{\partial u}{\partial z'} = -\frac{1}{2}(z'^2 - 1). \tag{10.23}$$

Since the depth-averaged value of u' is zero, i.e.,

$$\int_{-h}^{\eta} u(z)\mathrm{d}z = 0, \tag{10.24}$$

it follows that if we integrate (10.23) with respect to z,

$$u = -\frac{\tilde{u}z'}{2}\left(\frac{z'^2}{3} - 1\right). \tag{10.25}$$

Note that (10.25) and (10.20) show that, as predicted, the current u is inversely proportional to K_z.

If we use an eddy viscosity K_z which varies with z, the form of (10.25) would be changed to

$$u = -\frac{g(h/2)^3}{2\rho}\frac{\partial\rho}{\partial x}\int\frac{z'^2-1}{K_z}dz' \tag{10.26}$$

and is readily evaluated for any assumed from of K_z. For example, using (10.21) in the linear form for a log layer which we found in Chapter 7 for a rough bottom, i.e., $K_z = \lambda u^* h(z'+1)/2$,

$$u = -\frac{gh^2}{8\lambda\rho u^*}\frac{\partial\rho}{\partial x}\int(z'-1)dz' \tag{10.27}$$

where u^* is the stress velocity at the bottom and so, applying (10.24),

$$u = -\tilde{u}\left(z'^2 - 2z' -\frac{1}{3}\right) \tag{10.28}$$

where we have redefined $\tilde{u}$ as

$$\tilde{u} \equiv g\frac{(h/2)^3}{\rho\tilde{K}_z}\frac{\partial\rho}{\partial x} \tag{10.29}$$

in which

$$\tilde{K}_z \equiv 2\lambda u^* h. \tag{10.30}$$

Equation (10.25) or (10.28) can now be used in the Taylor model presented in the last section. For convection, the velocity shear is dependent on the density gradient so that the flux of tracer is proportional to the square of the density gradient times the gradient of the tracer concentration. This makes the dispersion coefficient proportional to the square of the density gradient. We have already noted this effect in (10.14) for the case in which the tracer is either heat or salt and which is responsible for the longitudinal density gradient.

For the sake of completeness, we now derive the dispersion for the case of convection from first principles. The transport of a tracer with concentration c is again analyzed. Writing the concentration as the sum of its depth-averaged value and the departure from that mean as before,

$$c = \bar{c} + c', \tag{10.31}$$

we obtain a tracer flux

$$uc = u(\bar{c} + c'). \tag{10.32}$$

In the steady state, to first order, we can assume a balance between the horizontal advective flux of tracer and its vertical mixing, i.e.,

$$\frac{\partial}{\partial z}K_z\frac{\partial c'}{\partial z} = u'\frac{\partial\bar{c}}{\partial x} \tag{10.33}$$

where K_z has been used for the vertical diffusivity of tracer. Hence, substituting (10.25) in (10.33) gives

$$\left(\frac{2}{h}\right)^2 \frac{\partial}{\partial z'} K_z \frac{\partial c'}{\partial z'} = -\frac{\tilde{u}}{2}\left[\frac{z'^3}{3} - z'\right]\frac{\partial \bar{c}}{\partial x}.$$

(10.34)

Integrating (10.34) gives $K_z\partial c'/\partial z$ which is the vertical flux of tracer due to mixing and we note from (10.34) that this is an even function of z' and we can stipulate that this flux must vanish at the bottom and top of the water, at $z' = \pm 1$, since tracer cannot pass through either the surface or the bottom. So, defining the number

$$\nu \equiv \frac{g(h/2)^5}{\rho K_z^2}\frac{\partial \rho}{\partial x}\frac{\partial c}{\partial x}$$

(10.35)

the integration of (10.34) gives

$$\frac{\partial c'}{\partial z'} = -\frac{\nu}{2}\left(\frac{z'^4}{6} - z'^2 + \frac{5}{6}\right).$$

(10.36)

Noting that that the depth average of s' vanishes, (10.36) gives

$$c' = \frac{\nu z'}{6}\left(z'^2 - \frac{z'^4}{10} - \frac{5}{2}\right)$$

(10.37)

and integrating the product of (10.37) and (10.25) gives the tracer flux

$$\overline{c'u'} \equiv -\frac{\kappa g^2 h^8}{\rho^2 K_z^3}\left(\frac{\partial \rho}{\partial x}\right)^2 \frac{\partial c}{\partial x}$$

(10.38)

where

$$\kappa \equiv \frac{1}{2^8}\left\langle \frac{z'^2}{12}\left(\frac{z'^2}{3} - 1\right)\left(z'^2 - \frac{z'^4}{10} - \frac{5}{2}\right)\right\rangle.$$

(10.39)

Using (10.21), (10.39) becomes

$$\kappa = \frac{1}{2^9}\int_{-1}^{1} \frac{z'^2}{12}\left(\frac{z'^2}{3} - 1\right)\left(z'^2 - \frac{z'^4}{10} - \frac{5}{2}\right)dz'$$

(10.40)

and, evaluating the integral,

$$\kappa = \frac{1}{2^8}\left(\frac{1}{21} - \frac{1}{270} - \frac{1}{5} + \frac{1}{70}\right),$$

(10.41)

i.e., $\kappa \sim 1.7\ 10^{-4}$. Note in particular that the tracer flux is proportional to the concentration gradient and of opposite sign, which has the familiar form of a diffusive flux, i.e., it obeys Fick's law with a coefficient of diffusivity

$$K_x = \frac{\kappa g^2 h^8}{\rho^2 K_z^3}\left(\frac{\partial\rho}{\partial x}\right)^2.\tag{10.42}$$

This is essentially Taylor shear dispersion (see Section 7.2) for the case in which vertical shear is produced by the longitudinal density gradient.

The diffusion coefficient is proportional to the square of the longitudinal density gradient. If the tracer is replaced by salt or heat the diffusivity is itself a function of function of the gradient since heat and salt affect the density. We now have non-linear diffusion and recall that we met non-linear diffusion (of momentum) in Chapter 7 when we discussed the mixing length. In the present case the horizontal flux becomes proportional to the cube of the gradient. The simplest way to represent this process is to adopt a characteristic length denoted by l such that

$$\frac{1}{\rho}\frac{\partial\rho}{\partial x} \equiv -\frac{1}{l_\rho},\tag{10.43}$$

so for example if the density changes by $3\,\mathrm{kg\ m^{-1}}$ over a distance of $3\,\mathrm{km}$, $l_\rho \sim 1020\,\mathrm{km}$.

The rate of increase of density with salinity is usually written in terms of a coefficient denoted here by $\beta \sim 0.7^5$ and can be defined either in terms of the salt concentration S or in terms of the salinity s, as $\beta \equiv (1/\rho)\mathrm{d}\rho/\mathrm{d}s$, or $\beta \equiv \mathrm{d}\rho/\mathrm{d}S$ where we note that $S = s\rho$, $\Delta\rho = \beta\Delta S$ and for example if $\Delta S = 35\,\mathrm{kg\,m^{-3}}$, $\Delta S \sim 0.0035$, i.e., 35 ppt, $\Delta\rho \sim 24.5\,\mathrm{kg\,m^{-3}}$,

$$\frac{\beta}{\rho}\frac{\partial S}{\partial x} \equiv -\frac{1}{l_\rho}.\tag{10.44}$$

Using our expression for K_z obtained in Section 7.20,

$$K_z = Buh \qquad B = \gamma\sqrt{C_\mathrm{d}} \approx 0.01,\tag{10.45}$$

where B has been evaluated for the friction due to sand particles,

$$K_x = \frac{\kappa g^2 h^5}{L_p^2 B^3 u^3}\tag{10.46}$$

$$K_x = \frac{6.10^5 h^5}{L_p^2 u^3}\tag{10.47}$$

---

[5] Definitions of β differ between authors.

and so for a tidal current of $u \sim 0.5\,\mathrm{m\,s^{-1}}$, and $h = 10$, we get $K_x \sim 2\,\mathrm{m^2\,s^{-1}}$ but clearly it decreases very rapidly with h.

10.4 Convective transport due to lateral shear

In a basin of finite width, and some lateral variation of depth, a longitudinal density gradient will create lateral shear. This produces a topographic gyre in the depth-averaged current giving longitudinal basin transport. The situation is similar to the two types of wind-induced shear in a basin treated in Chapter 5.

Using the same coordinate system as in the last section, the steady state longitudinal momentum balance gives, with linear bottom friction,

$$0 = -gh(y)\frac{\partial \eta}{\partial x} - C_d\, u_f\, u - g\frac{h^2(y)}{2\rho}\frac{\partial \rho}{\partial x} \tag{10.48}$$

where C_d is the bottom drag coefficient (~ 0.0025) and u_f is a background velocity used to determine bottom friction. It is assumed here that there is no lateral pressure gradient or advection (along the y axis). Using angular brackets to denote a lateral average (over y) and primes for departures from the laterally average of the depth-averaged current and salinity,

$$\langle u'\rangle = 0 \tag{10.49}$$

$$u' = -G\big(h - \langle h^2\rangle/\langle h\rangle\big) \tag{10.50}$$

$$G \equiv \frac{gh}{2C_d u_b \rho}\frac{\partial \rho}{\partial x}. \tag{10.51}$$

This produces a lateral variation in salinity s' (and a similar analysis applies to temperature):

$$\frac{\partial}{\partial y}K_y\frac{\partial s'}{\partial y} = u'\frac{\partial \langle s\rangle}{\partial x}. \tag{10.52}$$

Analysis of the lateral problem will be made in terms of a symmetrical basin of half width b, with a parabolic depth variation, i.e., choosing the x axis ($y = 0$) to be at the center of the basin (where depth is maximal),

$$h = h_{max}\left[1 - py'^2\right] \tag{10.53}$$

where $-1 \le y' \le 1$ and $y' = y/b$ and p is the ratio of the total variation in depth to the maximum depth. Hence (10.48) to (10.53) give

$$u' = -\frac{Gph_{max}^2}{1+p/3}\left[(1+p/3)y'^4 + (1-p^2/5)y'^2 - (1/3+p/5)\right]. \tag{10.54}$$

Integrating (10.54),

$$\frac{\partial s'}{\partial y} = -\frac{Gpbh_{max}^2}{1+p/3}\frac{1}{K_y}\frac{\partial\langle s\rangle}{\partial x} * \left[\frac{p}{5}(1+p/3)y'^5 + \frac{1}{3}(1-p^2/5)y'^3 - (1/3+p/5)y'\right] \tag{10.55}$$

where the lateral salt flux across the sides of the basin has been set to zero. Using (10.52), and integrating by parts, gives

$$\frac{\langle u's'\rangle}{K_y}\frac{\partial\langle s\rangle}{\partial x} = \left\langle \left(\frac{\partial s'}{\partial y}\right)^2 \right\rangle \tag{10.56}$$

and so

$$\langle u's'\rangle = \frac{1}{C_d^2}\frac{2}{945}\frac{g^2 b^2 p^2 F(p)h^4}{p^2 u_f^2 K_y}\frac{\partial s}{\partial x}\left(\frac{\partial\rho}{\partial x}\right)^2 \tag{10.57}$$

where

$$F(p) \equiv 1 - 1.601p + 0.788p^2 - 0.116p^3 + 0.006p^4. \tag{10.58}$$

The maximum value of $F(p)$ is 1, when p is zero, and it decreases to about 0.08 when p reaches its maximum value of unity. Hence, using (10.57) the ratio of lateral to vertical transport R is

$$R = \frac{2*10^6 p^2 F(p)K_z^3}{h^4 u_f^2 K_y}b^2 \tag{10.59}$$

or if we use (10.47)

$$R = \frac{0.05up^2 F(p)}{hK_y}b^2 \tag{10.60}$$

where we have made the association $u_f \sim u$, since on the long timescales of salt diffusion the tidal current becomes an effective background flow. Taking the maximum value of $p^2 F(p)$, which is 0.1 (when $p \sim 0.7$), the ratio R in (10.59) becomes

$$R = \frac{0.005u}{hK_y}b^2. \tag{10.61}$$

In Tampa Bay and many other basins, we are always interested in the effect of navigational channels on salt transport. The value $K_y \sim 2 \text{ m}^2\text{s}^{-1}$ is perhaps not untypical, reasonably representative of channels of width several hundred meters or more. In that case $R \sim 1$, which implies that lateral shear is an important mechanism for salt transport along the channel.

10.5 Flow through tidal channels

To understand the action of the channel, we can construct a simple model based on linear friction in the channel. We can also assume that the seiching time of the basin is much less than the periodic time of the tide, so that the water level in the basin is essentially horizontal. Let us write the surface area of the basin as A, the cross-sectional area of the channel as a, the depth current through the channel as u, the elevation in the basin as η_b, and the ocean elevation as η_o. We assume that the current is related to the difference between these two elevations, with positive u directed into the basin, i.e.,

$$u = \Gamma(\eta_o - \eta_b). \tag{10.62}$$

The volume continuity equation is

$$au = A\frac{d\eta_b}{dt} \tag{10.63}$$

so if we assume a simple sinusoidal form for the ocean tide with amplitude, and use complex notation,

$$\eta_o = C_o e^{i\omega t}. \tag{10.64}$$

Hence (10.62) to (10.64) give

$$A\frac{d\eta_b}{dt} + a\Gamma\eta_b = a\Gamma C_o e^{i\omega t}. \tag{10.65}$$

If we ignore initial transients

$$\eta_b = C_b e^{i\omega t}, \tag{10.66}$$

(10.65) become in complex notation

$$i\omega A C_b + a\Gamma C_b = a\Gamma C_o, \tag{10.67}$$

and hence

$$C_b = \frac{1}{1 + i\omega A/a\Gamma}C_o \tag{10.68}$$

and in terms of amplitude and phase

$$C_b = \frac{C_o e^{i(\omega t - \phi)}}{1 + (\omega A/a\Gamma)^2} \tag{10.69}$$

$$\tan \phi = \omega A/a\Gamma \tag{10.70}$$

so that the basin tide lags the ocean tide by phase angle ϕ. When Γ is small, this phase lag tends to 90° and the basin amplitude tends to zero. When Γ is large, the phase lag tends to zero and basin amplitude tends to become the same as that of the ocean tide. One can see in (10.69) the same effect of changing the ratio of the areas a/A and more concisely, that the phase lag and reduction in tidal amplitude is controlled by $\omega A/a\Gamma$. The phase lag means that the ocean tide turns before the basin tide, and for example, for a semidiurnal tide the tide in the basin may still be rising for up to 3 h after high water has been reached in the ocean. If the flow through the channel is controlled by non-linear effects due to friction and inertia, the basin tide will not have a simple harmonic variation and the current in the channel (as noted in Section 4.5.6) tends to swing quite suddenly from positive to negative at the turn of the ocean tide. The tidal channel may greatly reduce the amplitude of the tide with the result that the basin tide may be microscopic, i.e., a few centimeters or less. In natural channels through sand, the cross-sectional area of the channel tends to adjust so that the maximum current is around 0.1 to 0.2 m s^{-1}, which is the region of the threshold for erosion and deposition of sand.

A very important parameter for coastal basins is their tidal prism $V_t = A\eta_b$, and for bar-built basins the magnitude of this tidal prism relative to the volume of the channel is critical for the flushing time of the basin. If we write the length of the channel as L, the ratio denoted by r is

$$r = \frac{aL}{V_t}. \tag{10.71}$$

If $r > 1$, the tidal prism of the basin can, in the absence of dispersion, move from the basin to the channel and back without entering the ocean. This means that there is effectively no advective exchange of basin water with the ocean, and exchange relies entirely on diffusion across a ocean–basin front that lies always in the channel and never enters either the ocean or the basin. If $r < 1$, the tidal prism does move from basin to ocean, and clearly the exchange time does depend critically on the value of r. Let us suppose that the basin has volume V and is initially filled with a tracer at concentration $c = c_0$. The ocean has zero concentration of that tracer. Assume further that time $t = 0$ corresponds to the basin high water. At low water in the basin, a volume of water equal to rV_t occupies the channel and the reminder $(1 - r)V_t$ has passed to the ocean. Assume that a fraction f_o of the water discharged to the ocean returns through the channel on the flood tide. If there is no mixing in the channel itself, the new tidal prism entering the basin has a concentration of tracer given by

$$c_1 = [r + f_o(1 - r)]c_0 \tag{10.72}$$

assuming that this mixes totally with water remaining in the basin at the end of the ebb tide. So the outgoing water has tracer concentration given by

$$c_2 = [r + f_o(1 - r)]s_t c_0 + (1 - s_t)c_0 \tag{10.73}$$

where $s_t = V_t/V$ where V is the volume of the basin at low water. Hence, in one tidal cycle the concentration of tracer is reduced by the fraction

$$\frac{\Delta c}{c} = -s_t(1 - r)(1 - f_o) \tag{10.74}$$

and assuming the left-hand side of (10.74) is small compared to unity,

$$\frac{dc}{dt} = -\frac{s_t c(1 - r)(1 - f_o)}{T_t} \tag{10.75}$$

corresponding to a flushing time of

$$T_f = \frac{T_t}{s_t(1 - r)(1 - f_o)} \tag{10.76}$$

which tends to infinity as r and f_o tend to unity, and is otherwise inversely proportional to the fractional volume of the tidal prism. The idea of complete mixing of the tidal prism with the water remaining in the basin at the end of the last ebb tide is, of course, a very coarse model. This can be overcome by considering the concentration in the basin at the end of each ebb tide and redefining s_t as

$$s_t \equiv \frac{\mu V_t}{V} \tag{10.77}$$

where μ is the fraction of the tidal prism which mixes during each flood tide.

10.6 Sub-classification of partially mixed basins

There is a sequence of classifications for coastal basins, and within the hierarchy of classifications based on physical properties of the water column, basins that are partially (or totally) mixed in the vertical are often subdivided according to their horizontal salinity gradient.

10.6.1 Salinity

All coastal basins can also be classified as either *fresh* or *hypersaline*. A fresh basin has lower salinity than the ocean and this is normally due to the freshwater runoff from a river, groundwater, or the land. A hypersaline basin has higher salinity than the ocean due to evaporation. Because the salinity of the ocean varies slightly in time, we need to be more specific about the definition of *higher*; we say that the mean difference over a season is greater than the standard deviation of the time series of ocean salinity. The hypersalinity is usually due to evaporation and requires that the exchange time of water between the basin and the open ocean is long. Very roughly the salinity difference between the basin and open ocean ΔS is given by a box model (Chapter 3) as

$$\Delta S = \frac{ES_o}{h} \tau_e \tag{10.78}$$

where E is the surface evaporative rate velocity, τ_e is the exchange time, S_o is the salinity of the open ocean, and h is the mean depth. This equation is simply based on the assumption that the water level in the basin remains the same as the level in the open ocean, so that as water is lost by evaporation, it is immediately replaced by a flow from the open ocean. Summer evaporative rates in *mediterranean climates* (hot dry summers) can reach several millimeters (even up to 1 cm) per day. The hypersalinity in a shallow basin of depth 1 m for $E = 1$ mm per day would be

$$\Delta S = \frac{0.001 S_o}{1} \tau_e|_{days} \tag{10.79}$$

where $\tau_e|_{days}$ is the exchange time expressed in days. If the exchange is 100 days (poor flushing) the hypersalinity would be 3.5 if $S_o = 35$. Some basins can achieve hypersalinities of 10 or 20, and this is usually due to the ratio E/h being larger than 0.001; for basins with long tidal channel in arid climates E/h can reach 0.01 giving hypersalinity of 35.

If the basin has seasonal freshwater flow, the salt balance will be altered and during periods of river flow E becomes the difference between the evaporation and river flow/ (unit area of basin). In some basins there is rarely any river flow. An example is Sharp Bay on the northern part of the west coast of Australia, which lies in a very arid region. Hypersalinity reaches very high values in the main basin and some more isolated pools become salt ponds; at very high salinity the evaporation decreases due to salt crusting. Many of the coastal lagoons become hypersaline during the dry seasons associated with either El Niño or La Niña years, but values are of hypersalinity are more typically 1 to 5.

10.6.2 Temperature

As we saw in Chapter 3, basins with fast exchange with the ocean experience only minor solar heating, simply because the water remains in the basin for only a limited time. But when the exchange time is long, for example 100 days, the heating is much greater. When the exchange time is short (few days) the temperature difference between the basin and open ocean ΔT increases in proportion to τ_e. The heat gain and heat loss become equal when the basin reaches an equilibrium temperature denoted by T_e. In shallow basins, the temperature is always close to T_e and if T_e is considerably greater than T_o, the basin is said to be *hyperthermal*. Otherwise, if the basin is close to T_o, it is said to be *thermal*. Basins can have $T_e < T_o$ and are then said to be *hypothermal*. The general situation in winter is that T_e is close to the mean air temperature (averaged over a 24-h period) since the heat provided by solar radiation is less than that lost through the heat exchange between the basin and air. If winds are strong, evaporation can be appreciable, even at the lower water temperatures encountered in winter and T_e may be less than mean air temperature. The necessary condition for hypothermia, or hyperthermia, is slow ocean exchange and a reasonably shallow basin.

10.6.3 Density

Basins can be classified as *classical* or *inverse* according to whether the density is less or greater than the ocean (respectively). A basin can be classical due to freshwater inflow or heating (both reduce density). *Inversion results from hypersalinity, but hypersalinity itself does not ensure inversion* because of the effect of heating: an increase in tempera-ture has the opposite effect to hypersalinity and reduces density. So a hypersaline basin may be inverse or classical depending on whether there is any hyperthermia and the degree of that hyperthermia. If the basin is cooled, i.e., hypothermal, it is almost certainly inverse, whereas if it is heated it may be classical or become inverse due to hypersalinity.

10.6.4 Buoyancy flux

In the steady state the sign of the density difference is the same as the sign of the net buoyancy flux to the basin. Buoyancy flux basically means the rate of density change times g (the acceleration due to gravity). So river flow provides positive buoyancy flux, heating provides positive buoyancy flux, and evaporation gives negative buoyancy. If the net buoyancy flux is positive the basin is said to be *positive*, otherwise it is *negative*. So in the steady state a classical basin is also positive and an inverse basin is always negative, but a classical basin may not be positive if it is removed from the steady state. Similarly, an inverse basin may not be negative. It must be noted that there are some differences in the definition of classical and inverse basins, with some authors using these terms for basins that are fresh or hypersaline and then using the terms positive and negative for basins that we call classical and inverse.

10.7 Dispersion and exchange rates in basins

The spreading of a tracer concentration distribution which initially has a step function form (see Chapter 2) is illustrated in Figure 10.2.

The effect of an advective velocity u is to move the profile a distance ut. The profile at time t is

$$c(x,t) = \frac{1}{2}\left[1 + erf\left(\frac{x - ut}{w}\right)\right] \qquad (10.80)$$

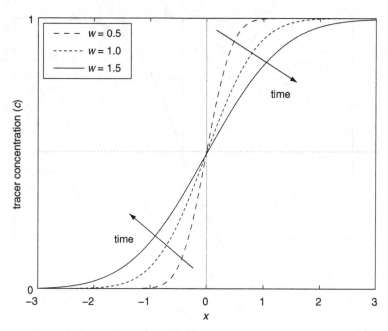

Figure 10.2 The spreading of a tracer front in a coastal basin. The initial distribution is a step function at $x = 0$. The spreading with time preserves the area under the curve, i.e., conserves the total quantity of tracer. The values of w show the half widths of the distribution as it increases with time; compare Figure 2.9.

where w is the width,

$$w = 2\sqrt{K_x t}, \qquad (10.81)$$

and erf is the error function shown in Figure 10.3:

$$\mathrm{erf}(a) = \frac{2}{\sqrt{\pi}} \int_0^a e^{-p^2} dp. \qquad (10.82)$$

Equation (10.80) is the solution of the advective diffusion equation in one dimension, and for example, describes the mixing of basin and ocean water across a front in a coastal basin. From Chapter 2 and Chapter 7, the flux of a tracer of concentration c is

$$Q_c = uc - K_x \frac{\partial c}{\partial x} \qquad (10.83)$$

and the continuity equation is

$$\frac{\partial c}{\partial t} = -\frac{\partial Q_c}{\partial x} \qquad (10.84)$$

and so

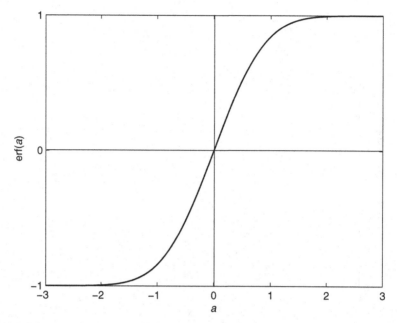

Figure 10.3 The error function erf(a) plotted as a function of a.

$$\frac{\partial c}{\partial t} = -u\frac{\partial c}{\partial x} + K_c\frac{\partial^2 c}{\partial x^2}. \tag{10.85}$$

Other useful solutions of the advection diffusion equation were treated in Chapter 2.

Suppose that a one-dimensional basin that initially has zero tracer concentration is subject to intrusion of ocean water, assuming that the process is described by the advection diffusion equation with horizontal diffusion coefficient of $10\,\mathrm{m^2\,s^{-1}}$ for salt and zero advection (Figure 10.4). After 1 day, the spreading is $2\sqrt{K_x t}$, which is $1860\,\mathrm{m}$ into both the ocean and basin water. So we would *estimate* the time t for an acceptably near uniform concentration of a tracer to occur over a basin of L as $L = 2\sqrt{K_x t}$. As an example, if $L = 20\,\mathrm{km}$, and $K = 10\,\mathrm{m^2\,s^{-1}}$ we find $t = 115$ days.

The flushing of a basin can be viewed as involving two processes. The first involves mixing and dispersion within the basin of a tracer introduced into the basin (such as fresh water or a nutrient), and the second is exchange of basin and ocean water. This was discussed in general terms in Chapter 3 based on a number of different processes. A traditional method of quantifying tidal exchange processes is based on the tidal prism. We often define a tidal volumetric flushing time as the time taken for the tidal prism to fill the basin. Firstly we define the relative tidal prism r by

$$r = \frac{\text{tidal prism}}{\text{basin volume at low water}} \tag{10.86}$$

so that

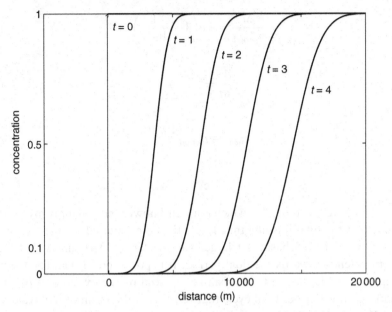

Figure 10.4 An advecting front of tracer with concentration c which is widening as it moves to the right (time t in hours). The advective velocity $u = 0.2\,\mathrm{m\,s^{-1}}$, and diffusivity $K_x = 20\,\mathrm{m^2\,s^{-1}}$.

$$r + 1 = \frac{\text{basin volume at high water}}{\text{basin volume at low water}} \qquad (10.87)$$

and note $r < 1$. If we had *total mixing* in the basin, so that the entire tidal prism totally mixed with the water of the basin, the tracer concentration would be uniform throughout the basin. Let us assume that the concentration of tracer is initially c_0 in the basin and this corresponds to low water. The concentration has a value of C in the ocean. At the next high water a volume of water equivalent to the tidal prism has entered the basin, and so the concentration c_1 is given by

$$c_1(1 + r) = c_0 + rC \qquad (10.88)$$

and generally

$$c_n(1 + r) = c_{n-1} + rC \qquad (10.89)$$

so that putting $C = 0$, i.e., the ocean has zero concentration of the tracer,

$$c_n = \frac{c_{n-1}}{(1 + r)}, \qquad (10.90)$$

and finally putting $c_0 = 1$,

$$c_n = \frac{1}{(1 + r)^n} \qquad (10.91)$$

$$\frac{\Delta c}{t_{\text{period}}} = -c\left(1 - \frac{1}{1+r}\right) \tag{10.92}$$

$$\frac{dc}{dt} = -\frac{c}{\tau_{\text{tidal}}} \tag{10.93}$$

$$\tau_{\text{tidal}} \equiv t_{\text{period}}\frac{1+r}{r} \tag{10.94}$$

$$c = e^{(-t/\tau_{\text{tidal}})} \tag{10.95}$$

when $r \ll 1$, τ_{tidal} tends to t_{period}/r which is a well-known (but generally invalid) result, and means that the tidal flushing time is just the time required for the tidal prism to replace the water in the basin. At larger values of r, (10.94) shows that the tidal flushing time tends towards the tidal period. It is possible to generalize the present treatment by considering the tidal prism be replaced by the *exchange prism*, i.e., the volume of ocean water per tidal cycle that remains in the basin or, equivalently, the volume of basin water that remains in the ocean.

The volumetric tidal flushing time fails badly in shallow partially mixed basins because the dynamics are quite different from those viewed from the perspective of the tidal prism. The reason is that we intuitively consider the tidal prism to sit either beneath or on top of the water remaining in the basin at the previous low water. In a shallow coastal basin this does not happen. Instead, ocean water advances into the basin from the mouth and the existing water tends to be pushed up towards the head of the basin. So, the mixing of the tidal basin tends to occur in the horizontal and not in the vertical. This means that the extent of mixing depends on the value of K_x, and it is possible to envisage a tidal process which just involves advection without any mixing. Tidal processes do contribute to mixing through many of the processes discussed in Chapter 7. In shallow basins, the process which is of greatest fundamental importance to ocean exchange is mixing through the basin. This means that there is no well-defined *exchange time* or *flushing time* and the quantity that appears in box models (Chapter 3) is only an average taken over the basin as a whole.

10.8 Age of particles

Determining an exchange time either analytically, or numerically, has exercised the skills of modelers for some decades and clearly this parameter is one of the most important for most applications of our understanding of basin dynamics to problems of ecology and basin management. Most modelers consider that moving beyond the ideas of simple box models, and perhaps one-dimensional models (which were discussed in Chapter 3), involves some sort of particle tracing technique. The most recent of these techniques at the time of writing is to determine the so-called *age of particles*. This involves releasing model particles into the basin and measuring their age while they still remain in the basin. The easiest way to assess age is to use an *internal clock*, which can be set at a rate most convenient to the basin under consideration and to the point of release of the particles. The internal clock is rather like an old-fashioned hourglass or egg-timer.

Within our model we have a large number of particles, so we cannot literally determine the age of each particle. We can, however, count within one spatial (and temporal) cell the number of particles that have age with a certain range of ages. This is akin to a demographic survey of the age range of a human population living with a certain region. The individuals that make up that population will have arrived in that region via a large number of routes, and their journey through life will have been motivated by many different reasons and aspirations. The same is true of particles that we introduce as traces into a coastal basin: the particles are moved by advection, mixing, and dispersion and can have visited many parts of the basin. Because we are using a basically Eulerian approach we cannot do more than find the average age of particles in any region of the basin, but we *can* conceptualize that a distribution of ages exists.

Every particle is treated as a pair of twins. These twins are not inseparable, but they do obey the same laws of motion. The difference between them is the twin labeled 1 has a decreasing mass, of which the concentration c_1 obeys the law

$$\frac{\partial c_1}{\partial t} = -\frac{c_1}{\tau_{\text{lifetime}}}. \tag{10.96}$$

This is exactly analogous to the sand running out of the top container of the hourglass, except that we never allow the container to empty simply by the device of making the flow proportional to mass in the container. The other twin has a constant mass and is labeled 0, i.e.,

$$\frac{\partial c_0}{\partial t} = 0 \tag{10.97}$$

and if we assume that at the time $t = 0$ of creation of a pair of twins $c_1 = c_0$, then over the lifetime of that pair

$$c_1 = c_0 e^{-t/\tau_{\text{lifetime}}}. \tag{10.98}$$

Twin 0 is like an hourglass without flow of sand. Because of the similar dynamics of the two twins, their spatial distribution is identical apart from the effects of the decay that we have introduced. So if we could measure both c_1 and c_0 for that pair we could determine the time t at which the measurement is made from

$$e^{t/\tau_{\text{lifetime}}} = \frac{c_0}{c_1}. \tag{10.99}$$

The age τ of the particles in any cell of the model is then given by (10.99) which we can write as

$$\tau = \tau_{\text{lifetime}} \log\left(\frac{c_0}{c_1}\right). \tag{10.100}$$

The age of particles is a tool that can be used in a variety of different ways to understand the fate of particles introduced into the basin. Some of the choices for use of the tool are whether to place a source on the open boundaries of the model, the river mouths, or in the bulk of the coastal basin. On the boundaries or the model, the age of particles provide some measure of the time taken for particles to move into the basin and reach the upper reaches of the basin. So, for example, we might find younger particles of age just a few days near the mouth and older particles of age over 100 days near the head of the basin, and in small side arms of the basin. If we release particles from the river mouths, this provides considerable information on the fate of river water inside the basin and the time taken to pass out through the mouth.

Let us consider how the age concept works inside a box model of the type introduced in Chapter 3. For the decaying concentration c_1 is controlled by

$$\frac{\partial c_1}{\partial t} = -\frac{c_1}{\tau_{\text{lifetime}}} - \frac{c_1}{\tau_{\text{flush}}} + S \tag{10.101}$$

where S is the rate of creation of both particles and τ_{flush} is the flushing time. In a steady state:

$$S = \frac{c_1}{\tau} \tag{10.102}$$

$$\frac{1}{\tau_{\text{total}}} \equiv \frac{1}{\tau_{\text{lifetime}}} + \frac{1}{\tau_{\text{flush}}}. \tag{10.103}$$

For the stable particle

$$\frac{\partial c_0}{\partial t} = -\frac{c_0}{\tau_{\text{flush}}} + S \tag{10.104}$$

so that in steady state

$$\frac{c_0}{c_1} = \frac{\tau_{\text{flush}}}{\tau_{\text{total}}} \tag{10.105}$$

and using (10.100), the steady state age of particles is

$$\tau = \tau_{\text{lifetime}} \log\left(1 + \frac{\tau_{\text{flush}}}{\tau_{\text{lifetime}}}\right). \tag{10.106}$$

When flushing is much faster than exchange, i.e., $\tau_{\text{flush}} / \tau_{\text{lifetime}} \ll 1$,

$$\tau \to \tau_{\text{flush}} \tag{10.107}$$

but otherwise the value of τ_{lifetime} does affect the steady state value of τ. The reason that this occurs is that we are really measuring the average age from a distribution of ages.

If the particles are released as a pulse at time $t = 0$, then the source has the form of a delta function (Chapter 2) in time and injects particles at a rate

$$S = \delta(t) \tag{10.108}$$

$$c_1 = U(t)e^{-t/\tau_{\text{total}}} \tag{10.109}$$

where U is the unit function defined in Section 2.6 and ensures that when we differentiate (10.109) in (10.101), the source term is reproduced. Using (10.109) in (10.104) gives

$$c_0 = U(t)e^{-t/\tau_{\text{flush}}} \tag{10.110}$$

so that

$$\frac{c_0}{c_1} = e^{t/\tau_{\text{lifetime}}} \tag{10.111}$$

and using (10.100)

$$\tau = t \tag{10.112}$$

as we would expect for particles born simultaneously. In this case there is no problem of averaging over a distribution.

Let us imagine that we can look at the particles in a particular cell of a model, denoted by position r, at time t, and describe their age distribution as $c_0(r, t, \tau)$ and $c_1(r, t, \tau)$. These are related by

$$c_1(r, t, \tau) = e^{-\tau/t_{\text{lifetime}}} c_0(r, t, \tau). \tag{10.113}$$

Generally we know only the sum over all ages, i.e.,

$$C_i(r, t) \equiv \int_0^\infty c_i(r, t, \tau)d\tau \quad i = 0, 1 \tag{10.114}$$

but we do not know the individual age distributions. For example, there could be particles that have recently arrived from the source via advection, and also particles that have reached the point over much longer times via more circuitous routes involving advection and diffusion.

The mean age is

$$\overline{\tau} \equiv \int_0^\infty c_0(r, t, \tau)\tau d\tau \bigg/ \int_0^\infty c_0(r, t, \tau)d\tau. \tag{10.115}$$

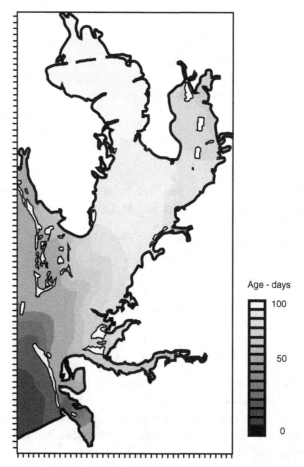

Figure 10.5 Residence times in Tampa Bay, on the west coast of Florida, obtained by the particle age technique. The model uses the grid shown in Figure 6.24.

If one applies the relation (10.113), we can define another mean age $\tilde{\tau}$,

$$e^{-\tilde{\tau}/\tau_{\text{lifetime}}} \equiv \int\limits_0^\infty c_1(r, t, \tau)\mathrm{d}\tau \bigg/ \int\limits_0^\infty c_0(r, t, \tau)\mathrm{d}\tau \tag{10.116}$$

and it is clear that this is affected by our choice of lifetime.

As an example, this technique was applied in Tampa Bay and the results are shown in Figure 10.5.

As another example of the use of the particle age technique consider a short basin of length only 1000 m connected to a freshwater reservoir at one end and closed at the other. The model is two-dimensional in the vertical and implicitly assumes that the basin is narrow in relation to its length and can be applied to a creek entering a larger basin (and that basin would form the freshwater reservoir). The model uses a z

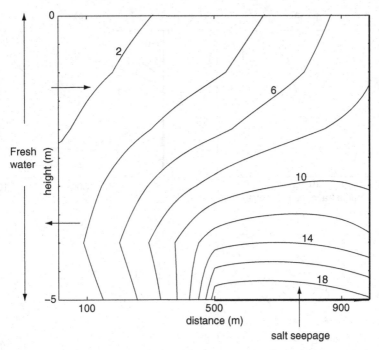

Figure 10.6 Contours of salinity shown in the vertical plane along the axis of the basin modeled in this section. The solid line on the right indicates that the basin is closed at its head. The solid line at the bottom right-hand half of the basin, and accompanying vertical arrow, shows the region of bottom seepage of salt. The horizontal arrows on the left indicate the directions of advection of water from, and to, the freshwater reservoir which forms the left-hand boundary.

coordinate system (compare Chapter 6) with five layers and cells of length 10 m in the horizontal. The basin is 5 m deep.

The model is driven by a seepage of salt into the bottom of the basin over that half of its length which is closer to its head. The model is initiated with the basin filled with fresh water and then run to a steady state. This process of salt seepage is common in regions where groundwater contains appreciable concentrations of salt often due to rising water tables following land clearing for agriculture. The model injects salt into the bottom of the water column at a specified rate. This increases the salinity of the bottom of the water column at the head of the basin and will drive a density driven (or negative buoyancy driven flow). The salinity contours (isohalines) are shown in Figure 10.6 (note that an increase in the number of layers would produce smoother contours) after the basin has reached a steady state with fresh water entering from the reservoir in the upper water column at the mouth and brackish water, i.e., water of salinity a few kilograms per cubic meter, leaving the basin in the lower water column. For the rate of salt seepage used in the model, the maximum values of salinity occur at the bottom of the basin near the head where salinities reach $19 \, kg \, m^{-3}$. This model does not include any dependence of vertical mixing rate on Richardson number as discussed in Chapter 9 although it is

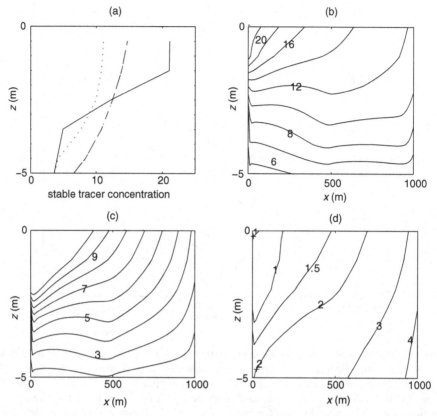

Figure 10.7 The top left panel (a) shows the steady state vertical profile of the stable tracer concentration near the mouth (solid line), half way along the basin, $x = 500$ m (dashed), and at the head $x = 1000$ m (dotted). The top right (b) and bottom left (c) panels show steady state contours of concentration of stable and decaying particles (respectively); concentration in the freshwater reservoir is 35 (arbitrary units). The bottom right (d) panel shows the contours of steady state age (in hours).

relatively easy to include that process. The model uses a constant coefficient of vertical mixing for momentum and the same constant value is also used for vertical mixing of salt. There is a set-up of the water surface from the head to the mouth which produces a pressure gradient that ensures that the depth-averaged current is zero in the steady state (since the basin is closed at the head). The convective motion is produced by the longitudinal density gradient with the total pressure being highest in the bottom of the water column at the head and the top of the water column at the mouth. At the mouth of the basin, the model takes the salinity to be zero and uses simple up-winding in the advective terms. The surface elevation in the freshwater reservoir is taken as zero.

Figure 10.7 shows contours of the two types of particle in which we have referred to the particles as stable and decaying. We set the concentration of both types of particle as 35 in the freshwater reservoir; the units of this concentration are kg m^{-3}.

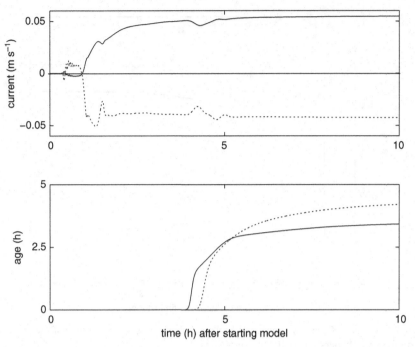

Figure 10.8 Top panel shows the time evolution (starting from the initial state at time $t = 0$) of currents at the head of the basin for the water surface (full line) and bottom of the basin (dotted line). Bottom panel shows the evolution of particle age at those same locations. Notice that the age is undefined for a time of about 4 h whilst particles reach the head of the basin from the mouth.

We have chosen the value in the reservoir to provide an easy comparison of the stable particles with salinity. Particles enter the basin with the inflowing water in the top of the water column at the mouth and are advected out with the bottom water at the mouth. During their time in the basin the particles are advected and diffuse into the main body of the basin. The lifetime in (10.96) has been chosen as 5 h after some experimentation to find the maximum age of particles in the basin which can be seen in Figure 10.7 to be about 4 h in the bottom of the water column at the head. We show in Figure 10.8 the time evolution (starting from the initial state at time $t = 0$) of the surface and bottom currents at the head of the basin and also the evolution of particle age at those same locations. The time for a hydrodynamic steady state to be reached is slightly over 5 h with the currents reaching about 0.05 m s^{-1} which gives a transit time from mouth to head of 5.6 h corresponding to the spin-up time. This time is fundamentally related to the rate of seepage of salt. After the hydrodynamic steady state is reached, the particle age comes into equilibrium after a further 3–4 h and Figure 10.8 shows that at the bottom the steady state age at the surface is 3.5 h and at the surface it is 3.5 h. Notice that these ages are undetermined for the first 4 h corresponding to the advective time for particles to reach the head and then increase

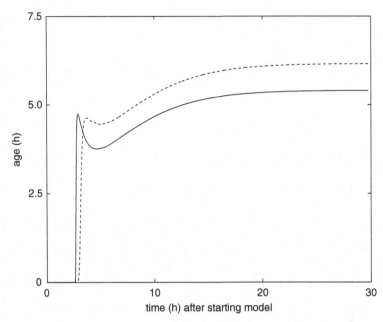

Figure 10.9 Repeat of the lower panel of Figure 10.8 showing the evolution of the particle age for a model with dispersion coefficient of $50\,\mathrm{m^2\,s^{-1}}$.

very rapidly to values comparable with their steady state values. The age of particles depends on the horizontal dispersion coefficient and Figure 10.9 shows the effect of increasing the horizontal dispersion of salt and tracer particles to $50\,\mathrm{m^2\,s^{-1}}$. This shows that the age of particles takes longer to settle to a steady state consistent with the greater importance of dispersion compared to advection and the age becomes rather greater.

10.9 Large-scale climate cycles

In Section 10.8 we mentioned briefly the influence of the *El Niño cycle* on the hypersalinity of coastal basins. El Niño is a large-scale climatic and oceanographic change that was originally identified on the western coast of South America, particularly Peru. It is an example (albeit probably the most important example) of the large-scale climatic cycles that affect all of modern civilization, including coastal basins. The influence of these cycles is important in many diverse ways, but we have chosen to include discussion in the present chapter because of relevancy to some of the cases presented here. Broadly, the effect of climatic cycles is seen through changes in local environmental forcing and ocean boundary conditions. Environmental forcing includes changing in river flow and wind conditions, and ocean boundary conditions are changed by major variations in near coastal water temperatures (themselves moderated by wind conditions).

The *El Niño cycle* is connected to the *Southern Oscillation* which is a large-scale oscillation in time-averaged atmospheric pressure that was identified by meteorologists and observed by them to produce an oscillation in the weather of the western Pacific. It was realized in the 1970s that the Southern Oscillation and El Niño are two aspects of the same phenomenon, which affects the South Pacific and the South-East Trade Winds. The opposite side of the El Niño cycle is called *La Niña*. Furthermore, we have found over the last few decades that these processes affect not only the South Pacific, but also most of the Earth's weather.

El Niño has been recorded in European records since the beginning of European settlement of South America from the seventeenth century. The west coast of South America usually has cold coastal waters because of upwelling produced by the *Peru Current*. Peru itself has a climate similar to southern California, which is very dry. In most years, there is a change in weather in Peru just after Christmas, when rains appear and the ocean becomes much warmer. The usual cold water from upwelling tends to produce high atmospheric pressure; lows are associated with warm oceans causing air to rise and highs with cold water causing air to descend. This is also the reason that southern California is so dry (all the water for Los Angeles is pumped from the Colorado River). So when the ocean warms after Christmas, atmospheric pressure drops producing rain. This is a great gift for the Peruvian community, and because of its proximity to Christmas it is called the "Boy Child" or El Niño. Upwelling is important to the Peru community because it creates nutrient-rich coastal waters, and the early Spanish explorers found an abundance of anchovies (a small silvery-green fish maximum length about 20 cm known since the times of antiquity).

The anchovies feed on plankton which feed on the nutrients brought to the surface by the upwelling. The anchovies were traditionally the basis of the Peru economy, and used for a variety of food sources for animals and humans. Salt is easily available in Peru because of the dry climate, and is produced by evaporation ponds. So, anchovies are traditionally preserved by using salt and hence canned anchovies are very salty, but this is not true of the fresh fish. Anchovies were additionally also used for making *fish paste* which was a cheap source of protein. When the El Niño arrives the anchovies disappear, because they have a very short life cycle. Hence, the fishers mended their nets and waited for the El Niño and the rains to disappear.

The annual El Niño sometimes gets much more intense, and lasts for the whole year (and even several years). It is this longer-lasting phenomenon that the whole world now calls El Niño, although strictly we should say *major* El Niño. There was a suggestion in the 1970s that the (major) El Niños occurred about every 7 years, each with only temporary effects over a period of about a year. However, there are historical records of much more frequent El Niños with horrendous rains in South America, and droughts in Australia and the Western Pacific. During the middle and late twentieth century, the anchovy fishery in Peru grew to huge proportions and anchovies were exported worldwide and were a major food source. During the 1970s concern was felt that the anchovies were being over-fished. A study was commissioned and catch limits were set, which turned out to be too high and consequently the anchovy fishery collapsed from 12 000 to 4000 tonnes per annum after the severe El Niño of the early 1970s; fortunately, much of the fishery returned in the middle 1990s. The reason for the cessation of upwelling is that the South-East Trade Winds weaken since the upwelling is wind-driven (the Peru Current is an *Eastern Boundary Current*). The Southern Oscillation shows that this is associated with a change in atmospheric pressure. The Trade Winds are basically caused by the convective cell set up by differences in solar heating.

There is a mass of warm water in the Western Pacific and this warm *surface pool* is often the warmest water in the Earth's deep oceans. It is held in place in the Western Pacific by surface currents, i.e., the South Equatorial Current and by the South-East Trade Winds. That warm pool of water is responsible for maintaining much of the low atmospheric pressure that brings rain to Australia. The strength of the Trades is maintained by the low pressure associated with the warm pool, combined with the high pressure due to upwelling of Peru but paradoxically, it is the Trades that maintain the warm water pool and the upwelling. So we can have two situations: strong Trades plus warm pool and upwelling *or* weaker Trades, no upwelling and the warm water dispersed across the Pacific from Australia to the coast of Peru. The second situation is the El Niño. When the Trades are especially strong we have La Niña (the "Girl Child").

10.10 Stommel transitions

10.10.1 Role of convection

Stommel transitions in coastal basins are a feature of buoyancy processes i.e. buoyancy input through the surface and buoyancy forcing of circulation (which we have called convection and is also commonly called *gravitational circulation*).[6] The vital ingredient of Stommel transitions is the dependence of convection on the horizontal density gradient. Stommel transitions can appear in models which contain both barotropic (for example, wind or tidal forced) circulation but *only* if convection (density forced) processes are also present. Stommel transitions are seen in box models and also in laboratory realizations[7] of box models.[8]

10.10.2 Density gradient

As we saw earlier in this chapter, there are several ways of classifying partially mixed coastal basins according to their horizontal temperature, salinity, and density

---

[6] The transitions are named after Henry Stommel (1920–92) who was one of the most original and innovative thinkers in physical oceanography of the twentieth century; see Stommel (1961).

[7] Whitehead (1995, 1998).

[8] Stommel considered a series of models based on interconnected boxes; our analysis corresponds to a generalization of one of these and its application to coastal basins.

gradients. Increases in temperature and salinity have opposite effects on density, and so a hypersaline, hyperthermal, basin may be either inverse (density increases towards the head) or classical (density decreases towards the head) or (in principle) *neutral* (density of basin is same as adjoining ocean despite being hypersaline and heated). The actual state of such a basin can be characterized by the ratio of the hyperthermal temperature induced *reduction* in density to the hypersaline *increase* in density. This is described as the compensation ratio. If the ratio is small, we have a standard inverse hypersaline basin, whilst if the ratio greatly exceeds unity, the basin is classical. Values of the compensation ratio close to unity create a nearly neutral basin.

10.10.3 Buoyancy fluxes

The horizontal dispersion coefficient denoted by Kx. As we saw in Section 10.7, the effective ocean exchange time for a basin of length L is

$$\frac{1}{\tau} = \frac{2K_x}{L^2}.$$ (10.117)

For a partially mixed basin, the exchange time is assumed to be the same for both heat and salt. As shown in Chapter 3, the difference between the average temperature of the basin T and that of the ocean, T_o, denoted by ΔT, is given by

$$\frac{dT}{dt} = \frac{Q}{\rho C_p h} - \frac{\Delta T}{\tau}$$ (10.118)

where Q is the net rate of surface heating, i.e., the difference between surface heating and cooling which we discussed in Chapter 3, C_p is the specific heat, and ρ the density, of the basin water, and h is the mean depth of the basin. Heating and cooling become equal when T reaches the so-called equilibrium temperature, T_e, so that when T is not too far removed from T_e:

$$Q = \rho C h \frac{T_e - T}{t_e}$$ (10.119)

where t_e is the characteristic time for T to relax to T_e and defined in Chapter 3.

Following the models developed in Chapter 3, the difference between the basin and ocean salinity, Δs, is given by

$$\frac{d\Delta s}{dt} = \frac{Es_o}{h} - \frac{\Delta s}{\tau}$$ (10.120)

where E is the surface evaporative velocity and s_o the ocean salinity. The difference between the density of the ocean and the basin, $\Delta \sigma$, is given by

$$\Delta \sigma = \alpha \Delta T - \beta \Delta s$$ (10.121)

where α and β are approximately constant coefficients. Consequently, (10.118), (10.120), and (10.121) show that the overall density (or *buoyancy*) dynamics are simply controlled by the surface buoyancy flux denoted by w_b,

$$\frac{d\Delta\sigma}{dt} = \frac{w_b}{h} - \frac{\Delta\sigma}{\tau}$$

(10.122)

$$w_b \equiv \frac{\alpha Q}{\rho C} - \beta E s_o$$

where both Q and E are functions of basin temperature T. It is convenient to work with the ratio b of w_b to the buoyancy flux w_b^o evaluated at the ocean temperature:

$$b \equiv w_b / w_b^o.$$

(10.123)

The value of b decreases with temperature due to increased cooling and evaporation (both of which give a negative surface buoyancy flux) and vanishes at a temperature $T_b < T_e$. Hence, the evaporative number Γ defined as

$$\Gamma^2 \equiv \frac{T_e - T_b}{T_b - T_o} ; \quad \Gamma^2 \geq 0$$

(10.124)

is a measure of the strength of the evaporative contribution. To a good approximation Q, and also the reduced buoyancy flux b, can be considered to decrease linearly with basin temperature T, so that

$$b = 1 - \frac{T - T_o}{T_b - T_o}$$

(10.125)

which enables temperature T to be expressed in term of b, and hence using (10.119) and (10.125), the heat balance (10.118) becomes

$$-\frac{db}{dt'} = \Gamma^2 + b - (1 - b)/\tau'$$

(10.126)

where the primes indicate dimensionless times normalized to t_e. Finally, using (10.123), the buoyancy balance (10.122) can be written

$$\frac{d\Delta\sigma}{dt'} = \sigma_e b - \frac{\Delta\sigma}{\tau'}$$

(10.127)

where σ_e is a convenient reference density difference,

$$\sigma_e \equiv \frac{t_e w_b^o}{h}.$$

(10.128)

10.10.4 Exchange time

The dispersion coefficient K_x generally consists of a term which is independent of the density difference $\Delta\sigma$ and another proportional to the square of the density difference

$$K_x = K_x^{(a)} + \kappa\Delta\sigma^2 \tag{10.129}$$

where κ is a constant. The first, density-independent, contribution in (10.129), $K_x^{(a)}$, is the barotropic term and this includes many processes due to circulation and mixing arising from the wind and tides (for example tidal stirring). The second term in (10.129) describes the convective processes discussed in Section 10.3. Equations (10.129) and (10.117) lead directly to the ocean exchange time τ provided that $\Delta\sigma$ is known. The barotropic diffusion length, along the longitudinal axis of the lagoon, over the characteristic time t_e, is denoted by L_a and given by

$$L_a^2 \equiv 2K_x^{(a)}t_e. \tag{10.130}$$

It is useful to express all properties that are functions of the lagoon length L in terms of a normalized inverse length λ defined as

$$\lambda \equiv L_a/L \tag{10.131}$$

in which the subscript and superscript a are used here to denote *barotropic* (density independent). Convective exchange, produces a normalized exchange rate represented by r^2 so that

$$\frac{1}{\tau'} = \lambda^2 + r^2; \quad r \equiv \left(\frac{L_b}{L}\right)^2 \frac{\Delta\sigma}{\sigma_e} \tag{10.132}$$

where r is the convective exchange number and L_b the *buoyancy length* of the lagoon which increases with the strength of convective processes; these decrease as vertical mixing increases and increase as the depth of the lagoon h increases. The purely barotropic lagoon has vanishing L_b and r. The relative importance of convection compared to barotropic processes is measured by a parameter θ,

$$\theta \equiv L_b/L_a. \tag{10.133}$$

A standard basin[9] of depth 5 m under summer drought conditions in a mediterranean climate (see Chapter 3) has $L_b = 10$ to 20 km; however note from Section 10.2 that L_b depends on the strength of vertical mixing and the depth h. For the expected range of the barotropic dispersion coefficient, $1 < K_x^{(a)} < 100\,\mathrm{m^2\,s^{-1}}$, L_a varies from 1 to 10 km so that θ is likely to be in the range 1 to 20.

---

[9] Hearn (1998).

10.10.5 Model equations

Substituting (10.132) in (10.126) and (10.127) gives, with the aid of (10.133), the pair of differential equations

$$-\frac{db}{dt'} = \Gamma^2 + b - (1 - b)(\lambda^2 + r^2)$$

(10.134)

$$\frac{dr}{dt'} = \theta^2 \lambda^2 b - (\lambda^2 + r^2)r$$

(10.135)

which represent the Stommel model of a basin based on barotropic and convective processes.[10]

The state of the basin is represented by the two parameters r, and b, which are solutions of (10.134) and (10.135). Setting their time dependencies to zero gives the basin steady states and we note that in the steady state r, and b, have the same sign as $\Delta\sigma$. Hence the lagoon is *classical* if $r > 0$, *neutral* if $r = 0$, and *inverse* if $r < 0$. It follows from (10.134) and (10.135) that the steady state exchange time of the neutral lagoon is τ_n is

$$\tau_n \equiv t_e / \Gamma^2$$

(10.136)

which is independent of the details of the physical processes in the lagoon. A typical value of t_e is about 5 days, and so for the value $\Gamma \sim 0.3$, $\tau_n \sim 50$ days. It is useful to define a steady state compensation ratio c

$$c \equiv \frac{\alpha \Delta T}{\beta \Delta s}$$

(10.137)

which is exactly unity for a neutral lagoon.

10.10.6 Phase plane

A good way of analyzing the model equations (10.134) and (10.135) is to construct a phase plane[11] based on the variables r and b. Phase planes were discussed in Section 4.5 and are a very powerful tool in understanding the properties of models controlled by a pair of non-linear ordinary differential equations. If we describe the state of a basin by a point (r, b) in the phase plane, the model dynamical equations tells us how that point moves in time, i.e., how that state of the basin evolves with time. We can take any point in the phase plane, and once chosen, this point defines a trajectory for the basin. Figures 10.11 to 10.14 show some examples of phase planes for specific values of θ, Γ and λ. The Matlab code for the model equations is shown in Table 10.1.

---

[10] Hearn and Sidhu (1999).
[11] The use of a phase plane for this model is due to Stommel although our analysis uses slightly different variables.

Table 10.1 *Matlab code for Stommel model*

```
theta = 5; Gamma = 0.3; dt = 0.005; lambda = 0.15; figure;
A = (theta^2)*lambda^2;
for side = 1:4;
        if side = = 1;rs = -3;bs = [-3:1:3];end;
        if side = = 2;rs = 3;bs = [-3:1:3];end;
        if side = = 3;bs = -3;rs = [-3:0.5:3];end;
        if side = = 4;bs = 3;rs = [-3:0.5:3];end;
        for r0 = rs;
            for b0 = bs;
            t = 0;
            clear bseries; clear rseries;
            bseries = b0; rseries = r0; r = r0; b = b0;
            while t< 30;
                    t = t+dt;
                    rs = lambda^2 +(r^2);
                    dboverdt = -( Gamma^2+b-(1-b)*rs);
                    droverdt = A*b-rs*r;
                    b = b+dboverdt*dt;
                    r = r+droverdt*dt;
                    bseries = [bseries, b]; rseries = [rseries, r];
            end;
            figure; plot(rseries, bseries); hold on;
        end;
    end;
```

The important features of phase planes are the so-called *singular* or *equilibrium* points and these correspond exactly to the *steady state* solutions of the dynamical equations (10.134) and (10.135). If the steady state solutions are *stable* the trajectories approach these equilibrium points asymptotically, i.e., they never quite reach the steady state points but instead have an exponential decay toward the steady state. A steady state solution could be *unstable* which means that the trajectories diverge away from the corresponding equilibrium point. Close to an equilibrium point, we can approximate the non-linear terms in the dynamical equations by linear terms in the differences δr and δb between r and b and their respective values at the equilibrium point. In the neighborhood of the equilibrium point, these linearized equations can be solved in terms of exponentials of form

$$\delta r \sim e^{-\Omega t}$$
$$\delta b \sim e^{-\Omega t} . \tag{10.138}$$

Because the dynamical equations consist of a pair of first order differential equations, there are two roots for the exponent Ω. These are called the solutions of the *eigenvalue*

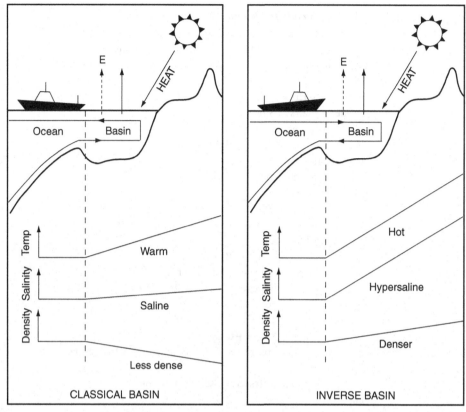

Figure 10.10 Schematic of a classical and inverse coastal basin showing the different directions of the convection cells due to the different sign of the longitudinal density gradient (the ship is moving downstream in both cases). The diagram illustrates temperature and salinity gradients. For a classical basin, the temperature-induced positive buoyancy dominates salinity-induced negative buoyancy. For an inverse basin, the negative buoyancy from hypersalinity has the greater effect.

equation or the *eigenvalues*[12] of Ω. which are either both real, or complex conjugates. If the eigenvalues are both real, (10.138) shows that if the equilibrium point is stable (note that we have used a negative sign in the exponent) both eigenvalues must be positive. This is called a *stable node*. If the eigenvalues are both negative, (10.138) shows that both δr and δb diverge with time and we have then an *unstable node*. If the real eigenvalues have opposite signs the trajectories tend to converge on the equilibrium point in one direction of the phase plane and diverge in the orthogonal direction and the equilibrium point is called a *saddle*. If the eigenvalues are complex conjugates with real parts that are zero, we have a *center* in which the trajectories orbit the equilibrium point. If the real part of the complex conjugates are non-zero, the

---

[12] *Eigenvalue* just means the *allowed* values, i.e., the values that come from the solution of the dynamical equations which describe the model.

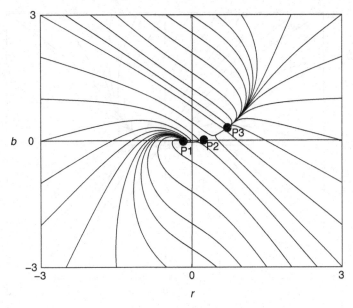

Figure 10.11 The phase plane of the Stommel model. The axes are r (related to density difference between ocean and basin) and b (related to surface buoyancy flux). The lines are trajectories which show the changing values of r and b as the basin evolves towards one of the steady state equilibrium points represented by P1 (inverse basin) or P3 (classical basin). P2 is an unstable saddle point. A Stommel transition involves the basin being perturbed in some way so that it switches between the neighborhoods of P1 and P3. The model parameters for this phase plane are $\theta = 5$; $\Gamma = 0.3$; $\lambda = 0.23$. The Matlab model code is given in Table 10.1. The coordinates for P1 are $(-0.16, -0.012)$ and for P3 $(0.72, 0.31)$.

equilibrium point is a *focus* which means that the trajectories either spiral into the point or out from the point according to the sign of the real part of the eigenvalues.

The equilibrium points of the Stommel model given by (10.134) and (10.135) correspond to the roots of $V(r) = 0$ where V is defined as the quartic

$$V(r) \equiv \left(\lambda^2 + r^2\right)\left[\theta^2\lambda^2 - \left(\lambda^2 + r^2\right)r - r\right] - \theta^2\lambda^2\Gamma^2. \qquad (10.139)$$

There are either one or three real roots of r depending on the values of the parameters θ, λ, and Γ and the corresponding values of b are

$$b = \frac{\left(\lambda^2 + r^2\right)r}{\theta^2\lambda^2}. \qquad (10.140)$$

If there is one root for r, the phase plane has only one equilibrium point and this corresponds to either an inverse or classical state of the basin (depending on the sign of the root of r). If there are three roots for r, the phase plane has two stable equilibrium points, one of which is classical and the other inverse; the third equilibrium point is unstable. This is illustrated in Figures 10.11 to 10.14 in which the value of λ is varied

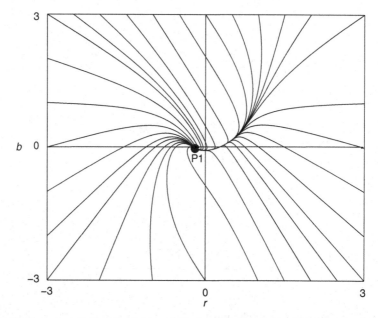

Figure 10.12 A repeat of Figure 10.11 with λ *decreased* to 0.15. This parameter is inversely proportional to the length of the basin and the figure shows that the 53% increase in length has removed the classical state and the basin may only be inverse. The coordinates of P1 are $(-0.2, -0.02)$.

between 0.15 and 0.35. The trajectories in these figures can be plotted using the code in Table 10.1 where the starting points of the trajectories are taken as equally spaced along the sides of the figures. In the figures presented here some of these trajectories have been removed for clarity.

In Figure 10.11 there are three equilibrium points: a stable focus P1 (inverse basin), the stable node P3 (which is a classical state of the basin), and the unstable saddle P2. Decreasing λ corresponds to increasing the length of the basin, and in Figure 10.12 the basin has one stable state shown by the stable focus P1 which is an inverse state of the basin. The other two figures (Figures 10.13 and 10.14) correspond to increasing λ, i.e., decreasing length of the basin and this first causes the inverse state to become *quasi-neutral* and then to disappear completely so that for the shortest basin only the classical state exists.

Note that as the trajectories in Figures 10.11 to 10.14 approach either P1 or P3, they tend to cluster around a curve on which b is quadratic in r. In terms of the mathematics of the phase plane this means that one of the eigenvalues is much larger than the other meaning that there is a fast and a slow process operating within the model. For the present choice of the controlling parameters of the model, the faster process is the basin coming into near equilibrium with the heat loss and gain through the surface; the slower process is exchange with the ocean.

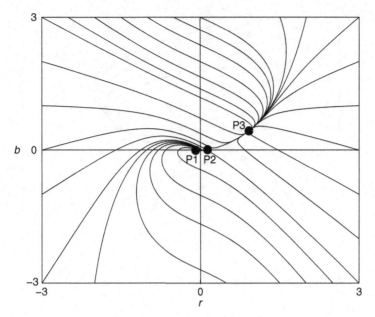

Figure 10.13 A repeat of Figure 10.11 but with λ *increased* to 0.28 corresponding to a basin that is 18% shorter than in Figure 10.11. The coordinates for P1 are $(-0.09, -0.004)$ and for P3 $(0.92, 0.43)$ so that P1 is approaching quasi-neutrality with almost exact compensation between the density effects of temperature and salinity (the truly neutral state is unstable).

The transition from a classical to inverse hypersaline basin may occur very quickly, and in a model box occurs as a rapid jump called a *Stommel transition*. Stommel discovered the transitions in box models of the thermohaline circulation of the deep oceans. They have since been seen in numerical models of the deep ocean.[13] Hearn and Sidhu[14] are responsible for a series of papers on the theory Stommel transitions in shallow coastal basins whilst Whitehead[15] and his co-workers have demonstrated the existence of the transitions in laboratory simulations.

Stommel transitions can occur within a particular basin provided that both P1 and P3 are present; it is envisaged that some perturbation causes the state of the basin to jump between the neighborhoods of P1 and P3. If we start with a basin at some arbitrary initial point in the phase plane, the basin will follow the trajectory that passes that point and finally arrive at either P1 or P3 depending simply on our starting point. In a sense each of the equilibrium points has its own *catchment area* and rather like a raindrop falling on the Earth, the final fate of the basin is decided by its starting point. If there is only one equilibrium point, say P3, then all trajectories will lead to P3. Otherwise the entire phase plane is divided between the catchment areas for P1 and P3.

---

[13] Rahmstorf (1995), Rahmstorf and Ganopolski (1999), and Stocker and Wright (1991).
[14] Hearn (1998), Hearn and Sidhu (1999).   [15] Whitehead (1995, 1998).

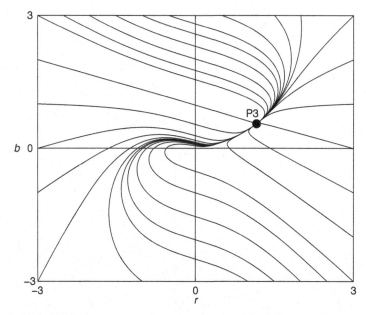

Figure 10.14 A repeat of Figure 10.11 but with λ further *increased* to 0.35 corresponding to a basin that is 35% shorter than in Figure 10.11. P1 no longer exists and the basin is necessarily classical. The coordinates for P3 are $(1.16, 0.56)$. For short basins, such as that illustrated in this figure, the exchange time is too short for the occurrence of hypersalinity.

The relative areas of these catchment areas is called their *attractiveness*. If we alter one of the basin parameters, for example by altering the length of the basin, we alter the *topography of the phase plane*, which means that an equilibrium point may disappear or a new equilibrium point may be born. This means that the attractiveness of an equilibrium point varies with the basin parameters. Changing the topography can cause a basin that is close to one equilibrium point, and therefore near to steady state, to become time dependent and hence to move to another stable state. This is another type of Stommel transition and occurs almost discontinuously as the parameters of the coastal basin are changed.

10.10.7 Tomales Bay

The hypothetical neutral basin is unstable in our model and instead the usual situation found in many basins is a quasi-neutral inverse states. This is illustrated in Figure 10.13 where the equilibrium inverse state represented by P1 is almost neutral. The first observation of a *quasi-neutral inverse state* was made in Tomales Bay (shown in Figure 2.3) by a team lead by Stephen V. Smith,[16] a scientist who has

---

[16] Smith, Hollibaugh, Dollar, and Vink (1991a, 1991b), Hearn and Largier (1997).

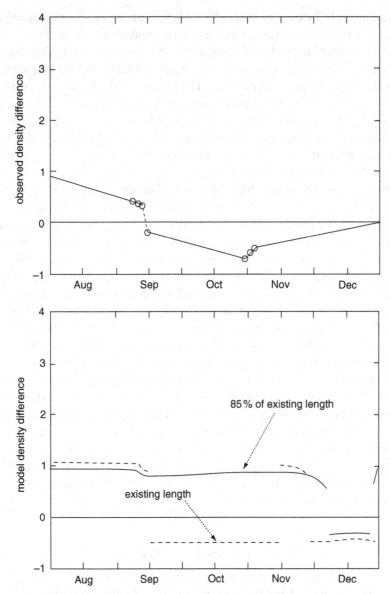

Figure 10.15 Upper panel shows a time series of the observed density difference (measured in kg m^{-3}) in Tomales Bay during summer and fall 1988 based on eight data points (open circles). The lines are linear interpolations between these data points. (Data from Smith *et al.*, 1991a.) The lower panel shows two corresponding time series obtained from a three-dimensional numerical model (Hearn and Sidhu, 2003) using both the 1988 length of the basin (broken line) and a basin 85% of that length (full line). For the 1988 length, there are indications of sudden jumps with a Stommel transition in late summer from a classical to inverse basin; the model suggests that transition would not be present if the basin were reduced to 85% of its 1988 length.

had a leading role in our understanding of hydrodynamics in the chemical and biological dynamics of coastal basins. Such quasi-neutral inverse states occur naturally in late summer for shallow basins in *mediterranean climates*; such basins are generally called *mediterranean basins* (see Chapter 3). During the summer such basins are typically hyperthermal and classical. The salinity of the basin may then start to increase, which moves the basin in the direction of a neutral basin. Near neutrality is very advantageous for the creation of hypersalinity, because it has very long exchange times due to the lack of density-driven (convective) flow and the longer is the exchange time, the greater the hypersalinity. This is simply a consequence of water particles spending longer in the basin.

The upper panel of Figure 10.15 shows a possible observation of a Stommel transition in Tomales Bay, based on approximately monthly data for the year 1988. The panel shows the difference between the density of the ocean and the basin. The sign of this density difference is such that it is positive when the basin is classical. Notice that as the basin becomes more hypersaline in late summer, the density difference decreases but stays positive (classical) until it quite suddenly jumps to a negative value, i.e., the basin jumps from classical to inverse. The basin remains inverse until late October when it increases and has the appearance of becoming positive again although there are no further data until the end of December. These changes in the basin are driven by monthly changes in the equilibrium temperature T_e discussed earlier in this chapter and changes in the net evaporative velocity E. The Stommel model shows that transitions are largely controlled by a parameter Γ defined in (10.124). Indications from the Stommel model of Tampa Bay is that the transitions shown in the upper panel of Figure 10.15, occurs at a measured value of $\Gamma \sim 0.6$. The changes of Γ with time are due to changes of the equilibrium and buoyancy temperatures which are a result of changing atmospheric, ocean, and river flow. It is possible to reproduce Stommel transitions in a three-dimensional numerical model as shown in the lower panel of Figure 10.15 for Tomales Bay. This also illustrates the critical dependence on the length of the basin.

10.11 Further reading

See further reading from earlier chapters. Additional texts for this chapter are:

Batchelor, G. K. (1996). *The Life and Legacy of G. I. Taylor*. Cambridge, UK: Cambridge University Press.

Caviedes, C. N. (2001). *El Niño in History: Storming Through the Ages*. Gainesville, FL: University of Florida Press.

Dyer, K. R. (1973). *Estuaries*. Chichester, UK: John Wiley.

Fischer, H. B. (1981). *Transport Models for Inland and Coastal Waters: Proceedings of a Symposium on Predictive Ability*. New York: Academic Press.

Fischer, H., J. E. List, C. R. Koh, J. Imberger, and N. H. Brooks (1979). *Mixing in Inland and Coastal Waters*. New York: Academic Press.

Philander, S. G. (2004). *Our Affair with El Niño: How We Transformed an Enchanting Peruvian Current into a Global Climate Hazard*. Princeton, NJ: Princeton University Press.

Tomczak, M. (1983). *Synthesis and Modeling of Intermittent Estuaries*. New York: Springer.

Turner, J. S. (1979). *Buoyancy Effects in Fluids*. Cambridge, UK: Cambridge University Press.

11

Roughness in coastal basins

11.1 Introduction

Roughness or *rugosity* has a major role in models of coastal basins at a variety of spatial scales. It is defined in the most general sense as a stochastic variation in the profile of the surface that defines the interface between two media. An example is the roughness of the seabed, and another is the roughness of a coastline. Both of the terms *roughness* and *rugosity* are in common use, although the latter does have a quantitative definition as the ratio of the *true* surface area to the *geometric area* of the interface. For example, the geometric area of the Earth is that of a sphere (or more precisely an oblate spheroid), whereas the true surface area involves the Earth's complex topography. In reality, the value obtained for the surface area depends on the spatial scale that we use, and this is discussed in more detail later in this chapter. So, rugosity is well defined if we specify the spatial scale at which we measure the true area. This idea can also be used in one dimension and, for example, a chain can be placed over a coral reef to measure the length its surface and this is then compared with the horizontal distance between the points measured with a surveyors tape. The result clearly depends on the length of the links in the chain but is a traditional survey tool for measuring rugosity. Light Detection and Ranging (LIDAR) methods are now used to provide more advanced measures of rugosity. LIDAR has the capacity to measure depths up to very small spatial scales, and so has potential to provide considerable information on many aspects of roughness.

Our perception of roughness is that a coral reef is much rougher than, say, the shell of a chicken's egg. Yet under a microscope we find that eggshell is not really smooth. So to be more quantitative about the comparative roughness of eggshell and a reef, we might consider the size of the roughness elements on the shell relative to the size of the egg, and make a comparison with roughness elements on a reef compared to its size. However, we do not know whether that sort of a comparison is a valid one until we can understand the effects of roughness. Most of that understanding comes originally from chemistry, and chemists have long known that the ability of a material to absorb a fluid, or to interact in any way through its surface, depends on the area of its surface per unit volume of the material. Roughness is important in any process that involves surfaces because of the very nature of solids and fluids. Although we shall talk about *smooth* surfaces, we recognize that at a molecular level all surfaces are rough.

11.2 Skin and form drag

In almost all fluid flows, friction is balanced either partially or completely by pressure gradients. That is evident in any situation as we watch water flow through a channel: the set-up of the surface as water flows past the support of a bridge structure, or even the rising level of the water in front of our legs as we wade into the ocean for a swim. The cause of this *friction* is the disturbance in the flow produced by some agency, which might be the body that impedes the flow. In aerodynamics it is usual to distinguish between a number of distinct contributions to friction, of which *skin friction* and *form friction* are amongst the most important. This distinction is partially a matter of spatial scale. *Skin friction* is drag associated with the stress between the surface of a body and the moving fluid. It is basically the friction which we discussed in Chapter 7 and is reduced by decreasing the surface roughness leading, in the limit of virtually no roughness, to the idea of the completely smooth surface. *Form friction* is a drag which is associated with the need for air to flow past over a body as a whole and its name comes from its dependence on the shape, or *form*, of the body. Form friction also depends on the skin roughness of the body and may be paradoxically reduced by increasing the surface roughness.

The discovery of form friction is really due to a paradox named after the French mathematician and physicist Jean le Rond d'Alembert (1717–83). The solution of a paradox has the potential for great advances in science and that was certainly the case for the solution of *d'Alembert's paradox*. It is easy to show that a theoretically perfectly smooth sphere should offer no drag to passing air, or at least what drag is present should be gradually reduced by continual polishing of the surface. So, we might imagine a ball with a surface like glass passing through air without friction. This result does not agree with experiment, and it is found that the smoothest sphere that we can produce shows drag of a significant magnitude. The reason for this discrepancy is that there is no such thing as a frictionless surface. Very close to any surface the flow speed must be zero and this produces a finite skin friction. Consequently, there is a velocity gradient that stretches out through the fluid from the surface. A pressure gradient is required around the sphere in order to accelerate and decelerate the particles transversely, as the fluid streamlines bend around the sphere. Consider a plane through the center of the sphere so that the plane is parallel to the flow. Let us denote by θ the angle between the direction of motion of the sphere and a point on the circumference of this circle. The pressure p can be determined at any point of the surface and can be considered a function of θ. In the absence of skin friction we would assume that there is no net force on the sphere, i.e., no drag. For this to be true, it is necessary that the pressure field does not itself produce drag and this means that the pressure is the same at $\theta = 0°$ and $\theta = 180°$. This requires completely symmetric flow around the sphere. For such a flow, a problem arises with the stability of the flow in the region between $\theta = 90°$ and $\theta = 180°$, where the pressure is increasing in order to slow the speed of the fluid. The flow simply breaks away from the surface, and we have *flow separation*. This instability is associated with the laminar flow region inside the viscous boundary layer. The flow separation is responsible for the *form* drag and requires higher pressure at $\theta = 0°$ than at $\theta = 180°$. This pressure gradient is required to supply energy to the turbulent field produced behind the sphere.

Even the non-golfers of the world are familiar with the shape of a golf ball, and have probably wondered why its surface is covered with small indentations, i.e., it is made deliberately rough. This seems strange, since golfers are interested in the ball moving as far as possible after being struck. The roughness on a golf ball does increase skin friction, but reduces form drag by suppressing flow separation associated with the viscous boundary layer. At a critical Reynolds number the drag decreases as the flow becomes turbulent, and this impedes (but does not curtail) the flow separation. The dimples cause the sphere to enter the turbulent regime at low Reynolds number, so that for a range of Reynolds numbers the drag of the dimpled sphere is lower than a sphere without dimples. Another way of reducing form drag is to *streamline* the body. These streamlined bodies have a *teardrop shape* that reduces the pressure change along the streamline so that the flow remains attached until very near the trailing edge; the resulting wake is therefore smaller with less form drag.

In coastal basins we deal with roughness that exists on spatial scales from fractions of a millimeter, through tens of centimeters, to many meters on complex structures such as coral reefs. The roughness elements in all of these surfaces experience drag that lies within the realm of stochastic processes. For small systems, the drag on such surfaces can be measured in a tank by driving water across the surface with a known pressure gradient and measure the flow speed. This technique is the same as that used in the earliest experiments in the nineteenth century. The drag can also be inferred from field measurements and model calibration.

11.3 Scales of spatial variability

If we look at an image of the seafloor it is often difficult to judge the spatial scale of the image unless a familiar object is kindly placed in view. We have the same problem in looking at images of clouds. In the same way, if we look at turbulent motions in water we see the same sorts of patterns at spatial scales from hundreds of kilometers down to centimeters. We summarize this situation by saying that spatial variability exists on all scales and is superficially similar. Figure 11.1 shows two pictures of the coastal bathymetry off Monterey on the Californian coast of North America. They are shown at different spatial scales, and the reader may agree that they do look very similar, and moreover the reader might find it difficult to identify the larger of the spatial scales. The idea of self-similarity is due to Gottfried Wilhelm Leibniz (1646–1716). It possible to

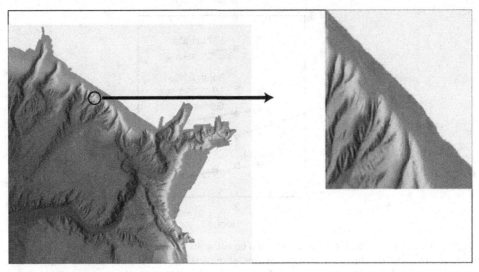

Figure 11.1 The bathymetry of Monterey canyon off the west coast of California drawn at two spatial scales. The right-hand image is a zoom of a small part of the left-hand image illustrating that similar structures exist at both spatial scales.

create a mathematical object that exactly captures this property, and is called a *fractal*. The development of the theory of fractals has a long history in mathematics, although the name fractal is due to Benoît B. Mandelbrot (b. 1924). The fractal is an extension of the ideas of traditional geometry to non-integer dimensions, and is a mathematical way of representing natural objects such as plants, trees, rocks, and coral reefs which are *not smooth* in the manner of the objects of Euclidean geometry, such as a sphere. The theory of fractals accepts that there is no small spatial scale at which natural objects become smooth.

One of the earliest contributions to the theory of roughness came from Lewis Fry Richardson (1881–1953). Richardson considered the length of a coastal as a function of the size of the ruler used to make the measurement (Figure 11.2). The length of a coast increases as one plots it on an increasingly fine scale, and this has been well known since the earliest production of maps. Richardson found that the scale dependence was approximately linear on a log–log plot, and the slope of the line clearly serves as a measure of roughness:

$$\log L = -(\sigma - 1)\log s + b \tag{11.1}$$

where σ is the *fractal dimension*. A straight line has a fractal dimension of 1 but for all non-Euclidean objects the fractal dimension is fractional. As a coast becomes smoother, so σ decreases towards 1. For Great Britain, $\sigma = 1.24$, a fractional value, whilst the coastline of South Africa is much smoother, virtually an arc of a circle, and $\sigma = 1.04$. Suppose we measure the length L of a coast with a chain having links of

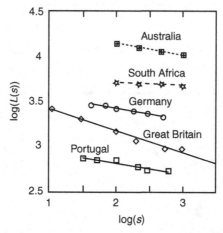

Figure 11.2 Richardson plot of the length of a coastline as a function of the length s of the ruler used to make the measurement. The slope fractal dimension σ is given by adding unity to the negative slope of the lines: Australia $\sigma = 1.13$, South Africa $\sigma = 1.04$, Germany $\sigma = 1.12$, Great Britain $\sigma = 1.24$, Portugal $\sigma = 1.12$. For a geometric coastline the dimension would unity; for these data, Great Britain has the roughest coast and South Africa the smoothest.

length s, and let s_0 be the total length of the chain, then it follows from (11.1) and noting that $L = s_0$ when $s = s_0$

$$\log(L/s_0) = -(\sigma - 1)\log(s/s_0).\tag{11.2}$$

The ratio of L to s_0 is called the rugosity r so that (11.2) becomes

$$\log_{10}(r) = -(\sigma - 1)\log_{10}(s/s_0)\tag{11.3}$$

or

$$r = (s_0/s)^{\sigma-1}.\tag{11.4}$$

So, for example, if we use a tape of length $s_0 = 10\,\text{m}$, and a chain with links $s = 1\,\text{m}$, (11.4) gives

$$r = 10^{(\sigma-1)}\tag{11.5}$$

and so if the rugosity is 1.5, fractal dimension is 1.2. The parameter $\sigma - 1$ is a measure of how well the fractal structure fills all available space. So for a straight line $\sigma - 1$ is zero, since a line has zero width and when σ approaches 2, the structure has effectively become two-dimensional and fills the whole of a two-dimensional space with $\sigma - 1 = 1$. Fractals in nature occur in three-dimensional space, and their dimension varies from $\sigma = 2$ for a plane surface, to $\sigma = 3$ for a structure that fills all of three-dimensional space (and is just a solid body). In either two- or three-dimensional space,

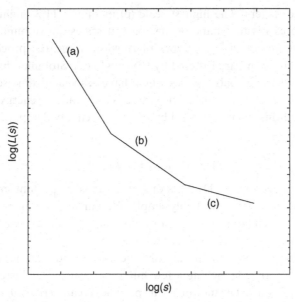

Figure 11.3 Hypothetical generalized Richardson plot. The fractal dimension, although constant over several orders of magnitude, does change with regions of spatial scale. In this example, (a) could be the roughness regime of individual corals, (b) the roughness of the entire reef, and (c) the roughness of the coastal region in which the reef is located.

the departure of the fractal dimension from an integer, i.e. σ, is a measure of roughness. The method used by Richardson is usually known as a Richardson plot and although there are now many better methods of accessing fractal dimension, it remains a good conception foundation stone of the theory of fractals.

The fractal has perfect *self-similarity* at all spatial scales which means that its form is exactly similar at all spatial scales. For natural objects, this self-similarity only exists over a limited range of scales and more generally we find that the roughness changes with scale. Figure 11.3 illustrates this idea for the case in which the fractal dimension increases as we move to smaller spatial scales, i.e., the coast becomes rougher. We call the plot in Figure 11.3a *generalized Richardson plot* because it extends the original idea of the Richardson plot to a wider range of spatial scales over which the fractal dimension varies with spatial regime. This variation of fractal dimension with the spatial regime is considered to be a consequence of the changing nature of roughness with regime. Figure 11.3 is based on a hypothetical coastline for which we have identified three regimes. The first (a) occurs at the smallest scales and we might call the *local* roughness regime and would correspond to the roughness of (say) the individual elements of a coral reef or perhaps sediment grains or small sand waves. The next highest regime (b) has been labeled *structural* and involves the roughness of individual structures on the coast such as complete reefs or individual rock formations. Typically, we would expect the change of regime from (a) to (b) to occur at scales of centimeters to meters. The highest scale (c) in Figure 11.3 might be called *geo-morphic* and represents the shape of the coast at scales above hundreds of meters. These regimes represent different types of roughness and the processes that both create that roughness and are fostered by the roughness. Stommel[1] has discussed the interconnection of these spatial scales which have associated time scales as we discussed in Chapter 7. As with turbulence there is a cascade of energy from larger to smaller scales and this is characterized by the rate of energy dissipation ε.

11.4 Models of reef growth

Diffusion limited aggregation (DLA) is one of the most important models of fractal growth. The growth rule is remarkably simple. We start with an immobile seed on the plane. A walker is then launched from a random position far away and is allowed to diffuse. If it touches the seed, it is immobilized instantly and becomes part of the aggregate. We then launch similar particles one by one and each of them stops upon hitting the cluster. After launching a few hundred particles, a cluster with intricate branch structures results. The motion of the particles is called *random walk*, and is said to resemble the motion of a walker taking each step in a random direction.

An example of a *DLA model* coded in Matlab is shown in Table 11.1. The seed for the growth is placed at the center of a mesh which we have taken as 100 by 100 cells. This can be enlarged as the user wishes. We introduce an occupation array which tells us whether each cell is occupied (occupation = 1) and empty (occupation = 0). This array is initially set to zero everywhere except at the position of the seed. We then release particles from a random point on the boundary, and they diffuse until they reach a cell which has an occupied nearest neighbor. The particles then stick, just as a coral larva might settle on a coral structure. Once that happens, we release a new particle and it randomly finds a place on the existing structure to stick. The types of structure that result is shown in Figure 11.4. Clearly the resulting structures are quite random in their form and we will not obtain the same structure after each run of the model (unless we code the model to produce the same set of random numbers for each run). The structures tend to have the form of arms stretching outward from the central original seed. The reason for this form is that as the structure grows, particles find it increasing difficult to penetrate to its inner region because they tend to stick to the outlying arms before they can diffuse to its center. Let us just look through some of the details of the code in Table 11.1 and we have annotated the code with letters A to F.

A: *rand(2)* is a 2 * 2 matrix with random entries, chosen from a uniform distribution on the interval (0.0, 1.0).

B: this line of code finds the position of the maximum random element generated by *rand(2)* and the subsequent lines use this position to find which of the four boundaries of the model

[1] Stommel (1963).

Table 11.1. *Matlab code for a simple diffusion limited aggregation (DLA) model of reef growth;* comments A to F are detailed in text

```
figure(1); N = 100; M = 100; P = 4*(N + M);
nseed = N/2; mseed = M/2; occ(1:N, 1:M) = 0; occ(nseed, mseed) = 1;
for r = 1:P;
        start = rand(2);                                          %A
        [Y,I] = max(start);                                       %B
        if I = = [1 1]; j = 1; end;
        if I = = [1 2]; j = 2; end;
        if I = = [2 1]; j = 3; end;
        if I = = [2 2]; j = 4; end;
        if j = = 1; n = 1; m = ceil(M*rand); end;               %C
        if j = = 2; n = N; m = ceil(M*rand); end;
        if j = = 3; m = 1; n = ceil(N*rand); end;
        if j = = 4; m = M; n = ceil(N*rand); end;
        check = 0; jumpmore = 1; count = 0;
        nseries = nseed; mseries = mseed;
        while jumpmore = 1;                                      %D
            while check = = 0;
                    jump = rand(2);                             %E
                    [Y,I] = max(jump);
                    if I = = [1 1]; j = 1; n = min(n + 1,N); end;
                    if I = = [1 2]; j = 2; n = max(n-1,1); end;
                    if I = = [2 1]; j = 3; m = min(m + 1,M); end;
                    if I = = [2 2]; j = 4; m = max(1,m-1); end;
                    if occ(n,m) = = 0; check = 1; end;
            end;
            stick = 0;                                          %F
            if occ(min(N,n + 1),m) = = 1; stick = 1; end;
            if occ(max(1,n-1),m) = = 1; stick = 1; end;
            if occ(n,min(M,m + 1)) = = 1; stick = 1; end;
            if occ(n,max(1,m-1)) = = 1; stick = 1; end;
            if stick = = 1;
                    occ(n,m) = 1;
                    jumpmore = 0;
                    nseries = [nseries,n]; mseries = [mseries,m];
                    figure(1); plot(nseries,mseries,'*'); hold on;
                    axis([1 N 1 M]); pause(0.1);
            end;
            check = 0;
        end;
end;
```

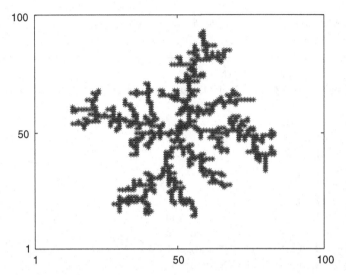

Figure 11.4 The diffusion limited aggregation (DLA) model of a fractal structure. The model contains 100 * 100 cells and 800 particles.

will be used for the release of the next particle. *P* such particles are released and this model has taken a value of *P* given as the sum of the number of cells along all four sides of the boundaries.

C: *m* is being chosen randomly as a position along a boundary.

D: *jumpmore* is an index held at 1 while the particle remains free to diffuse in the model, i.e., jump between cells. When the particle sticks, *jumpmore* becomes 0.

E: *jump* is a 2 * 2 random array that tells a particle which of its four nearest neighbor cells are to be the end point of its next jump. Note that the particle is only jumping to nearest neighbors and not next nearest neighbors; it is easy to extend the code so that particles do move further on each jump.

F: *stick* is an index that is initially zero but will be raised to unity if any of the nearest neighbor cells is occupied.

The particles typically take NM/4 jumps before they stick to the existing structure although that number decreases as the structure grows. To obtain structures in three dimensions at fine resolution requires considerable computing time.

Notice that in Figure 11.4 only 8% of the mesh is occupied, but its area greatly exceeds that fraction, and its line surface area per unit cross-sectional area is close to 75%. The value of μ from the previous section looking downwards on this structure is therefore If the same particles were placed in a layer at the bottom of the square in Figure 11.4, they would form a block of length 100 and height 8 with a line surface area that is 14.5% of the cross-sectional area. In two dimensions, structures such as that shown in Figure 11.4 have a major blocking effect on flow in the two-dimensional plane that is presented, but that effect is much smaller for a three-dimensional

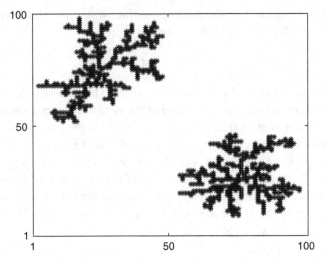

Figure 11.5 A repeat of the diffusion limited aggregation (DLA) model in Figure 11.4, but with two seed particles (the model again has 100 * 100 cells and contains a total of 800 particles).

structure. If two seed particles initiate the growth, we obtain a much more open structure, as illustrated with the same number of particles in Figure 11.5. This emphasizes that (as with all models) the result is dependent on starting conditions. Simulation of such structures should take into account dominant current flows, and will then produce a structure that has a more open structure (in that the structure interacts with the flow), and we obtain a boundary layer near the growth center with slower advection and preferential growth.

11.5 Nutrient uptake

11.5.1 Mass transfer limited processes

We saw in Chapter 7 that there exists around the surface of submerged aquatic vegetation (SAV) a *diffusive boundary layer* in which the transport of nutrients can only occur through molecular processes. We say that the supply of nutrients through this boundary layer is *mass transfer limited*. If A is the *active surface area of a system* such as a coral reef or seagrass bed, we have a total uptake rate R under mass transfer limitation

$$R = \beta A. \tag{11.5}$$

If the concentration of nutrients near the outer surface of the boundary layer is C we can suppose that

$$\beta = \sigma C \tag{11.6}$$

where β is a coefficient of transfer from the surface water into the system. Denoting R/C as the volume flux per unit time w, we can write

$$w = \sigma A \tag{11.7}$$

and note that w has the units of $m^3 s^{-1}$ and σ is a speed. The model implied by (11.6) takes the concentration of nutrients at the surface of the system to be zero which is consistent with our assumption of mass transfer limitation in that it gives the maximum transport through the boundary layer. Equation (11.6) implies that if we were able to load the water column with more nutrients, C would increase, and the uptake rate would increase proportionally. This is accepted to be true provided that the supply of nutrients is limited by transfer across the boundary. If that is not true and the demand for nutrients is less than the rate given by (11.5), the concentration at the inner surface of the layer will rise and could eventually reach a value which would totally cut off supply to the system.

In principle, a living organism can increase its area A so as to increase the value of w, and it is important to emphasize that A is the area of the *entire* system. So if σ has already reached a maximum, we can always increase w by increasing A. This means that such systems will never be truly *mass transfer limited* until they reach a maximum value of A that is consistent with other criteria affecting the system. This result is well known in the chemistry of uptake of ions in the chemical engineering industry. The maximum value of A increases with the fractal dimension of the system although increasing the area can also reduce water flow past, or through, the system. Increasing fractal dimension, or roughness, will not only block flow but also increase surface stress in relation to mechanical strength. This means that a model of some sort of optimum value for A or fractal dimension, is quite difficult. Furthermore, some consideration must be given to the spatial scale at which A is measured.

By contrast with the problems implicit in calculating A, it is possible to propose a simple model for the coefficient of transfer from the surface water σ into the system. However, such a model can only be verified by measurements of the total rate w and therefore verifying values of σ still involves determining A. A short cut around this problem is to measure w relative to the horizontal area of the basin. This only has real meaning on a large system such as an extended seagrass bed or a coral reef. Nevertheless, this type of measurement has been made in a *flume* either placed over part of a reef, or seagrass bed, or with pieces of living coral distributed on top of a sealed bottom. The rate of nutrient uptake per unit horizontal area is denoted by S which is called the *uptake constant* and S, like σ, has the units of a speed. To relate S to σ requires a factor which we call μ defined as the surface area of the system per unit horizontal surface area subtended by the system. Then

$$S = \mu\sigma \tag{11.8}$$

so that if the entire system, subtends a surface area of a the uptake rate is aSC which is identical to $a\mu\sigma C$, i.e, $A\sigma C$. For example, if the system was a set of perfect hemispheres arranged on the bottom of the basin $\mu = 2\pi r^2/\pi r^2 = 2$. The value of μ needs to account for the distances separating individual parts of the system. The problems of defining A are carried over into determining a value for μ. Recall that there is a possible control of the value of μ in that terms of impeding the flow or producing excess stress on the surfaces of the system but true mass transfer limitation will only occur once μ has reached that maximum.

11.5.2 Uptake dissipation law

The *uptake dissipation law* (UDL) relates ion transfer rates through the diffusive boundary layer to the energy dissipation rate ϵ. It has the form

$$\sigma = \frac{D}{\delta} \tag{11.9}$$

where σ is the uptake constant introduced in the last section, D is the molecular diffusivity of the ion, and δ is the thickness of the diffusive boundary layer. The thickness of the diffusive boundary layer is related to the *Batchelor length* that we discussed in Chapter 7.

Let us consider an eddy inside a turbulent fluid. Suppose that the eddy has mean radius r and speed u. The rate of damping of the eddy by inertial (non-linear) forces is proportional to u^3/r, and this is the rate of transfer of energy to smaller eddies. Hence, in a steady state the rate of energy cascade is $\varepsilon = au^3/r$, where a is a universal constant (a number). So, the eddy kinetic energy/mass $E\,(=u^2/2)$ is

$$E = A(\varepsilon r)^{2/3} \tag{11.10}$$

and $A \equiv a^{2/3}/2$ and is now known to be about 1.5 by measurements of turbulence spectra. The rate of dissipation by heat is

$$v(u/r)^2 \tag{11.11}$$

where v is the molecular viscosity. So the ratio of energy loss by viscosity to cascade damping is $2A/(r/r_K)$ where r_K is the *Kolmogorov length* defined by

$$r_K = (v^3/\varepsilon)^{1/4} \tag{11.12}$$

As the effect of viscosity increases so there is less energy within the cascade, and hence u and E decrease rapidly at small r. When $r = 2Ar_K$ the ratio is 1. When $r = r_K$ the ratio is about 3 and there are few remaining eddies.

George Batchelor (1920–2000) was a pioneer in the application of the ideas of Andrei Nikolaevich Kolmogorov (1903–87), who formulated a possible equilibrium theory of homogeneous turbulence. These ideas also exercised the creative talents of G. I. Taylor, who was Batchelor's original mentor at Cambridge. Batchelor pointed out that even the tail of the turbulent cascade produces more ionic diffusion than molecular ionic diffusion, because diffusivity associated with the latter is several orders of magnitude smaller than viscosity. Batchelor derived the form of the tail of the energy distribution of the eddies of radius r as

$$E = \frac{u^2}{2} = A(\varepsilon r)^{2/3}(r/r_K)^{4/3} \quad \text{for} \quad r \leq r_K. \tag{11.13}$$

Fluctuations in ion concentration (and temperature) of these small eddies exist down to a radius at which the molecular diffusion across the eddy is greater than the advection by the eddy, i.e.,

$$u = \frac{D}{2r} \tag{11.14}$$

or using the Batchelor length

$$r_D \equiv \left(\frac{D^2 \nu}{\varepsilon}\right)^{1/4} \tag{11.15}$$

we obtain

$$2r = \frac{r_D}{(A/2)^{1/4}} \tag{11.16}$$

where $(A/2)^{-1/4} \approx 1.07$ and so the distance $2r$ is about equal to the Batchelor length. The variation of ion diffusivity with r is

$$K = D\left[1 + (r/r_H)^2\right] \tag{11.17}$$

$$r_H \equiv r_D \big/ (8A)^{1/4} \tag{11.18}$$

For a diffusive boundary condition, we apply the boundary condition $C = 0$ at the inner edge of the layer $z = 0$, and let $C \to C_0$ as $z \to \infty$, so that

$$C(z) = 2r_H \int\limits_0^{z/r_H} \frac{B \, dz'}{D\left[1 + (z')^2\right]} = \frac{C_0}{\delta} r_H \tan^{-1}(z/r_H) \tag{11.19}$$

with

$$\delta = \frac{\pi}{(8A)^{1/4}} r_D. \tag{11.20}$$

Figure 11.6 compares the form (11.19) with the simple concentration formed derived in Chapter 7 showing that the diffusive boundary is broader than the Batchelor length.

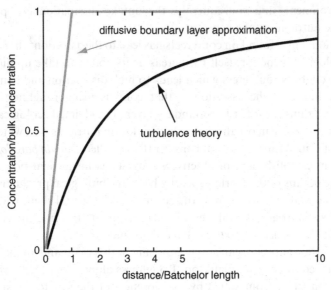

Figure 11.6 The variation of concentration of a nutrient with distance from an absorbing surface (where the concentration is zero). The heavier line is the prediction of turbulence theory and the lighter line shows the assumption of a discrete finite thickness diffusive boundary layer discussed in Chapter 7 (compare Figure 7.9).

The uptake dissipation law then becomes

$$r = D \left/ \frac{\pi}{(8A)^{1/4}} r_D \right.$$ (11.21)

or if we take the value of the surface area constant μ as 2 (corresponding to hemispheres),

$$S = \left(\frac{D^2 \varepsilon}{\nu} \right)^{1/4}.$$ (11.22)

11.5.3 Energy production and dissipation on coral reefs

As we shall see later in this chapter, bottom friction is the dominant force in the hydrodynamics of flow over coral reefs and this is also true of seagrass beds. As we saw in Chapter 1, friction is usually a dominant force in coastal basins for a number of reasons and additionally reefs probably have some of the roughest surfaces which we find near the coast. This is due to both the small-scale roughness that we would find if we tried to handle coral but also due to the larger-scale roughness associated with the individual coral structures. This means that very considerable forcing is necessary to drive currents over reefs such as set-up of the water surface by breaking waves. As a consequence, energy dissipation on reefs is high compared to the usual values in the marine environment, and consequently, the uptake dissipation law predicts high values for the nutrient uptake constant S.

The high values of S found on coral reefs has lead to the assertion[2] that coral creates a special ecological niche for itself by increasing its ability to take up nutrients from the ocean through its roughness which leads to high dissipation and hence increased values of S. Basically the assertion is that coral is using roughness to produce turbulence that thins the diffusive boundary layer. The claimed ecological advantage is that coral can survive in ocean waters that are low in nutrients, i.e., waters that have a low value of C in (11.6). This would answer the so-called *Darwin paradox*[3] as to how coral reefs can establish high productivity ecosystems in nutrient-poor waters. We recall that the oceans of the world generally have low biological productivity with the exception of special regions, such as the coast and continental shelf, where there are local increases in nutrients. Coral reefs are an exception to this rule and can be likened to an *oasis in the nutrient desert* and have the same productivity as a tropical rain-forest.[4] Coral are complex organisms comprising both a plant and animal component and they can feed directly on plankton. Roughness alone cannot achieve high productivity and must be accompanied by the closeness of the reef to the surface giving high light penetration; this very shallow environment also ensures wave breaking which aids energy dissipation on the reef and a low-energy environment in the lagoon behind the reef.

Coral reefs consist of much more than just coral and much of the ecosystem exists in the lagoon behind the reef and comprises a large trophic range of microbes, plants, and animals. Reefs also recycle nutrients and are certainly not entirely nutrient limited. An added argument to the importance of high value of S to coral reefs is that reefs generally seem adapted to nutrient poor waters. Corals are usually harmed by the effects of low salinity, or warm, water and seem perfectly adapted to an environment which is removed from terrestrial influences but is sufficiently shallow for reefs to be able to grow close to the surface and receive sufficient sunlight.

Mass transfer limitation is usually based on some form of restriction of transport of nutrients. Usually, nutrients are present in the ocean but plants are simply unable to access them, i.e., nutrients cannot be transported sufficiently rapidly enough across the plant membranes or across the diffusive boundary layer. Increasing nutrient concentration always helps that transport but that effect should not be confused with a limitation in the *quantity* of nutrients within a basin. Even in the presence of nutrient limitation, the fluxes of nutrients into and out of the basin are far greater than the uptake flux. The limited uptake can be likened to a man with poor fishing skills trying to catch fish from the bank of a fast-running river filled with fish. Unless nutrient fluxes across the boundaries of a coral reef system are measured with extreme

---

[2] *New Scientist*, September 21, 2002, 17.

[3] Most paradoxes are named after a person whose work has, in retrospect, highlighted the paradox and the use of Darwin's name here signifies his comments on reef productivity without implying his use of the actual word *paradox* in his writings.

[4] Hatcher (1990).

accuracy, this might lead to the erroneous conclusion that there is no net uptake by the system, since budgets based on measured water fluxes are rarely accurate enough to distinguish differences or a few percent. The key point for the reader to note is that nutrient uptake is primarily controlled by the concentration C (although the actual current does also increase dissipation and hence uptake). Nevertheless, the uptake by coral reefs can be sufficient for the horizontal flux to be significantly depleted.[5]

11.5.4 Verification of the uptake dissipation law

Detailed justification of the UDL depends on assessment of the energy dissipation rate ε. There are basically two approaches to that assessment. The first method is to attempt to derive the dissipation directly from measurements of the spectra of turbulence using acoustic Doppler velocimeters (ADVs). This method relies the identification of an inertial sub-range of the spectrum which may not always be easy to identify. The second and most dependable method is to deduce a value for the dissipation based on the rate of energy production and assume all that energy is dissipated in the system.

A series of such estimates have been made by several authors.[6] All of these estimates are based on turbulent energy production by bottom friction. A comparison of the UDL and measurements in Tampa Bay (Figure 11.7), on the west coast of Florida, is shown in Figure 11.8. Measurement of S involves the time rate of decay of nutrients inside a closed tank either placed over the benthos. Figure 11.8 gives uptake rate constants for ammonium, nitrate, and phosphate over a range of hydrodynamic conditions for three types of benthos (algal mat over sand, *Thallassia testudinum*, and a sparsely vegetated carbonate hard bottom). Comparison of measured and expected uptake rate constants for ions with diffusivities that vary up to a factor of three. Variation of dissipation rate ϵ involves changing the rate of flow of water over the benthos either by the use of pumps or through a variation of surface level from one end of the tank to the other. For the latter method, finding the dissipation rate ε involves measuring the change in the height of water, together with the flow velocity. In the former case, using pumps, the dissipation rate ϵ is estimated by measuring the spectrum of turbulence with an ADV.

11.6 Hydrodynamics of coral reefs

Coral reefs appear to be some of the roughest surfaces in the ocean and harness the turbulent energy from breaking waves for ecological purposes. They maintain their roughness by a growth process that compensates for the very high rates of erosion induced by their extreme roughness with roughness lengths often being an appreciable fraction of their water depth. Typical mean water depth over a nearshore reef is 1 or 2 m.

---

[5] Baird, Roughan, Brander, Middleton, and Nippard (2004).
[6] Hearn, Atkinson, and Falter (2001), Brander, Kench, and Hart (2004).

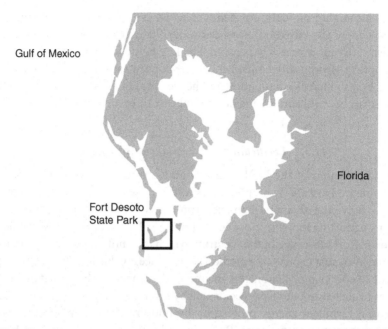

Figure 11.7 Site of the measurements of the nutrient uptake rate in seagrass meadows in Tampa Bay on the west coast of Florida (compare Figure 11.8).

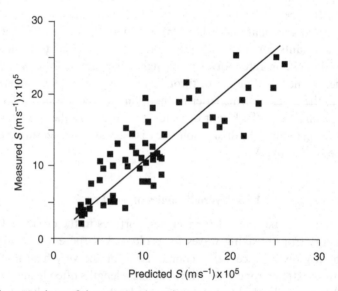

Figure 11.8 A comparison of the measured nutrient uptake constant S with predictions of the uptake dissipation law (UDL) for three types of nutrient ion by various types of benthos (compare Figure 11.7). (Data kindly supplied by Dr. Florence Thomas of the University of Hawaii.)

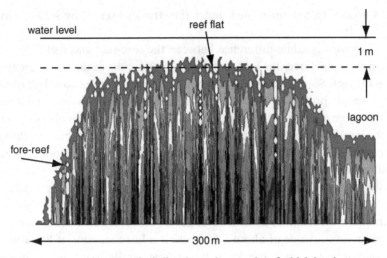

Figure 11.9 A two-dimensional vertical slice through a coral reef which has been generated as a surface of fractal dimension 2.64. The various regions of the reef are indicated (compare Figure 11.18). The roughness (0.64) is high because the surface forms interconnecting channels and cavities which allow it to occupy a large fraction of the three-dimensional space. Fractal realizations of coral reefs are becoming used in hydrodynamic models (Hearn, 2001).

11.6.1 Bathymetry of reefs

Figure 11.9 shows the basic topographic, or bathymetric, structure of a reef consisting of a *fore-reef* behind which is a *reef flat* (and *back reef*). Behind the reef flat is usually a lagoon. The elemental hydrodynamics of a reef consist of waves breaking on the fore-reef, and pumping water across the fore-reef and reef flat into the lagoon. This process is the basis of reef hydrodynamics, although it is clearly augmented by wind forcing of currents and modulated by changes in tidal height. Most coral reefs have a height that allows them to be exposed for some hours near low water during spring tides. This exposure is thought to be the ecological price paid by coral reefs for the benefit of being very close to the surface and therefore maximizing the amount of sunlight that reaches the coral. This is important to the coral because it helps photosynthesis of the *zooxanthellae*, which are the symbiotic algae that live within hard or stony corals. The symbiotic relation is based on the coral's inability to capture sufficient amounts of food from the ocean, and the algae's ability for photosynthesis and converting chemical elements into energy. The proximity to the surface also produces wave breaking, which has the negative impact on the coral of high erosion, and the positive impact of the very high rates of energy dissipation which aid the zooxanthellae through the uptake dissipation law. Living in such high energy dissipation environments is also advantageous to other submerged aquatic vegetation for similar reasons, despite the price of the mechanically strong structures needed to remain rooted under the stresses exerted by strong waves and currents.[7]

The basic hydrodynamic difference between the fore-reef and reef flat is that the water changes rapidly over the fore-reef and so causes most of the waves breaking that occurs on the reef. Some breaking does occur on the reef flat but to a first approximation waves simply propagate over the reef flat. So when we look down on a reef from an aircraft we see a line of white breakers on the fore-reef. This is also called the *surf zone* in the vocabulary of the science of *nearshore processes* and occurs through the same wave breaking processes that produces the surf zone on a sloping beach. The major difference between the fore-reef and most beaches is that a fore-reef has greater bottom slope so that the surf zone is usually more sharply defined for a coral reef.

There is an approximate criterion that waves will break if their height becomes greater than some fraction of the still water depth. This fraction is usually denoted by γ and taken to have a value of about 0.8. There is a huge amount of literature on the precise conditions for wave breaking, but it is generally agreed that the resulting spatial change in wave amplitude produces a force known as the *radiation stress*.[8] As with all of the forces that we have encountered in coastal basins, the immediate response of the basin is to produce a surface set-up that tries to counter this stress and bring the system into a steady state. This is rarely totally successful, because surface set-up just produces a pressure gradient and the effectiveness of such a gradient is dependent on the total depth of the water column. The radiation stress on the fore-reef therefore produces a major increase in surface level over the *reef crest* (the back of the fore-reef) so that the resulting pressure gradient almost balances the radiation stress. On the other side of the reef crest is the reef flat and this is much shallower, and much longer, than the downward-sloping fore-reef. The elevated water level at the reef crest therefore drops down along the reef flat and comes back to (almost) ocean level in the lagoon. This downward slope of the surface along the reef flat produces a current, which is mainly a balance between pressure gradient and bottom friction.

By our normal standards of surface slope in coastal basins, this downward surface slope along the reef flat is quite high, and without the extreme friction that we find on reefs, the current would be extremely high, and most of the potential energy lost by the water as it falls from the reef crest to the lagoon would be converted to kinetic energy in a jet of water issuing into the lagoon. Instead, most of that energy is converted by friction into turbulent energy and becomes available for thinning of the diffusive boundary layer. But this high friction does involve very high erosion and tests the ability of corals to grow. In reality, a coral reef is an assemblage of species of which coral is only one part but is instrumental in establishing a structure with high roughness. Waves continue to propagate across the reef flat and there is some limited wave breaking on the reef flat as the surface decreases in height with distance.

[7] A good introduction to the role of water flow in marine ecosystems is Mann and Lazier (2005).

[8] Radiation stress theory was developed in a seminal paper as recently as the early 1960s: Longuet-Higgins and Stewart (1964).

11.6.2 Wave–current interaction

Waves on the reef flat do also contribute to friction. This wave contribution to friction clearly also exists on the fore-reef, but friction is not the dominant force on the fore-reef and so our concern with the role of waves in friction is focused on the reef flat. The process responsible for waves affecting friction is called the *wave–current interaction*.[9] It is always important in shallow water, but the process is much stronger on reefs because of the greater roughness. The wave–current interaction is basically a consequence of bottom friction being quadratic in total current. Hence, the amount of friction that is produced by two currents each of the same strength is four times (not twice) the friction we would obtain from either of the two currents acting separately. The process is therefore *non-linear* and represents an interaction between the two current systems, or in our case, between the waves and current. The reason why the friction is quadratic is that the current does two different things when it runs along the bottom. The first is that it creates turbulence and the second is that it interacts with that turbulence. If the friction were merely due to molecular viscosity, the current would simply produce a viscous drag and the friction would be linear in the current strength. So, the wave–current interaction is due to the waves producing turbulence with which the current interacts.

For a seabed of normal roughness, the turbulence produced by waves is confined to a very thin layer at the bottom called the *wave boundary layer*. The reason that this layer is so thin is that the bottom stress produced by the waves does not penetrate far into the water column, and so the turbulence is also very confined. This is a consequence of the oscillatory nature of waves. Any oscillatory force applied to the surface of a system has a finite distance of propagation, which is controlled by the ability of the medium to react to the rate of change of the force. This wave boundary layer serves almost as the equivalent of a roughness layer, and within it the stress due to the current varies linearly with distance. The situation for a very rough bed may be somewhat more complex and is considered in some detail here.

Quadratic friction for reef flat has the form

$$\tau_b = \rho C_d \, \bar{u}^2 \tag{11.23}$$

where $\bar{u}$ is the depth-averaged (or *bulk*) current speed, and C_d the bottom drag coefficient. For a pressure-driven flow, the relation between C_d and roughness length, d, assuming that all turbulence is produced by drag on reef is

$$C_d \sim \left[\lambda \Big/ \ln\left(\frac{h}{d}\right) \right]^2 \tag{11.24}$$

---

[9] The seminal paper on wave–current interactions in shallow water is Grant and Madsen (1979).

where λ is the von Kármán constant (0.41). Bottom stress from waves penetrates only a finite distance into water column, over the *wave boundary layer* (wbl), which is of thickness some multiple of $u*/\omega$ where $u*$ is the bottom stress speed and ω the angular frequency of the waves. For normal coastal situations wbl $\ll h$. Reefs are so rough that wbl $\sim h$, and so the turbulence created by the interaction of the waves with the rough bottom affects most of the water column. *Inside the wave boundary layer*, the effect of the waves is maximal and the usual quadratic friction law is replaced by a linear relation

$$\tau_b = \rho C_d u_w \bar{u} \qquad (11.25)$$

where u_w is the *wave particle velocity* defined as some multiple of the wave particle speed. This linear form is a consequence of the interaction of the bulk residual current $\bar{u}$ with the turbulence due to the waves passing over the rough bottom.

Figure 11.10 shows the variation of bottom stress with the depth-averaged current speed over the reef as determined by the model described in Table 11.2 There are two points in each cell of the model: the z points at $z = 0$, dz, $2dz$, etc. up to $(N - 1)dz$ and the ztau points which are the midpoints of each of these cells with the uppermost at $z = (N - 1/2)dz$. The ztau points are used to calculate the stress τ. The roughness length is taken as 0.1 m which is typical of coral reefs. The broken curve in Figure 11.8 shows the best fit to the model points in the absence of waves and provides a quadratic relation like (11.23) with $C_d = 0.09$ which corresponds to typical values of C_d measured in flumes and on reefs. The full line in Figure 11.10 shows the best fit to the variation of bottom stress with depth-averaged current speed in the presence of waves of height 0.8 m and period 15 s as determined by the model. The model uses waves with peak particle speed of 0.35 m s^{-1}. The best fit is a straight line corresponding to (11.25) with $u_w = 0.51$ m s^{-1}, which implies that u_w is a factor of 1.4 times the peak particle velocity or twice the root-mean-square wave particle velocity. When $\bar{u}$ becomes greater than the peak particle speed associated with the wave (0.35 m s^{-1}), the bottom stress in Figure 11.10 departs from the *linear wave friction law* in (11.25), with the line gradually tending towards the quadratic law given by (11.23). Notice that the points in Figure 11.10 occur in pairs for model runs with, and without, the presence of waves. Every pair of points demonstrates that for the same pressure gradient the current is lower in the presence of waves, but the shear stress is higher.

Figure 11.11 is a comparative plot to Figure 11.10 and shows the variation of bottom stress with the depth-averaged current speed over the reef when the roughness length is taken as 0.01 m which is typical of a relatively rough coastal environment but smaller than for coral reefs. The broken curve again shows the best fit to the model points in the absence of waves and provides a quadratic relation with $C_d = 0.014$ which is six times greater than the usual value for sand (0.0025) indicating a fairly rough bottom. The full line shows a fit to stresses derived in the presence of waves for current speed $\bar{u} < 0.35$ m s^{-1}. This is again quadratic, allowing use to define a *wave enhanced* value of bottom drag coefficient which is a standard technique used to determine the

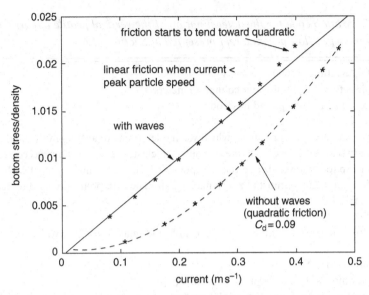

Figure 11.10 Average bottom shear-stress/density plotted against depth-averaged current (time-averaged over a wave cycle) for a roughness length of 0.1 m and currents from waves of height 0.8 m, period 15 s, for a water column of height 1 m.

current wave interaction under typical conditions of low coastal roughness; in the present case $C_d = 0.024$. So, the linear law only occurs in high roughness conditions with corresponding high penetration of wbl into the water column.

Figure 11.12 shows snapshots of the stress (at specific times in the wave cycle) for roughness lengths of 0.1 m, 0.01 m, and 0.001 m. The thickness of the wave boundary layer (wbl) increases to 0.4 m when the roughness length reaches 0.1 m and occupies a significant part of the water column. Figure 11.13 shows a repeat of Figure 11.12, but including the stress profiles in the absence of waves. The profiles are linear in the absence of waves and reflect the constant pressure gradient down through the water column, which arises on a reef flat from the set up at the reef crest. The stress is zero at the surface of the water column.

Linear friction due to waves implies a linear relation between wave height and current as observed at Ningaloo Reef on the north-west coast of Australia and shown in Figure 11.14. This is of major importance for the flushing of reef lagoons. It is to be emphasized that the waves act in two ways on coral reefs. Firstly, they produce a wave set-up that creates a pressure gradient that drives current over the reef flat. Secondly, they propagate over the reef flat to produce turbulence that increases friction. Wave friction on the reef flat does not vary with incoming wave height, provided that waves break on the fore-reef, because u_w is dependent only on the height of waves propagating over the reef flat, which is controlled by the depth of reef flat. The linear relation in Figure 11.14 implies linearity in both the friction law and the relationship between the pressure gradient and height of incoming waves.

Table 11.2. *Matlab code for the determination of the vertical profile of current in the*
numerical model of flow across a reef presented in Section 11.6.4

```
Nmax = 200; maxtime = 2000
h = 1; %this is total height of water column in meters
period = 15; %this is wave period in seconds
zr = 0.1;
%Establish z coordinates of points at which we are determine quantities in the model
%There are two points in each cell: the 'z' points which start at z = 0 and proceed z = dz,
%z = 2dz etc up to (Nmax-1)dz and the ztau points which are the midpoints of each of
%these cells with the uppermost at z = (Nmax-1/2)dz. The ztau points are used to %calculate
      the stress tau.
for n = 1:Nmax
             z(n) = (n − 1)*dz; ztau(n) = (0.5 + (n-1))*dz;taumax(n) = 0;umax(n) = 0;
end
% At the surface n = Nmax;
A = 6e-5; % which is the steady pressure gradient
g = 9.81; lambda = 0.41; %von Karman's constant
u(1: Nmax) = 0; % u(n) represents vertical profile of velocity and is initially set to zero
tau(Nmax) = 0; %tau is shear stress in the water column taken as zero at surface
height = 0.8*h; this is height of waves on reef flat
dz = h/(Nmax-1); dz is vertical increment;
kzmax = lambda*(h + zr)*ustarmax
% kzmaz is max value of eddy viscosity at surface calculated using mixing length
dt = dz**2/maxk; %this is dt for stability
speed = sqrt(g*h); %speed of gravity waves
wavelength = speed*period; %wavelength of gravity waves
wavepres = g*height/wavelength; used in determining the pressure from gravity waves
while time. < .maxtime
      time = time + dt; ustar = lambda*zr*u(2)/dz;ustarsum = ustarsum + dt*tau(1);
      for n = 1,Nmax-1
             if k(n).> maxk; maxk = k(n); end;
             tau(n) = k(n)*(u(n + 1)-u(n))/dz
             if abs(tau(n)). > .taumax(n)); taumax(n) = abs(tau(n)); end
             if (abs(u(n)).gt.umax(n)) then; umax(n) = abs(u(n)); end
             epsi(n) = tau(n)*(u(n + 1)-u(n))
      end;
      presgrad = A + wavepres*(1-exp(-time/tstartup))*sin(2*pi*time/period)
      do n = 2,Nmax
             u(n) = u(n) + dt*((tau(n)-tau(n-1))/dz + presgrad)
      end
      dt = 0.3*dz**2/maxk
end
```

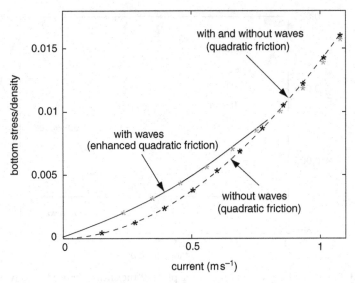

Figure 11.11 Repeat of Figure 11.10 for a roughness length of 0.01 m.

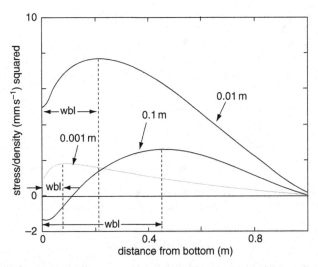

Figure 11.12 Snapshots of shear stress/density against distance from the bottom of the basin for model runs identical to Figures 11.10 and 11.11, with three different values of bottom roughness. The figure shows progressive penetration of the wave boundary layer (wbl) as the roughness length increases.

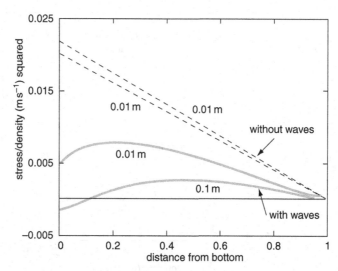

Figure 11.13 Repeat of Figure 11.12 including plots of the shear stress when waves are absent (stress decreases linearly with height above the bottom to become zero at the surface).

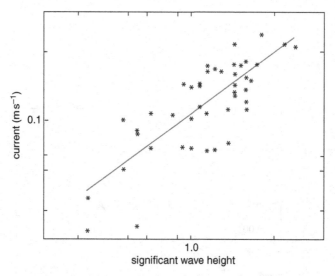

Figure 11.14 Observations of the current across a coral reef compared with the significant height of incoming waves (log–log plot). (Details of data collection: Hearn and Panker, 1988.)

11.6.3 Wave set-up

The wave set-up on a reef, or the head difference across the reef flat, can be determined from the height of incoming waves. The maximum height of surface gravity waves that can propagate through a water column of depth h is usually taken to be some fraction γ of h, i.e., so that the maximum wave height is γh. In reality, the value of γ depends on the slope of the bottom and decreases as the slope decreases, i.e., a rapidly upward sloping bottom can support higher waves than a very flat bottom of the same depth. However, for the present purposes, it is sufficient to take γ as a constant of value about 0.8. Within this approximate model, it is useful to define an *excess wave height* by $\Omega = H_w - \gamma h$, where H_w is the wave height, so that wave breaking only occurs if $\Omega > 0$. The steady state wave set-up at the front of a coral reef is proportional to this excess wave height, and this known as the *radiation stress law*:

$$\Delta \eta = \Omega / Y \qquad (11.26)$$

where

$$Y \equiv \gamma [1 + (8/3\gamma^2)] \qquad (11.27)$$

where Y is called the wave set-up coefficient[10] and using $\gamma = 0.8$, $Y = 4.1$. Equation (11.26) is a remarkably simple model: for a wave that has a an excess height of 1 m, i.e., it is 1 m higher than the minimum height for waves to break, will cause a set-up on the reef crest of about 0.25 m. This will then stream water back across the reef flat, and it is the high roughness of the reef flat that dampens the speed of water across the reef flat; water free-falling a distance of 0.25 m would reach a speed of over $2 \, \mathrm{m \, s^{-1}}$. Equation (11.26) is based on a balance of radiation stress and pressure gradient, which involves the linear and non-linear terms in the continuity equation on the fore-reef (and should *not* be linearized). There is, however, an approximation here which assumes that the fore-reef is very steep so that some *set-down* of the surface in front of the *surf zone* (the zone of breaking waves on the fore-reef) can be ignored; this set-down slightly increases the value of Y. The set-down is literally the opposite of the surface set-up that occurs on the reef crest and is due to the current that flows on the fore-reef. This current is very much smaller than the current on the reef flat because the fore-reef is so much deeper than the reef flat and by volume continuity the volume flux across the fore reef is the same (in a steady state) as the volume flux across the reef flat. The depth over the fore-reef in front of the surf zone is perhaps 10 m of more and so the current is just one-tenth of that on the reef flat. This current increases as the depth decreases and must be forced by a pressure gradient, i.e., a downward slope of the surface in the direction of the current. This slightly reduces the set-up caused by the breaking waves and produces a set-down in the region in front of the surf zone. The magnitude of the set-down is dependent on friction. The set-down is present also in the surf zone, although it is overwhelmed by the set-up due to breaking waves.

---

[10] This law is easy to prove and is originally due to Tait (1972).

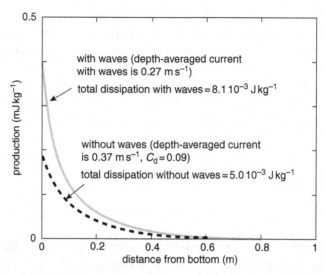

Figure 11.15 Modeled effect of waves on turbulent energy production and dissipation on a coral reef under moderate current conditions for roughness length of 0.1 m.

Over the reef flat the bottom stress must match the pressure gradient due to this set up, and so for a reef flat that stretches a distance a from the reef crest to the lagoon,

$$\tau_b = \frac{\rho g h}{a Y} \Omega \qquad (11.28)$$

which gives

$$\bar{u} = \frac{g h \Gamma}{C_d u_w a Y} \Omega. \qquad (11.29)$$

Figure 11.15 shows the depth profile of the production of turbulent kinetic energy due to shear stress for a roughness length of 0.1, assuming that total production equals dissipation. This figure represents a moderate depth-averaged current over the reef of $0.27 \, \mathrm{m \, s^{-1}}$ (currents on a reef can exceed $0.5 \, \mathrm{m \, s^{-1}}$). The figure shows the increase of production towards the rough bottom which is characteristic of a logarithmic layer. In this figure, waves are enhancing turbulence production by a factor of about 60%. Notice that the very high levels of dissipation due to the high current and roughness, greatly exceed values elsewhere in the coastal ocean. The model is run with and without waves, using the same time-averaged bottom stress. The increase in production and dissipation in the presence of waves is due to wave dissipation in the wave boundary layer (which is a significant part of water column). This dissipation comes directly from the attenuation of waves passing over the reef, and has magnitude $3 \, 10^{-3} \, \mathrm{J \, kg^{-1}}$.

Figure 11.16 shows turbulence production and dissipation for low current conditions and roughness of 0.1 m. Dissipation is almost entirely due to waves at $3 \, 10^{-3} \, \mathrm{J \, kg^{-1}}$, but

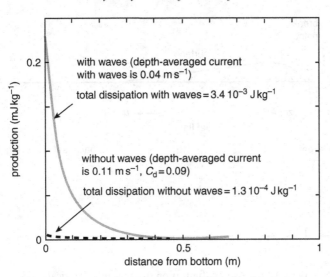

Figure 11.16 Repeat of Figure 11.15 for low current conditions with the same roughness length of 0.1 m.

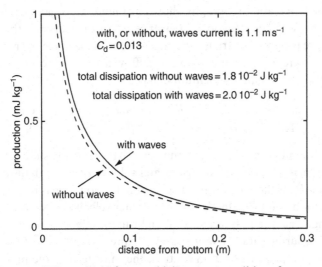

Figure 11.17 Repeat of Figure 11.15 for very high current conditions for a smaller roughness length of 0.01 m.

again very high for the marine environment due to roughness length being a significant fraction of water depth. At the smaller roughness length of 0.01 (Figure 11.17) but necessarily higher current, the wave dissipation drops to $5 \ 10^{-4} \ \mathrm{J \ kg^{-1}}$ consistent with the lower drag coefficient, $C_d = 0.014$.

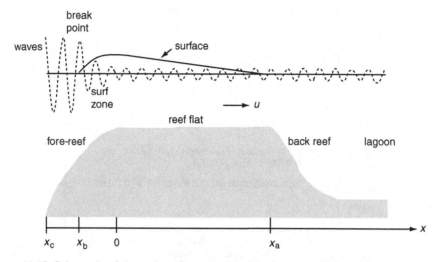

Figure 11.18 Schematic of the reef configuration used in the numerical model.

11.6.4 Numerical model of flow across a reef

We will discuss a one-dimensional hydrodynamic model of an idealized reef corresponding to the schematic topography shown in Figure 11.18, and consists of a fore-reef, reef flat, back reef, and lagoon.[11] Distance across the reef from the open ocean is in terms of increasing x. The transition points from the fore-reef to reef flat and reef flat to back reef are at $x = 0$ and $x = x_a$. The flow across the reef is modeled by the depth-averaged steady state momentum balance

$$\frac{\partial u}{\partial t} = -\frac{3g}{16(h+\eta)}\frac{\partial H_w^2}{\partial x} - \frac{F_z(h+\eta, u, z_0, H_w)}{\rho(h+\eta)} - g\frac{d\eta}{dx} - u\frac{du}{dx} \qquad (11.30)$$

where h is the water depth below the datum level, η is the steady state surface elevation above the datum, u is the depth-averaged water velocity (positive in the direction of x), g is gravity, and H_w is the *wave height*. The bottom friction term is represented by $F_{z=0}$ which is a function of total water depth, mean water velocity, wave height, and the reef roughness length z_0. The first term on the right-hand side is the *radiation stress term* due to breaking surface gravity waves, and is expressed in terms of wave height H_w. The second term represents bottom friction, the third term is the pressure gradient, and the fourth is the momentum advection term. Waves are not modeled explicitly, but parameterized by wave height H_w. The incident wave height at the ocean is prescribed and wave height is not permitted to increase in magnitude across the reef:

$$H_w \le \gamma(h+\eta) \qquad (11.31)$$

---

[11] See also Hearn and Schulz (in press).

where the wave-breaking ratio γ is taken as 0.8. Notice that we use the total height of the water column in (11.31), and this is found to be significant on the reef flat where η may be a significant fraction of h; the decreasing value of η across the reef flat causes some very minor wave breaking and consequently a radiation stress on the reef flat.

The continuity equation is used to determine η:

$$\frac{\partial \eta}{\partial t} = -\frac{\partial q}{\partial x}$$ (11.32)

where the volume flux is

$$q \equiv u(h + \eta).$$ (11.33)

The bottom friction term is determined from a non-linear formulation which allows for a non-constant vertical stress profile. This term is calculated with the separate one-dimensional vertical model outlined in Table 11.2 for various combinations of $h + \eta$, u, z_0, and H_w which are held constant throughout the model run. The least-squares quadratic best fit between $F_{z=0}$ and u is determined for a range of water depths, roughness lengths, and wave heights. This information is stored in a table. The shear stress is

$$\tau(z) = \rho K_z \frac{\partial u_z}{\partial z}$$ (11.34)

where $u(z)$ is the water velocity at a height z above the reef surface and the vertical eddy viscosity is

$$K_z = l_z^2 \left| \frac{\partial u_z}{\partial z} \right|$$ (11.35)

where the mixing length l_z is positive and has the form

$$l_z = \lambda \begin{cases} z_0 & z \le z_0 \\ z_0 + z & z > z_0 \end{cases}$$ (11.36)

where $\lambda = 0.4$ is the von Kármán constant and the turbulent length scale is determined as $l_z = \min[h + \eta, z_0 + (h + \eta)/2]$. This formulation of l_z needs to recognize enhanced mixing within the roughness layer since z_0 is a significant fraction of water depth. Equations (11.34) to (11.36) are solved and then

$$\frac{\partial u(z)}{\partial t} = \frac{\partial(\tau_z/\rho)}{\partial z} + g\frac{\partial \eta}{\partial x} + \frac{gH}{T\sqrt{g(h+\eta)}}\sin\left(\frac{2\pi t}{T}\right)$$ (11.37)

where T is the wave period and t is time. The second term on the right-hand side is constant and represents a steady pressure gradient, and the third term represents an oscillating pressure gradient due to waves.

The total production of turbulent kinetic energy (TKE) is also calculated:

$$P = \sum_{z=0}^{h+\eta} \frac{\tau_z}{\rho} \Delta u_z \tag{11.38}$$

where Δu_z is the vertical shear in the horizontal velocity. A turbulent kinetic energy model is included in the one-dimensional hydrodynamic reef model based on Section 7.9:

$$\frac{\partial E}{\partial t} = P - u \frac{\partial E}{\partial x} - \varepsilon \tag{11.39}$$

where E is turbulent kinetic energy, P in (11.38) is the total production from currents and wave motion, and ε is the dissipation parameterized as

$$\varepsilon = \frac{(2E)^{3/2}}{16.6 l_z}. \tag{11.40}$$

The model code is shown in Table 11.3. With the parameters given in the table, the model simulates 40 min of real time and the wave forcing is ramped up over the first 10 min. The model rapidly reaches a steady state after the ramp expires. The model domain is $x = -153$ to 347 m with $dx = 1$ m. The regions of the reef are located as follows: fore-reef: $x_c = -153$ m to $x = 0$; reef flat: $x = 0$ to $x_a = 73$ m; back reef: $x_a = 73$ m to $x = 250$ m; lagoon: $x > 250$ m.

The water elevation η is zero in the absence of waves. The depth at the beginning of the fore-reef $h(x_c)$ is 10 m, the depth on the reef flat $h_r = 1$ m, and the depth of water in the lagoon is 10 m. The shape of the fore-reef, reef flat, and lagoon are specified as a series of linear segments, which are then smoothed by fitting a tenth order polynomial with the reef flat being forced to be constant. The reef roughness length is kept constant throughout the fore-reef and reef flat at $z_0 = 0.1$ m, and assigned a smoother value of 1×10^{-3} m on the back reef and lagoon.

In Figure 11.19, the wave height (panel a) has a constant value of 1 m at $x = -153$ m until the fore-reef is shallow enough for (11.31) to take effect. At the reef crest the wave height is 0.83 m and reduces to 0.80 m at $x = 75$ m. Panel b shows the smoothed depth profile of the reef. The vertically averaged current obtained once the model has reached a steady state is shown in panel c. The maximum current is $0.23 \, \mathrm{m \, s^{-1}}$ at the end of the reef flat. The surface elevation η, shown in panel d, displays a small set-down in the fore-reef region with a decrease of 0.005 m over 133 m. The set-up begins 19 m before the reef flat and has a magnitude of 0.043 m over a distance of 20 m. The set-up gradually declines over the reef flat and reaches zero around at $x = 100$ m. The bottom frictional shear stress, shown in panel e, is largest at the end of the reef flat, and is mainly controlled by the value of the current. The momentum advection, $-u \, du/dx$ (panel f), indicates that the water accelerates as it approaches the reef, and then slows once past the shallowest point on the reef flat. This term is an order of magnitude smaller than bottom friction.

Table 11.3. *Matlab code for the determination of the across-reef current in the numerical model presented in Section 11.6.4*

```
ampocean = 1/2;period = 10; cd = 0.1; z0 = 0.1;dx = 1;
hoc = 10; %depth of ocean at front end of model
sealevel = 0; hr =1% depth of water on reef flat
hlag = hoc; n = 500/dx; % n is number of cells in model
g = 9.81; gamma = 0.4; wave breaking amplitude/depth criterion
radconstant = 3*g/(4*dx); % constant multiplying the radiation stress/wave breaking
c = sqrt(hmax*g); dt = 0.5*dx/c; timemax = 40*60; % run time;
ntime = round(timemax/dt); %number of time step in run
for i = 1:nfr; hh(i) = hr + (hoc-hr)*(nfr + 1-i)/nfr; end
for i = nfr + 1:nbf; hh(i) = hr; end
for i = nbf + 1:1:nlag;hh(i) = hr + (hlag-hr)*(i-nbf-1)/(nlag-nbf); end
for i = nlag + 1:1:n-1;hh(i) = hlag; end; time = 0;
%TKE equation
kappa = 0.4; B = 16.6;% length scale and TKE dissipation
for ncount = 1:ntime
        time = time + dt; ht = h + eta;
        amp(1) = ampocean; % this is amplitude of wave = 0.5 height at cell 1
        amp(2:n) = gamma*ht(2:n); amp0 = amp;
% amplitude of waves
        x = find(diff(amp(2:n)) > 0);size(x);
        if ~isempty(x);
            for i = 2:1:n;if amp(i) > amp(i-1);amp(i) = amp(i-1);end;end
        end
        rad(1:n-1) = -radconstant*diff(amp.*amp);
% radiation stress;
        hu(1:n-1) = ht(1:n-1) + 0.5*diff(ht);
        %find hu at right edge by interpolation
        ux(2:n-2) = (u(3:n-1)-u(1:n-3))/(2*dx);
        ux(n-1) = (uend-u(n-2))/(2*dx);
        ux(1) = (u(2)-uend)/(2*dx);% ux is gradient of u
% forward (or downwind), differencing scheme
        ux1(2:n-1) = (u(2:n-1)-u(1:n-2))/dx; ux1(1) = (u(2)-uend)/dx;
        px(1:n-1) = (diff(e))/dx;
        uf(1:n-1) = 2*pi*amp(1:n-1)/period;
        adv = -u.*ux; adv1 = -u.*ux1; press = -g*px; rad1 = rad./hu; frict1 = frict./hu;
% TKE budget
        dtkedx(2:n-1) = (tke(2:n-1)-tke(1:n-2))/dx; % horizontal grad in tke
% horizontal diffusion of TKE
        tls = min([hu; z0 + hu/2]);% turbulent length scale
        wave = cd.*((uf).^3)./hu;% input from wave orbital velocity
        prod = -frict.*u./hu; trans = -u.*dtkedx; diss = ((2*tke).^(3/2))./(B*kappa*tls);
```

Table 11.3. (*cont.*)

```
tke = tke + dt*(tke_prod - diss + trans );
xx = find(tke < 0);if ~isempty(xx),tke(xx) = 0;end % remove negative tke
u = u + dt*(rad1 + frict1 + press + adv1); flux = hu.*u; ediff = diff(flux);
% continuity equation to time step elevation
eta(2:n-1) = e(2:n-1)-dt*ediff(1:n-2)/dx;
dtf(1:n-1) = abs(e(1:n-1) + h(1:n-1))./(uf(1:n-1).*cd(1:n-1));
end;
```

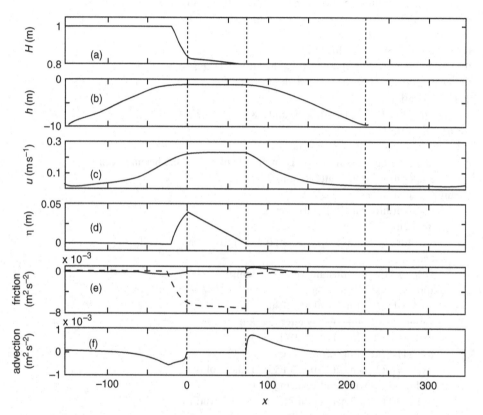

Figure 11.19 Model variables as a function of *x*; wave height (panel a), water depth (panel b), current *u* (panel c), surface elevation (panel d), friction (dashed) and advection (panel e), and advection (panel f). The location of the beginning and end of the reef flat and the beginning of the lagoon are indicated.

The balance of forces acting on the water column (Figure 11.20) varies with position along the reef. The major balance is between pressure and friction throughout the domain, except in the surf zone, where the balance involves radiation stress and pressure. The advective force is everywhere very small.

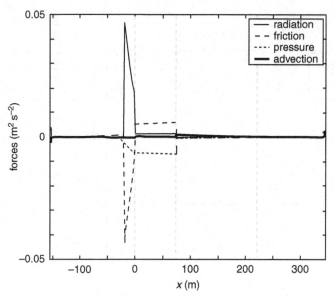

Figure 11.20 Forces on a coral reef as a function of x; radiation stress (unbroken), friction (dashed), pressure gradient (dash–dot), and momentum advection (thick unbroken).

The turbulent kinetic energy, energy dissipation ε, and nutrient uptake coefficient S are displayed as a function of distance along the reef for a reef flat roughness length $z_0 = 0.1$ m, with reef flat depths of $h_r = 0.76$ m and 1.24 m (Figure 11.21). The shallower reef displays larger TKE, ε, and S on the reef flat compared to the deeper case. We also examine the TKE production (P) versus dissipation (ε) along the reef flat (Figure 11.22) for the standard model run. Production balances dissipation, except in the surf zone where there is a 20% excess in production, and beyond the end of the reef flat where there is a large (60%) excess of dissipation, i.e., there must be significant transport of turbulence. It is expected that this process will increase with larger-scale spatial inhomogenuity, i.e., larger-scale roughness.

The Matlab code for the model used in this section is given in the tables and the reader can experiment with the hydrodynamics of flow across reefs. As always, the output routines are not given and the codes are written in the simplest forms so that they can be reconfigured in Fortran or other languages. The vertical model can of course be greatly simplified by the use of standard quadratic or linear friction. The model can also be cast into two dimensions in the horizontal which requires a two-dimensional form for the radiation stress. Two-dimensional models are essential if we are to resolve the alongshore currents parallel to the reef. Grid cells used for hydrodynamic models of coral reefs need to be smallest over the surf zone simply because that is usually the region in which depth varies most rapidly. This clearly invites the use of nested, or perhaps better, bathymetry following flexible meshes. More complete models of the hydrodynamics of coral reefs clearly require a precise

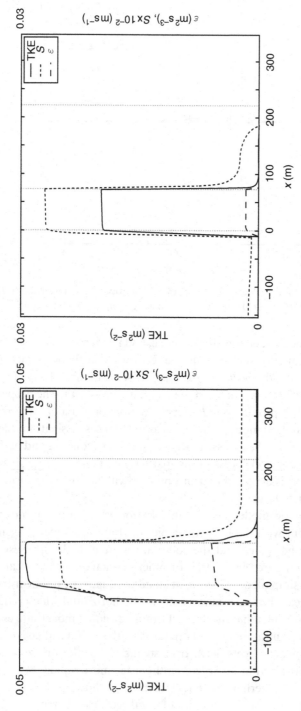

Figure 11.21 Turbulent kinetic energy (TKE), nutrient uptake coefficient (S), and TKE dissipation (ϵ) as a function of x. The reef flat has $h_r = 0.76$ m and $z_0 = 0.1$ m (left panel) and $h_r = 1.24$ m and $z_0 = 0.1$ m (right panel) respectively.

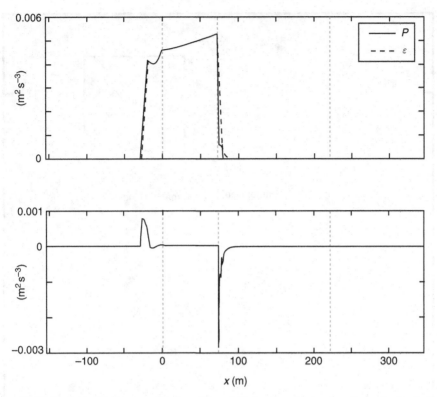

Figure 11.22 Turbulent energy production (P) and dissipation (ε) as a function of x (top), $P - \varepsilon$ (bottom). The reef flat has a depth of $1.0\,\text{m}$ and a roughness length of $z_0 = 0.1\,\text{m}$.

formulation of the dynamics of the gravity waves and such models are mentioned in Chapter 12.

11.6.5 Circulation model of a coastal lagoon

Figure 11.23 illustrates Kañe'ohe Bay (or Kaneohe Bay in standard US spelling) on the east coast of the Hawaiian island of Oahu.[12] It is a coral reef coastal lagoon of the type that we discussed in Chapter 1 and showed in more detail in Figure 1.5. Another example of such lagoons that has been mentioned in this book is Ningaloo Reef on the north-west coast of Australia. Such coastal coral reef lagoons are very common on the coasts of tropical islands. The formation of coral reefs is dependent on water temperature and usually only occurs in tropical regions. The dynamics of coastal lagoons is shown in Figure 1.5 with waves breaking on a central, or *barrier*, reef and driving water into the lagoon where it escapes back to the open ocean through gaps in the reef.

---

[12] An extensive review of Kaneohe Bay hydrodynamics is given by Smith, Kimmerer, Laws, Brock, and Walsh (1981).

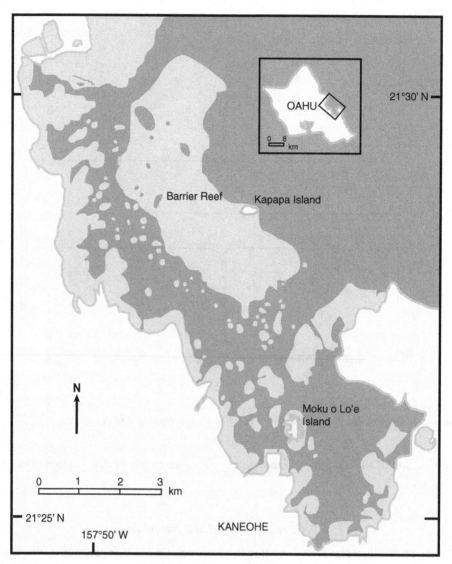

Figure 11.23 Kaneohe Bay on the east coast of the Hawaiian island of Oahu. Lighter-colored regions are coral reefs. The hydrodynamics of the bay is dominated by waves breaking on the main barrier reef. Note the two islands in the bay, one of which, Moku o Lo'e (Coconut Island), is home to the University of Hawaii's Institute of Marine Biology. The bay experiences the North-East Trade Winds and intense winter storms.

The barrier reef is evident in Figure 11.23 as are two gaps in the reef at the northern and southern ends of the bay. Those channels are natural but have been enhanced for navigational purposes. There are also extensive reefs on the shoreline that are physically separate from the barrier reef. Figure 11.24 shows a single snapshot of a model of

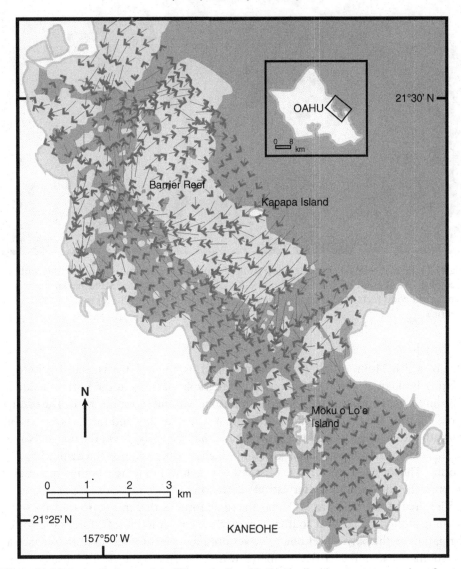

Figure 11.24 A circulation model of Kaneohe Bay. Notice the dominant wave pumping of water across the barrier reef and its exit back to the open ocean through the channels at the southern and northern ends of the barrier reef.

water flow in the lagoon which has been adapted from a three-dimensional numerical model called SPECIES[13] which includes the radiation stress from breaking waves. The model in Figure 11.24 is run with a grid size of 30 m although the current vectors are then averaged over much larger cells for ease of display. There are several islands in the

[13] Hearn (2000).

Figure 11.25 Kaneohe Bay looking north from Moku o Lo'e (Coconut Island) on a day without incoming ocean swell or wind waves. The Ko'olau Mountains are in the background and submerged coral reefs can be seen as lighter coloring of the water. (Photograph by Clifford J. Hearn.)

Bay of which Moku o Lo'e (or Coconut Island) is home to the Hawaii Institute of Marine Biology. Figure 11.25 shows a photograph taken from the northern side of Coconut Island during an early windless morning without incoming swell. The picture looks towards the tip of the northern edge of Kaneohe Bay and the reefs are visible beneath the water surface by the lighter shade that they produce in the image. Notice the Ko'olau Mountains in the background which produce major surface runoff during storms. The image shows a range of scales of roughness ranging from the mountains through the coral reef to the terrestrial vegetation.

The circulation in Figure 11.24 has the same form as that in Figure 1.5 with some seiching motion in the most southerly part of the bay. A notable feature is the vortex formation on the barrier reef due to waves producing greater set-up of the surface at the southern end of the barrier compared with the northern end. This in turn is due to the reef flat being slightly shallow in the south. The SPECIES model also uses a roughness length that varies over the reef according to a benthic classification of the types of coral and algae in cell of the spatial grid. Although this means that water flow across the reef is much more complex than we have assumed in Section 11.6.4, it remains true that here, as in all coastal models, the lower-dimensional model captures much of the physics of flow across a reef. Reduction in flow observed in models by local increases in roughness produces so-called *sticky water*.[14]

[14] Wolanski and Spagnol (2000).

11.6.6 Effect of changing water level on flow over reefs

Water flow over a coral reef varies considerably with the tidal level, especially so during spring tides when reefs are typically dry for some hours either side of low water, and may then become deep enough at high water to curtail wave breaking except under conditions of heavy swell. The low water situation may be one of severe environmental stress to the reef, often exacerbated by human exploitation. High water could also be stressful through lack of flushing, lowered light conditions, and turbulence. The tidal cycle clearly modulates the water flow across a reef, even if the incoming wave field and other environmental forcing remains constant. This is also true of other shallow parts of coastal basins, especially those in the intertidal regions. One of the effects of this process is that we see a new component in the tidal *current*, which is usually *in phase* with the tidal elevation. The usual situation in a coastal basin is that the tidal current is 90° out of phase with the amplitude. Strictly speaking, this describes a standing wave as we discussed in Chapter 6. It is a wave in which water level and current can be simply represented as products of space and time. As the tidal level rises in the ocean outside of a coastal basin, a pressure gradient is developed which forces water into the basin and as the tidal level falls, so the pressure gradient acts in the opposite direction. If there are no frictional forces present, the resulting phase difference between elevation and current is 90°, i.e., high water and low water correspond to slack currents.

Let us envision a coral reef that forms a *barrier* across a coastal basin rather like that in Figure 1.5. Such systems are very common types of coral structures, and the resultant coastal basin is a lagoon. Barrier reef topography always involves a number of gaps in the reef, so that water pumped over the reef by wave breaking can circulate through the lagoon and exit via the gaps. In the absence of gaps in the barrier reef, the water level would simply rise in the closed lagoon and block further wave driven flow. As the tide rises in the outside ocean, a tidal pressure gradient is created which forces water into the lagoon. Water will flow both over the reef and through the gaps, and so the purely wave-driven circulation is augmented by this tidal component: the across-reef and gap currents are altered in such a way as to supply the tidal flow into, and out of, the lagoon.

A key element of the tidal dynamics of a coral reef is that any change in water level affects the wave-driven flow over the reef. If we assume that the flow responses almost instantly to the change in water level, and we are able measure that flow as a function of time, we will see an oscillation basically at the tidal frequency, and it will be in phase with the tidal elevation. It is important to note that this observation would tell us whether the wave-driven current increases or decreases with water level. There are two opposing processes which control the sign of the change in flow relative to the change in water level. Firstly, the excess wave height Ω is reduced by an increase in water level, and that clearly reduces water set-up on the reef crest and hence, tends to reduce the current. Secondly, the increase in water level taken by itself (for constant surface set up on the reef) increases the pressure gradient over the reef flat, and this tends to increase current. There are similarly two opposing influences as we reduce water height, and certainly the current vanishes as the reef becomes dry.[15]

11.7 Coastal roughness and trapping

Coastal roughness is a complex issue which is only recently coming within the realm of modeling capabilities.[16] Coastlines, as we saw earlier in this chapter, contain intrinsic roughness at all spatial scales. A coastline is, therefore, a rough wall which forms the boundary of a coastal basin. It is suggested that aspects of the flow associated with the roughness may be responsible for trapping particles close the coast. This is envisaged to involve local areas of closed circulation, and genetic distributions do provide support for this suggestion. It is certainly possible to relate the transport of larva to features of circulation patterns, and many coastal basins do contain *closed gyres*. These gyres are almost always a result of topography in the broadest sense, but can be driven by any of the forcing functions that we have considered in this book; examples are wind, pressure gradients, and advection of momentum.

Whatever fixed or flexible mesh is used, some approximation must be made in the representation of any coastline. Although a flexible mesh appears to do a superior job of representing the coast, this merely takes the approximation down to smaller spatial

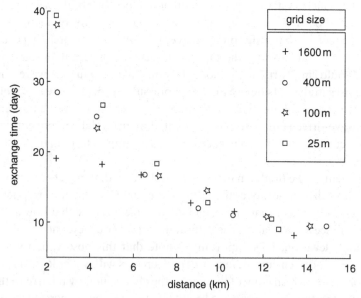

Figure 11.26 Dependence of ocean exchange time on distance from a coastline for a rectangular mesh of varying size.

---

[15] Hearn (1999).   [16] See, for example, Zeidberg and Hamner (2002).

scales. At whatever scale we use, we must eventually represent a coast by a series of straight lines, or smooth geometric curves. Figure 11.26 shows the ocean exchange calculated for particles released from points at varying distance from the coast within a coastal basin. There are two interesting aspects of this figure. The first is that the exchange time decreases rapidly with distance from the coast indicating that there is some degree of coastal trapping. The other important conclusion from this figure is that the increase in exchange time saturates as we approach the coastal at a value which increases as we reduce the grid size.[17]

11.8 Further reading

See further reading from earlier chapters. Additional texts for this chapter are:

Birkeland, C. (2003). *Life and Death of Coral Reefs*. New York: Springer.

Bowen, J. and M. Bowen (2002). *The Great Barrier Reef: History, Science, Heritage*. Cambridge, UK: Cambridge University Press.

Falconer, K. (2003). *Fractal Geometry: Mathematical Foundations and Applications*. New York: John Wiley.

Mandelbrot, B. B. (1982). *The Fractal Geometry of Nature*. San Francisco, CA: W. H. Freeman.

Rothschild, W. G. (1998). *Fractals in Chemistry*. New York: Wiley-Interscience.

---

[17] Hearn (2006).

12

Wave and sediment dynamics

12.1 Introduction

In this chapter, we briefly consider available wave models for coastal basins and the physical conditions which lead to the erosion, transport, and deposition of sediment. Sediments are moved about more in shallow water than in the deep sea, because surface waves can affect the seabed, and tidal currents are typically stronger than in the open ocean, because of increased tidal ranges. Sediment transport and deposition are also more easily studied in shallow water, but the principles governing these processes are as valid in the deep ocean as they are in an estuary, beach, or anywhere there is moving water.

12.2 Wave models

Wave modeling in coastal basins is a highly developed tool necessary for both studies of sediment transport and *nearshore processes*[1] including coastal engineering. As with hydrodynamic models, wave models can be run in both *hindcast, nowcast*, and *forecast* modes. Our interest within this chapter is directed towards the use of wave models in sediment transport within coastal basins. The study of all of the techniques of wave modeling is outside our present scope and establishing a wave model for a particular coastal basin requires not only skill, and a proper assessment of the type of predictions that are required from such a model, but also very good data sets to drive, calibrate and verify the model. As a generalization, wave models tend to be much more dependent on good data than are hydrodynamic models since they tend to contain more adjustable parameters. If the wave climate of the basin is dominated by local wind waves, it is very important to have good wind measurements inside the basin. If the coastal basin has a large component of ocean swell the model is dependent on wave data at its ocean boundary.

---

[1] Nearshore processes are those very close to the shore including the surf zone.

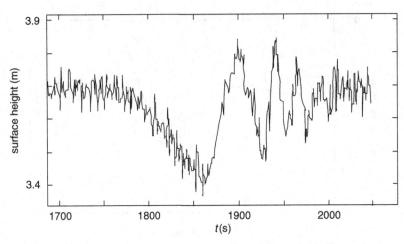

Figure 12.1 Measured water elevation (m) as a function of time t (in s) due to the passage of a ship. The main wave consists of a depression of the surface from about $t = 1750$ to 1900 s followed by secondary waves of higher frequency at larger t.

Sverdrup and Munk were significant pioneers in developing the wave forecast technique.[2] Wave models tend to divide into a number of different groups, some of which compute surface elevation and currents over the basin. Others work only with the wave spectrum within each cell of a model. The choice of wave model is highly dependent on the particular type of application, forcing, and scales that we are considering. For example, we might be interested in the effect of ship wakes on the stability of shorelines inside the coastal basin. In that case, we would need to model the actual wake from a ship and study its propagation across the basin.

12.2.1 Ship wakes

The wake from a ship consists of two components. These are also clearly visible if one stands on a beach and watches a large ship pass along the coast. The ship displaces water as it moves, and this produces a corresponding wave at the shore which has a duration about equivalent to the time taken by a ship to move past a point, i.e., if the ship is 300 m long and travels at a speed of $3\,\mathrm{m\,s^{-1}}$ this time is 100 s. This is seen as a depression of the water surface, and dynamically this is the *hole* left behind the ship into which water streams as the ship passes. This creates a surface depression across the coastal basin which moves with a speed similar to that of the ship. Figure 12.1 shows the recording from a pressure gauge set in 2 m of water near the shore of Tampa Bay on the west coast of Florida. It shows the surface depression of the surface after a

---

[2] Harald Sverdrup (1888–1957) was a Norwegian oceanographer and meteorologist who made a number of important theoretical discoveries in these fields. As director of California's Scripps Institution of Oceanography from 1936, he made significant contributions to oceanography. Walter Munk (b. 1917) has been a very major contributor to the field of physical oceanography and geophysics at Scripps Institution of Oceanography in La Jolla, California.

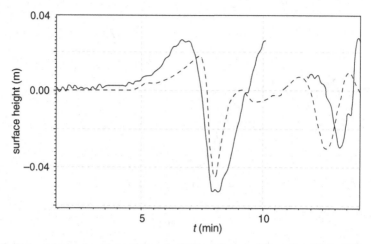

Figure 12.2 Comparison of a primary wave (solid line) from a ship wake off MacDill peninsular, Tampa Bay with the DHI model MIKE 2 (dashed line) (DHI, 2002).

ship has passed followed (at later times) by higher-frequency waves. As the depression moves into shallower water, it will become steeper (and deeper) as we saw in models of waves in Chapter 6. As such, the long period waves from ships are easily visible when walking on a beach. However, the waves that are often observed when aboard a boat are the higher-frequency secondary waves. Figure 12.2 shows a comparison of a primary ship wake with a two-dimensional numerical model in Tampa Bay. The model uses the ship as a moving surface depression and is dependent on accurate information on the ship's track. In Tampa Bay, the actual path is known accurately as the ship must move through the navigation channel illustrated in Chapter 6. Information on the position of large ships in coastal basins as a function of time is available from port authorities. The model does require quite accurate information on the shape of the ship's hull and its displacement, i.e., the height of the ship in the water. For the model shown in Figure 12.2, a flexible mesh was used to maximize spatial resolution near the ship.

Ship wakes can be categorized in terms of the Froude number Fr of the ship defined as the ratio of the vessel speed Vs to the speed of surface gravity waves (Figures 12.3 and 12.4). The *critical speed* of the vessel for which this Froude number becomes unity is clearly equal to $\sqrt{gh}$, which in a typical navigational lane is about 10 m s^{-1} or roughly 20 knots. A knot is 0.51 m s^{-1} and defined as 1 nautical mile per hour. A nautical mile is 1000 fathoms or 1852 m. The word *knot* comes from an old method of measuring a ship's speed by throwing a log over the stern and counting the passing out of knots on a rope tied to the log in a given time (also the origin of *log book* in which speed was recorded). Ships tend to maneuver in coastal basins at speeds in the vicinity of 10 to 15 knots and so should be just subcritical. These patterns change as the Froude number Fr changes from *subcritical* to *supercritical*. We can estimate the speed of ships in a coastal basin by considering these wave patterns. Physically, it is clear that

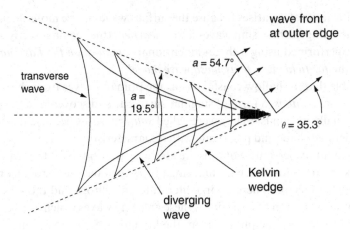

Figure 12.3 Schematic of secondary waves from a ship.

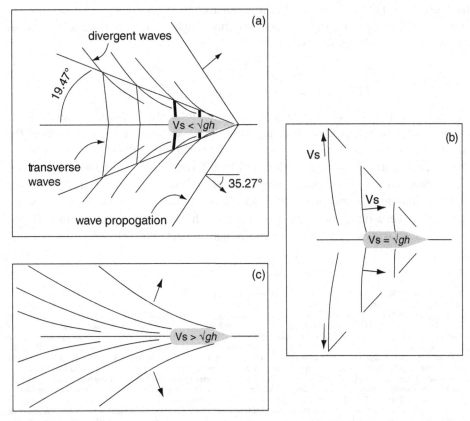

Figure 12.4 Schematic of waves from a ship traveling at three speeds: (a) subcritical, Fr < 1, (b) critical, Fr = 1, (c) supercritical, Fr > 1.

the supercritical situation arises because the surface waves move more slowly than the ship. The modeling of these ship waves in the *nearfield* (the region directly around the ship) is best performed using a three-dimensional *computation fluid dynamics* (CFD) model and the *far field* via a *Boussinesq wave model*.[3]

It is possible that in shallow coastal basins the primary waves from ships may form solitary waves or *solitons*, a form of singular surface first discovered in the nineteenth century originally in canals.[4] They were called *solitary waves* because they can maintain a well-defined shape and propagate with a characteristic speed over long distances without apparent dispersion.[5] Solitons are also easy to produce in the laboratory in a narrow tank. Two solitons have the amazing property of passing through one another without change of shape. There is large literature on solitons and the solution of the non-linear wave equation that produces these amazing waves which are now observed in a variety of macroscopic physics from the human body to fiber optical communication.[6] There are also *internal solitons* (which are basically the same phenomenon propagating on the pycnocline), and these can produce a surface manifestation of cold bottom reaching the surface which is visible as a thermal signal in infrared images from satellites. There is an amusing anecdotal account of solitons being used for high-speed canal transport in early nineteenth century England by C. S. Forester's fictional character Horatio Hornblower.[7]

12.2.2 Parametric and full spectrum wave models

For most of our work in coastal basins, we are not interested in such individual events as the passage of a ship. Our concern is with the wave climate in various parts of the basin, i.e., the basic properties of waves over time intervals of days, or weeks, months, seasons, and years. These time periods reflect different aspects of the wave dynamics. Yearly properties, or even decadal properties, show how the wave climate may change with the long-term atmospheric weather climate (compare Section 10.9). Seasonal changes in the wave climate are a response to the seasonal variations in weather. Properties of waves over shorter periods may reflect daily changes in wind (for

---

[3] Named for Valentin Boussinesq (1842–1929), a French mathematician who made major contributions to the theory of fluid flow.

[4] J. Scott Russell, "Report on waves", *Fourteenth Meeting of the British Association for the Advancement of Science*, 1844. "I believe I shall best introduce this phenomenon by describing the circumstances of my own first acquaintance with it. I was observing the motion of a boat which was rapidly drawn along a narrow channel by a pair of horses, when the boat suddenly stopped – not so the mass of water in the channel which it had put in motion; it accumulated round the prow of the vessel in a state of violent agitation; then suddenly leaving it behind, rolled forward with great velocity, assuming the form of a large solitary elevation, a rounded, smooth and well defined heap of water, which continued its course along the channel apparently without change of form or dimension of speed. I followed it on horseback, and overtook it still rolling on at a rate of some eight or nine miles an hour, preserving its original figure some thirty feet long and a foot to foot and half in height. Its height gradually diminished, and after a chase of one or two miles I lost it in the windings of the channel. Such, in the month of August 1834, was my first chance interview with that singular and beautiful phenomenon which I have called the Wave of Translation a name which it now very generally bears."

[5] Dispersion is the dependence of propagation speed on frequency found for most waves and leads to the spreading, or dispersion, of wave packets. It is the reason that messages from public address systems are so difficult to understand at a distance.

[6] Dauxois and Peyrard (2006).   [7] Forester (1953).

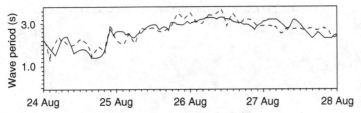

Figure 12.5 The peak wave period (s) in Tomales Bay, Florida, measured (broken line) and simulated (solid line) with a parametric wave model plotted against days in August, 2004. This period corresponds to the passage of Hurricane Frances (compare Figures 5.4 to 5.6). The wave model is based on the bathymetry following flexible mesh shown in Figure 6.25. The wave period is typical of local wind-generated waves. The station is close to the mouth of the bay but with limited penetration of swell waves from the Gulf of Mexico.

example a daily sea breeze as mention in Section 10.3) or some hazardous episode, such as a severe storm. Models of this type perform averages in each spatial cell of a model grid. Ideally, that grid should be identical to the grid used for the hydrodynamic models of the same basin in order to facilitate studies of the wave–current interaction (see Chapter 11) and sediment transport.

Our starting point is a class of wave models that determine a limited number of *parameters* of the waves, such as significant wave height (the average of the *highest one-third* of the waves), the mean wave period, and the mean direction. These are called *parametric wave models*. They are often based on a set of relationships published in the nearshore wave literature notably by the Coastal Engineering Research Center (CERC) of the US Army Corps of Engineers.[8] These are entirely empirical formulations, based on the collection and synthesis modeling of wave and wind data and must be solved for each cell in the model. For a flexible mesh (compare Section 6.7) this involves some complex technical coding of the model. Illustrations of results of a parametric wave model are shown in Figure 12.5.

The other major type of model (and there are many others) is the *full spectrum wave model*, which solves for the wave spectrum at as many points over the spectrum as may be chosen by the user. Full spectrum wave models are much more computationally intensive than parametric wave models. In small relatively sheltered coastal basins, such as Tampa Bay which is illustrated in Figures 12.6 and 12.7, much of the wave spectrum occurs at periods of a few seconds which implies that the waves are dominantly created by local winds; as we saw in Section 1.3 incoming ocean swell has typical periods of 10 to 15 s. Near the mouth of Tampa Bay there is penetration of swell from the Gulf of Mexico but within the bulk of the bay waves are locally wind driven. Furthermore the waves are *fetch limited* in that they have not reached a steady state under the local wind field. Fetch limitation plays a major part in the inhomogeneous nature of the wave climate in many coastal basins and this can be exacerbated by human modifications such as the islands of dredge spoil from dredging in Tampa Bay.

---

[8] A good starting point for this technical literature is the excellent text Dean and Dalrymple (2001).

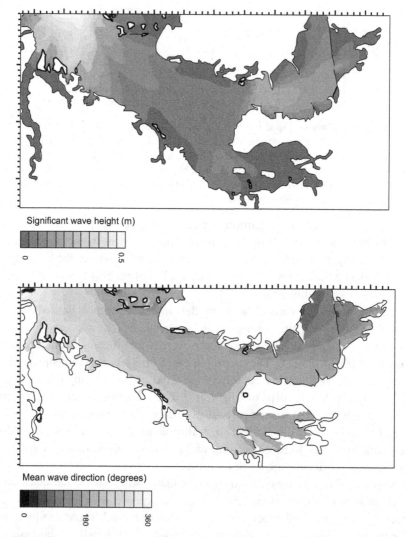

Figure 12.6 Full spectrum wave model simulation in Tampa Bay, Florida, based on the flexible mesh shown in Figure 12.7. The panels are snapshots of significant wave height and mean wave direction at 2.45 pm on August 13, 2004 corresponding to the passage of Hurricane Charley across Florida.

12.3 Sediment particle size

12.3.1 Size classification of particles

Modeling sediments in coastal basins requires some simple terminology for the types of sediment. There are several classification systems but the physics of sediment movement is mainly concerned with the size of sediment grains (and to a lesser extent their shape) and the tendency of grains to adhere to one another. The latter involves

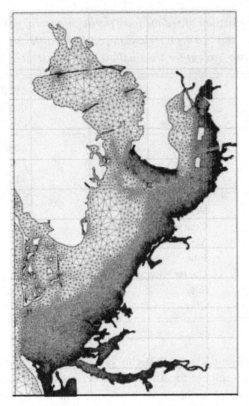

Figure 12.7 Flexible mesh used for the full spectrum wave model in Figure 12.6.

dividing sediments into those that are non-cohesive and those that are cohesive. Non-cohesive modeling is obviously much simpler than dealing with cohesive sediments and we have not attempted to treat cohesive sediments in this book. It is certainly true that no sediments are totally non-cohesive but the theory of non-cohesive sediments forms the basis of all sediment modeling and as usual in this book we take the view that simple models are justified as forming the foundation of understanding. Sediments can also be classified according to their origin: lithogenous (from rock), biogenous (from organisms), hydrogenous (from water), and cosmogenous (from space).

For grain size, we adopt the widely used classification system shown in Table 12.1. For each of these size classes there is one or more physical or chemical origins for the sediment. Referring to Table 12.1, *gravel* forms through physical weathering of rock. A piece of gravel is usually a rock fragment composed of one or more minerals such as quartz. *Sand* forms through the breakdown and disintegration of rocks. *Silt* originates from the chipping of coarser grains during sediment transport, or from the disintegration of fine-grained crystalline rocks. *Clay* originates primarily from chemical weathering of feldspars and other alumino-silicate minerals (those which contain aluminum and silicon). The size demarcation between silt and clay is variously defined. The real

Table 12.1. *The classification of sedimentary particles according to size from the Wentworth scale; note this is a geometric scale, each size differing from those below and above by a factor of two, except for the smallest and largest sizes*

Main class	Sub-class (if any)	Minimum diameter (mm)	Maximum diameter (mm)	phi (ϕ) (fineness)
Boulders		256	none	< -8
Gravel	Cobbles	64	256	-6 to -8
	Pebbles	4	64	-2 to -6
	Granule	2	4	-1 to -3
Sand	Very coarse	1	2	0 to -1
	Coarse	0.5	1	1 to 0
	Medium	0.25	0.5	2 to 1
	Fine	0.125	0.25	3 to 2
Fine Sediments or mud	Coarse silt	0.06	0.125	4 to 3
	Silt	0.002	0.06	6 to 4
	Clay	none	0.008	>8

difference between silt and clay is that clay is generally considered to be a cohesive sediment, i.e., particles adhere together on the bed or even in suspension. This is in contrast to a non-cohesive sediment such as sand. The term *mud* is also used, but is again somewhat ambiguous, and is used by modelers mainly for cohesive sediments.

The Greek lower-case ϕ (*phi*)[9] denotes a logarithmic size scale for sediment grains called *fineness* defined as

$$\phi = -\log_2(d/d_0) \tag{12.1}$$

where d_0 is the *standard* grain size equal to 1 mm and notice the base 2 for the log. The equivalent phi (ϕ) is given for the sediment groups in Table 12.1. The inverse of the above equation for converting back from ϕ to millimeters is

$$d = d_0\, 2^{-\phi}. \tag{12.2}$$

Coastal basins usually have layers of sediment forming their bed and also have sediments suspended within the water column. Like the sediments on the bed of the coastal basin, the sediments within the water column have a spectrum of sizes and types. Concentrations of *total suspended solids* (TSS) range from less than 1 mg l^{-1} (equivalent to 10^{-3} kg m^{-3}) in the deep ocean to hundreds or thousands of milligrams

---

[9] There are two ways of pronouncing the Greek letter *phi* in English; one of these is *fie* (which rhymes with *pie*) and this alliterates with *fineness* and is the reason for the choice of this symbol (the other pronunciation is *fee*).

per liter in coastal basins (although TSS can be below $100\,\mathrm{mg\,l^{-1}}$ in some basins). Note that the density of seawater is about $1030\,\mathrm{kg\,m^{-3}}$ and so the fractional density concentration of sediments is still usually less than 1%. However, the density of the total fluid is increased and, for example, water heavy with mud particles (which are slow to separate out) may produce a bottom layer of what is called *fluid mud*. It is a highly concentrated suspension of cohesive fine-grained particles, which can reduce the depth considerably, depending on its density, viscosity, and thickness. Echo sounders, used for the detection of water depths, are sometimes, depending on their operational frequency, not able to differentiate between fluid mud and the solid bottom, which can lead to problems in defining the navigability of coastal regions if fluid mud is considered *unnavigable*. We use the term *hyperconcentration* to describe concentration of sediments which exceed a fractional concentration by volume of a few percent and hence able to change the properties of the fluid. *Hyper* is used generally as a prefix for *high* in coastal oceanography, and *hypo* as *low*, hence we have hypersalinity and hyperthermal, hyposaline and hypothermal.

12.3.2 Role of grain size

Grain size, current, and wave energy are primary determinants of the behavior of a sediment in a coastal basin. From our own observations, we know that gentle waves breaking on a sandy beach are capable of washing sand grains up and down the beach but do not normally shift pebbles. There are four modes of sediment transport in water called *sliding, rolling, saltation*, and *suspension*. *Sliding* particles remain in continuous contact with the bed, merely tilting to and fro as they move. *Rolling* grains also remain in continuous contact with the bed, whereas *saltating* grains jump along the bed in a series of low trajectories (Latin *saltare* "to dance"). Sediment particles in these three categories collectively form the *bedload*. In contrast to the bedload is the *suspended load* which consists of particles in suspension, i.e., particles that follow long and irregular paths within the water and seldom come in contact with the bed, until they are deposited when the flow slackens. Sliding and rolling are prevalent in slower flows, saltation and suspension in faster flows.

The heart of understanding suspended sediment transport is Figure 12.8 (Hjulstrom's diagram) which illustrates the processes of suspension and settlement from the bed of the basin. We will work through an explanation of this diagram because it explains much of the physics of the way particles of material move in coastal basins. The *Hjulstrom diagram* plots two curves (note that this is a double logarithmic plot). The upper curve represents the minimum stream velocity (from current or waves) required to suspend sediments of varying sizes from the stream bed. The lower curve is the minimum velocity required to continue to transport sediments of varying sizes. Notice that for coarser sediments (sand and gravel) it takes a velocity that is just a little higher to initially suspend particles than it takes to continue to transport them. For small particles (clay and silt) considerably higher velocities are

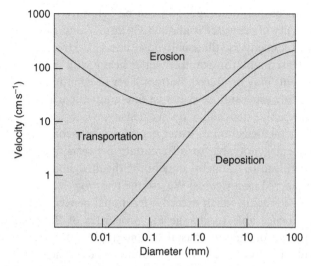

Figure 12.8 A Hjulstrom diagram illustrating the erosion, deposition, and transport regimes of particles of varying diameter in water of varying speed.

required for suspension than for transportation, because these finer particles are cohesive. The upper curve has a minimum for a grain size somewhere near a few tenths of a millimeter which means that this size is the easiest for a current to suspend: larger particles are heavier and smaller particles probably adhere together on the bed. There is no similar minimum in the lower curve so that the bigger the particle, the more difficult it is to keep in suspension. The Hjulstrom diagram is a major simplification and for example we find that small grains can adhere in the water column and so be deposited more rapidly. Also the bed will contain a spectrum of grain sizes and small grains may help prevent larger grains being suspended.

Transport within the water column involves suspension, vertical mixing, advection, diffusion, and deposition to the seabed. Sediment particles are normally denser that water and so sink according to *Stokes' law*, which involves a balance of their negative buoyancy and molecule viscosity. Sedimentologists measure the rate of sinking by simply putting the particles into a settling tank. Traditionally these tanks have been used in the laboratory but are now also deployed in the field with digital imagery. Larger particles are affected by turbulence created by their sinking velocity which provides extra frictional forces. Stokes' law refers to the movement of the particle relative to the ambient water mass which is itself affected by turbulence. This means that it is difficult to measure sinking speeds in the field.

12.3.3 Stokes' law

We briefly discussed Stokes' law in Chapter 7 as part of our consideration of molecular viscosity and its effects at small spatial scales. For all particles, the forces of friction and buoyancy have a different dependency on size. Buoyancy is controlled by the particle volume which varies as diameter cubed, while friction involves the cross-sectional surface area and so varies linearly with the diameter. This means that large particles are dominated by negative buoyancy and sink faster than small particles that are dominated by friction. Hence plankton which tend to be very small remain in the water column through the effects of friction. The speed with which particles descend to the bottom is called the *settlement speed* w_s. The value of w_s depends on whether the settling particles are affected only by molecular viscosity, or by a combination of molecular forces and turbulence due to the motion of the particle. This is determined by the Reynolds number Re for the particle:

$$\text{Re} = \frac{wd}{\nu} \tag{12.3}$$

Initially, let us assume that wd is small enough for $\text{Re} \ll 1$ so that only molecular forces are important. In this case, the drag is due to a laminar viscous force

$$F = 3\pi\nu dw \tag{12.4}$$

where ν is the molecular viscosity. Importantly, as noted above, this force varies linearly with the size (diameter d) of the particle. The buoyancy force on the particle is

$$g(\rho_s - \rho)4\pi\left(\frac{d}{2}\right)^3 \tag{12.5}$$

which varies only as the cube of d, i.e., is proportional to the volume of the particle. If we assume that the particle achieves a terminal velocity w_s (even in the presence of external turbulence) in which the drag F matches the negative buoyancy force,

$$w = c_s d^2 \tag{12.6}$$

$$c_s \equiv \frac{(\rho_s/\rho - 1)g}{18\nu} \tag{12.7}$$

where c_s is a constant of proportionality.[10] If we take a very fine sand grain, of size 0.1 mm, $\nu = 2*10^{-6}\,\text{m}^2\,\text{s}^{-1}$ and $\rho_s/\rho = 2$, Stokes' law gives $w_s = 0.003\,\text{m}\,\text{s}^{-1}$ or $10\,\text{m}\,\text{h}^{-1}$ which corresponds to $R = 0.14$ confirming laminar motion. As we reduce particle size and move to silt and clays, the settlement velocity becomes extremely slow. For example, a clay particle of diameter 0.01 mm and $\rho_s/\rho = 1.1$ has the settlement speed of $w_s = 3\ 10^{-6}\,\text{m}\,\text{s}^{-1}$ or 0.25 m per day.

Settlement in coastal basins is evidently very dependent on grain size. The reason is perhaps rather surprising to newcomers to the field of sediment dynamics, because coastal basins are fundamentally turbulent. Particles within the water column are

---

[10] Equation (12.6) is called Stokes' law after George Stokes (1819–1903), a brilliant mathematical physicist and one of the many great physicists at the University of Cambridge in the nineteenth century. His scientific contributions include fundamental studies of friction in moving fluids.

moved by any motion of the fluid. This is true of organic material (such as plankton) and non-organic (such as clay) particles. Friction keeps these particles tied to the local water mass, and we recall that *plankton* comes from the Greek word meaning "to wander" (just as a *planet's* position in the night sky appears to *wander* across the background of fixed stars). Plankton occupy an ecological niche made possible by the microscopic size of plankton, which means that their world is dominated by molecular viscosity. This viscosity may seem insignificant to an animal the size of a whale, but to a small creature of just a few micrometers in diameter the water is highly viscous. The reason is that we need to scale our picture of the motion down to the size of the particle and that means that viscosity effectively increases in proportion to the decrease in area. Turbulence in the water column does not penetrate down to spatial scales of the size of plankton because they are smaller than the *Kolmogorov length* (see Section 7.5) although this is an assumption that needs to be carefully considered in extremely high energy environments. The local environment of a small particle looks like very viscous water. If you watch a wave on the beach, you will see that it is often full of fine suspended particles that it carries around in the same way that it would carry a dissolved chemical. If we watch the motion of suspended particles with an underwater camera, we see that the particles are passive tracers of the larger-scale turbulent eddies and are very difficult to follow.

12.3.4 Measuring sediment size

Because of the comparative importance of bulk and surface forces for sediment particles, it is extremely important to know the size distribution of sediments in a coastal basin. This task divides into measurement of sediment samples collected from the bottom of the basin and samples within the water column. We will deal firstly with bottom samples. For gravel, grain size is normally assessed by direct measurement with calipers or very large mesh sieves (up to approximately 20 mm). For sand, we use nested sieves of progressively smaller mesh, from 2 mm to 63 μm. Normally, the sand is first washed with fresh water (to remove salt and finer sediment), then dried and shaken in a sieve stack. Sieves are graded in ϕ intervals of 0.25 to achieve an accurate ϕ distribution. Finer particles are measured by allowing them to settle through a column of still water, to obtain the terminal settling velocity w from which the *free-fall diameter d* of the particles is calculated using Stokes' settling law (12.6) or a modified form of Stokes' law. The free-fall velocity may be obtained by either the direct observation of particles, the measurement of the decrease of density in the settling tube as the sediment settles out, or measuring the weight of the settled sediment with time. Coarse silts can be measured using fine-meshed sieves (down to about 30–40 μm). Problems frequently occur with this method due to clogging of the mesh and cohesion. The Coulter counter is a superior method for finding relative size distribution. It measures the change in electrical conductivity as a particle passes through a small cavity. A small concentration of the sediment is suspended in an electrolyte, and a fixed volume of the electrolyte plus liquid is sucked through the cavity by a vacuum pump. The number of particles that give rise to

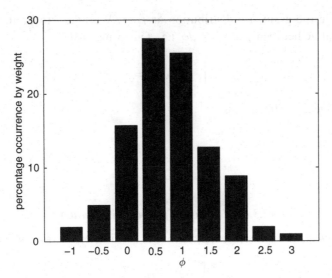

Figure 12.9 A typical histogram of grain size against fineness ϕ (phi). This has a peak at $\phi = 0.5$ which corresponds to coarse sand.

a change in resistance above a critical threshold is counted. The resistance threshold is then lowered and the process is repeated, which gives the grain size distribution of the sediment. Note that all clays and most silts (other than very coarse silts) exhibit cohesion, and are therefore difficult to analyze for grain size. The presence of fine particles may also cause cohesion in coarser sediments. Preparation of samples often requires the breaking up of cohesive aggregates using a chemical solution.

Measuring sediments suspended in the water column involves two techniques. The first is to collect water samples of a given volume which are prepared onboard a boat and then brought back to the laboratory and weighted. This gives total suspended solids (TSS) which includes non-organic and organic material. It is also possible to find TSS by in situ instruments using a variety of optical techniques. To find the size distribution in the water column, there is a technique based on the angular distribution of scattered laser light (see Section 12.3.6).

12.3.5 Size distribution of particles

All natural sediments contain a range of sediment sizes and the distribution of sizes is very important to all modeling. Several of the grain size analysis methods outlined in Section 12.3.4 give the grain size distribution, which is usually plotted as a histogram against fineness ϕ. Our reason for using ϕ rather than just diameter d is that we are naturally interested in the distribution over the many orders of magnitude of diameter that occur in a coastal basin. Interestingly, the distribution in terms of ϕ is often close to being Gaussian (the Gaussian is also called a *normal* distribution). Figure 12.9 shows a typical histogram. Remember that fineness increases as grain size decreases.

Let us write the grain size distribution as $f(\phi)$. We take f to be the fraction of sediments that lie between ϕ and $\phi + d\phi$, i.e., f is normalized:

$$\int_{-\infty}^{\infty} f(\phi)d\phi = 1 \tag{12.8}$$

and the average grain fineness is

$$\overline{\phi} = \int_{-\infty}^{\infty} \phi f(\phi)d\phi. \tag{12.9}$$

12.3.6 Moments of fineness distribution

We define *central*, or *standardized*, moments of the fineness distribution as

$$\mu_n = \int_{-\infty}^{\infty} f(\phi)(\phi - \overline{\phi})^n d\phi \tag{12.10}$$

where n is called the *order of the moment*. For a distribution which is symmetric about its mean (such as the Gaussian) all of the central moments whose order is an odd integer are zero. The *standard deviation* σ is the second order central moment, i.e., $n = 2$:

$$\sigma = \mu_2 \tag{12.11}$$

and so if $f(\phi)$ is Gaussian,

$$f(\phi) \rightarrow \frac{1}{\sigma\sqrt{2\pi}} \exp\left(-\frac{\phi - \overline{\phi}}{2\sigma}\right). \tag{12.12}$$

The cumulative distribution $F(\phi)$ is defined as the fraction of sediments which have fineness less than ϕ:

$$F(\phi) \equiv \int_{-\infty}^{\phi} f(\phi)d\phi \tag{12.13}$$

which for a Gaussian distribution is

$$F(\phi) \rightarrow \left[1 + \mathrm{erf}\left(\frac{\phi - \overline{\phi}}{\sigma\sqrt{2}}\right)\right] \tag{12.14}$$

where erf is the error function which we have met in Chapter 10. A cumulative distribution is shown in Figure 12.10.

Skewness of the fineness distribution is defined as

$$\gamma_1 \equiv \frac{\mu_3}{\mu_2^{3/2}} \tag{12.15}$$

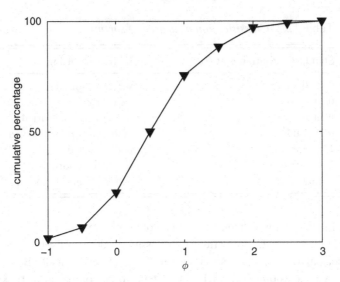

Figure 12.10 Example of a cumulative distribution based on the histogram of Figure 12.9.

and note that skewness is a number. Skewness is zero for *any* symmetric distribution (such as a Gaussian) and generally measures the extent to which the distribution is skewed towards higher values of ϕ. Negative skewness means that the fineness is skewed to lower values of ϕ, i.e., there are more coarse grains than one would expect for a symmetrical distribution. Skewness helps us to construct the life history of sediments that we may find at a particular site because fineness has a major control on suspension and settlement by water.

Kurtosis is a measure of the size of the tails of a distribution relative to its peaks (from the Greek word *kurtos* roughly meaning "convex, arched" and is defined as

$$\beta_2 \equiv \frac{\mu_4}{\mu_2^3}. \tag{12.16}$$

For a Gaussian distribution, the kurtosis $\beta_2 = 3$ and authors sometimes quote values of *excess kurtosis* $\beta_2 - 3$. Kurtosis measures peaking near the mean and low kurtosis indicates flatness.

If the distribution of sediments at a particular point in the basin is very peaked at one value of fineness, i.e., all particles are closely similar in size, we say that they are *well sorted*. Generally, sediments in the coastal environment have a wide distribution of grain sizes simply because they come from a variety of sources. So if we find just one size at a particular site, we may be able to make some judgment as to the relevant transport and settlement processes. Usually this involves the settling out of the sediment from transport within a water column and usually we find that there is a graduation of the size of the well-sorted sediments with distance. As an example, we might find that near a river mouth the deposited sediments are large and then gradually reduce in size with distance from the river mouth; this is the expected result of the progressive reduction in water with

Table 12.2. *Classification of degree of sorting*

Standard deviation σ of $f(\phi)$	Degree of sorting
0.00–0.35	very well sorted
0.35–0.50	well sorted
0.50–0.71	moderately well sorted
0.71–1.00	moderately poorly sorted
1.00–2.00	poorly sorted
2.00–4.00	very poorly sorted
> 4.00	extremely poorly sorted

distance from the mouth which allows particles of ever increasing fineness to settle out of the moving water. Table 12.2 is an attempt to classify the degree of sorting. Sorting is very evident in coastal basins, and especially on the continental shelf. Beaches that are exposed to large waves (*high-energy* beaches) tend to be composed of rocks, boulders, and pebbles, because only these particles are able to resist suspension by the waves. On such beaches a drop of wave energy, as for example often occurs with a change of season, will see the rocks covered with sand. This sand tends to be transported shoreward and then settle out and is not resuspended. Suspension of sand requires a speed of about $0.2 \, \mathrm{m \, s^{-1}}$, and the Hjulstrom diagram shows that currents of less than this speed will not suspend particles; they can *transport* sediments, but not suspend them.

Roundness is a measure of the removal of any corners of a sedimentary particle. As sediment is transported, it suffers abrasion by coming into contact with the bottom of a stream or basin, seafloor, or other grains of sediment. This abrasion tends to round off the sharp edges or corners. Rounding tends to vary with the size of the grains. Boulders tend to round much more quickly than sand grains because they strike each other with much greater force.

Sphericity refers to the particle shape, and can be described as high or low. According to this definition, a ball would have highly sphericity, but so would a cube (high sphericity, but low roundness). In contrast, a cigar would have low sphericity. A shoe-box would have both low sphericity and low roundness. Sand grains may have high or low sphericity. Some minerals may produce elongated or flattened grains, depending primarily on original crystal shape and cleavage. A well-rounded grain may, or may not, resemble a sphere. And a spherical grain may or may not be well rounded.

Texture is an indicator of energy levels in the deposition area (the place where sediment accumulates like a beach, a river bed, a lake, or river delta). High energy may be due to waves or currents. Quiet or *still water* (water without waves or currents) is considered to be a low-energy environment. As sediments experience the input of mechanical energy (the abrasive and sorting action of waves and currents), they pass through a series of textures from *immature* (unsorted sediment containing clay and/or silt) to *supermature* (well sorted and rounded sediment with no mud).

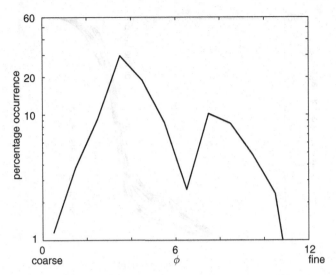

Figure 12.11 Example of bipolar distribution of grain size in bottom sediments. One peak lies just below $\phi = 4$ and the other, smaller, secondary peak just below $\phi = 8$.

Sediment distributions usually show several dominant peaks as in Figure 12.11, which has two peaks indicating two types of sediment, i.e., sand and silt/clay (mud). These two groups of sediment are likely to have different dynamics and so should be treated separately. This means that separate skewness and kurtosis are important to considering processes that differentially affect the two types of sediment. Commonly used laboratory programs use the cumulative distribution to determine the percentage (by volume) of clay, silt, and sand. They can also determine separate statistics for clay, silt, and sand by breaking the distribution into three separate ranges of ϕ. More importantly, it is possible to distinguish between the various ranges of sand grain size. This is important because coarse and fine sands do behave differently and have different roles in basin dynamics. Figure 12.12 shows bipolar distributions for a variety of sites within a single coastal basin. There is a variation in the relative fraction of silt, clay, and sand around the basin with sand becoming more prevalent near the mouth (as wave energy and tidal currents increase).

Figure 12.13 shows a corresponding size distribution for TSS mentioned earlier in this chapter and illustrates the effect of mechanically stirring the bottom sediments.

12.3.7 Generalization of Stokes' law

Stokes' law is based on spherical, non-interactive, settling particles affected only by molecular viscosity. In reality, the settlement speed does depend on particle shape. One of the characteristic numbers used to describe lack of *sphericity* is the *Corey shape factor*

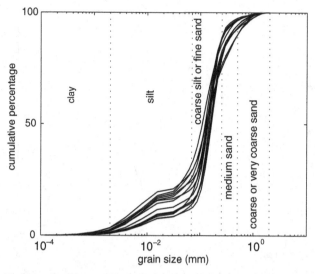

Figure 12.12 Cumulative bipolar distribution of grain size in bottom sediments for a variety of sites around a coastal basin (each line represents a different site). Notice the variation in the relative amounts of silt, clay, and sand between sites. Sand is usually more prevalent as one moves to higher-energy regions of a basin.

$$C = \frac{d_c}{\sqrt{d_a d_b}} \tag{12.17}$$

where d_a, d_b, and d_c are the major, minor, and intermediate axes of the particle. Provided that the particle has a simple geometric shape, the theoretical settling speed can be calculated by determining the frictional force. For a disc shape (often found for clay minerals), Stokes' law still applies (assuming only molecular drag) if we use an effective diameter

$$w_s = \frac{\bar{d}}{2k\nu d_c}(\rho_s - \rho)g\bar{d}^2$$
$$\bar{d} \equiv \frac{d_a + d_b}{2} \tag{12.18}$$

where the coefficient k has a theoretical value of 5.1 for *broadside settling* of infinitely thin particles. For spherical particles, $\bar{d} = d_c$, and note that $k = 9$ reproduces the Stokes' law.

It is important to check that the settling velocity does not produce a Reynolds number Re for the particle (12.3) in excess of 0.5. Otherwise, the flow around the particle will not be laminar (see Section 7.2). Taking the molecular viscosity ν as approximately $2\ 10^{-6}\ \mathrm{m^2\,s^{-1}}$, (12.3) indicates that the flow is only laminar if $wd < 10^{-6}\ \mathrm{m^2\,s^{-1}}$, and if we use (12.6) and (12.7) for w, we get $d < 10^{-4}$ m, i.e., the diameter must be less than 0.1 mm corresponding to $\phi > 3.5$. Note that the failure of Stokes' law is not

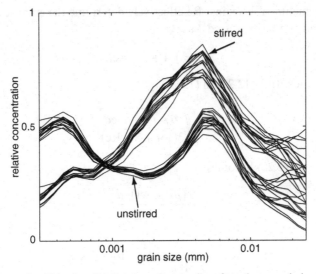

Figure 12.13 Example of bipolar distribution of grain size of total suspended solids (TSS). The figure shows that the manual stirring of bottom sediments brings coarser sediments into suspension.

due to ambient turbulence in the fluid, but turbulence produced by the particle itself. One would intuitively expect that the transition to turbulence would increase friction and hence decrease the settling speed. A similar reduction of speed would also be expected from departures of shape from the pure sphericity assumed in Stokes' law.

To obtain a more general expression for the settlement speed, w, which extends Stokes law to higher speeds, we first write the frictional force F as

$$F = \frac{1}{2} C_d w^2 A \tag{12.19}$$

where w is the speed of the particle, A is the cross-sectional area of the particle normal to the direction of its motion, and C_d is a coefficient of drag. If C_d were a constant (independent of d and w), the form of (12.19) would implicitly assume that the drag were due to turbulence generated by the motion of the particle. In fact, for small particles the friction is generated primarily by molecular viscosity in which case the motion is *laminar*, i.e., the friction depends *linearly* on d and w. For this to be true C_d must, at small values of the particle diameter d, vary inversely as dw. The effects of that turbulence relative to molecular viscosity are determined by the Reynolds number Re of the particle. If we again assume that the particle achieves a terminal velocity w in which the drag F given by (12.19) matches the negative buoyancy force (12.5), we have

$$w_s = \sqrt{\frac{4(\rho_s/\rho - 1)gd}{3C_d}}. \tag{12.20}$$

Hence measurements of w against Re effectively determine the empirical dependence of C_d on Re and can be expressed[11] as

$$C_d = 1.4 + 24/\text{Re}, \tag{12.21}$$

and if we now substitute (12.21) in (12.20),

$$w = \sqrt{\frac{4(\rho_s/\rho - 1)gd}{3(1.4 + 36/\text{Re})}}. \tag{12.22}$$

So when Re $\ll 1$, (12.22) becomes

$$w = \frac{(\rho_s/\rho - 1)gd^2}{18\nu} \tag{12.23}$$

$$w = \frac{(\rho_s/\rho - 1)gd^2}{18\nu} \quad \text{Re} < 0.5 \tag{12.24}$$

which reproduces Stokes' law, whilst at high Reynolds numbers

$$w = 0.98\sqrt{(\rho_s/\rho - 1)gd} \quad \text{Re} > 1. \tag{12.25}$$

So that in this limit, a particle of size $d = 0.002\,\text{m}$ would fall at a speed of about $0.17\,\text{m s}^{-1}$ compared to $3.6\,\text{m s}^{-1}$ by the original form of Stokes' law.

The settlement speed is also affected by the fractional volume concentration c (volume of sediment particles per unit volume of fluid) and clearly becomes zero as c tends to unity with as much as a 20% reduction in settlement speed when c is 10%.

12.4 Littoral drift and tidal channels

An interesting example of the effect of erosion and settlement is provided by a tidal channel leading from the continental shelf to an otherwise enclosed coastal basin. There is often a strong alongshore drift of sand caused by waves meeting the shoreline obliquely. This flux of sand along the shore is also called the *littoral drift* and was discussed in Chapter 2. If the waves hit exactly at right angles to the shore there is no littoral drift, but otherwise the drift lies in the direction at which the waves strike; waves from the south tend to create a flux from south to north. This littoral drift tends to fill in holes in the coast and, for example, if we dredge out an inlet, the littoral drift will tend to fill that inlet; as the littoral current enters regions that are out of the direct effect of the waves, sand tends to be deposited. Littoral drift is also responsible for erosion of the beach in areas where there are offshore structures that prevent the supply of sand from further upstream of the drift. This happens for

---

[11] Fredsøe and Diehards (1992).

example if we build a *groin*[12] outward from the beach. Such groins are intended to prevent erosion by storms and do have the effect of slowing the littoral drift. They have been used by coastal engineers for many centuries and are a familiar feature of many shorelines.

The need for protection of a coastline is often a product of human occupation of beach areas. Beaches are usually in a very dynamic state of balance between seasonal, annual, and episodic events of erosion and deposition. This includes the *beach* as we traditionally understand it, but also the *back beach* or regions of sand dunes which serve as a reservoir of sand used to supply the beach after extreme storm events. If there is construction of roads and buildings on this back beach, those constructions will be at extreme risk from erosion. The building of groins creates erosion upstream of the groins because the groins impede the littoral flux, so that sand moving down-stream of the groins is not replaced. There is similarly deposition of sand on the upstream side of the groin. These areas of erosion can be sufficient to effectively remove the whole of the beach near the groin. The deposition of sand can be sufficient to extend the beach out to the end of the groins so that the sand can effectively bypass the groin. Littoral drift tends to seal the mouth of estuaries and lagoons, unless the mouth is kept open by river outflow (and perhaps tides). The action of the latter must be to mobilize sand and transport it outward from the mouth, and that requires current speeds in excess of about $0.2\,\mathrm{m\,s^{-1}}$. Provided that the mouth remains open, the actual speed achieved by tidal outflow is approximately inversely proportional to the area of the mouth. As the littoral drift gradually seals the mouth, the current speed increases to the point where it can erode at the same rate as sand is deposited. While the mouth is open, tidal currents can maintain the mouth in the form of a tidal channel. The currents in the channel will then be of order a few tenths of a meter per second. If the currents become larger (due to closure of the channel) there will be erosion and the channel will be enlarged and if the current drops below about $0.2\,\mathrm{m\,s^{-1}}$ the channel tends to narrow and seal.

12.5 Coastal classification based on waves and shorelines

So far in this book we have looked at various types of classification for coastal basins. These include geomorphology and vertical stratification, and in Chapter 10 we also looked at classifications based on horizontal gradients of physical quantities such as salinity and density. When considering sediment dynamics, we are especially inter-ested in the wave conditions in basins and the form of the shoreline and any beaches. This does lead to another useful classification system which helps clarify the type of sediment processes to be considered and the models which are likely to be most useful. The cases are broadly as follows.[13]

---

[12] The alternative spelling of *groin* outside the United States is *groyne*.   [13] Mangor (2004).

12.5.1   Exposed littoral coast

This is the archetypal wide beach coastline that is highly valued for recreational purposes and attractive to coastal developers. It is generally associated with seasonal high wave energy. This is responsible for the littoral drift or alongshore drift of sediments, which contribute to the sculpting of beaches. Beaches are seasonally dependent on wave conditions and generally larger in summer than winter. All beaches are dynamic formations which are maintained by erosion and settlement, which is heavily influenced by stresses and currents due to waves. Exposed ocean beaches, in their natural state, are often (perhaps *usually*) backed by dunes that are wholly or partially vegetated. These dunes are part of the beach, although their sediment is only *mobilized* during storms of sufficient severity. The type of sediment is generally variable from sand to gravel, or boulders, with sand returning during calmer seasons. As we move from the ocean-fronting beaches to more protected beaches at the entrance to a coastal basin, the beach characteristics change from those associated with high energy to those found on lower energy beaches.

Exposed beaches may have many systems of offshore bars, and these typically exchange sediment with the beach. Beaches of this type can be extremely hazardous to swimmers. They are subject to *rip currents* formed by alongshore wave-driven currents swinging outward to sea in a narrow fast-flowing ribbon of water. The dynamics of rip currents involves a convergence zone in the alongshore wave-driven current which is able to produce a pressure gradient sufficient to overcome that due to the radiation stress of breaking waves (Section 11.6).

Sand movement on these beaches can be very large, up to 1 million m^3 per year (which is enough sand to create a beach 20 m wide and 2 m deep over a length of 25 km in one year). There are a variety of specialized models that deal with such littoral drift and its dependence on wave conditions. In Chapter 2, we noted that these models are based on the continuity equation for sediment, plus the essential ingredient of a dynamical component that determines the drift as function of wave conditions. Many exposed beaches are a lesson in the science of littoral drift.

12.5.2   Moderately exposed beaches

This type of shoreline is often found inside the entrances to coastal basins in regions where ocean swell and storm waves are effective. They usually have one offshore bar and a narrow beach.

12.5.3   Protected or marshy beach

Found in low-energy areas with either no sandy beach or a very small beach and no offshore bars. This type of shoreline commonly contains wetlands and marshy areas and is very common in tropical and subtropical regions, usually facing small coastal basins.

12.5.4 Tidal flat coast

A common shoreline on coasts with tide range much greater than wave height. These coasts usually have very wide beaches at low water.

12.5.5 Monsoon coast

Characterized by persistent waves of about 1 m in height. If there is an adequate supply of mixed sediments, these are found to be well sorted at the coast.

12.5.6 Muddy coast with mangroves

Dominated by mud with little sand at the shoreline. Usually found in tropical regions where rivers supply abundant fine terrestrial material and often accompanied by mangroves and wetlands.

12.5.7 Coral coast

Tropical coast dominated by corals with nutrient-poor water (see Section 11.5) with low suspended solids. Carbonate beaches are dominated by coral and coral debris and waves usually break on the reef.

12.6 Critical shear stress

The effect of water flow over a sediment bed is to dislodge individual particles and force them into motion. There is a critical shear stress at which particles in the bed start to move and this clearly depends on the size and density of sediment grains, and we would like to have a simple model that allows us to find the critical stress from these two parameters. However, some sediments are cohesive in character, and this has a significant effect on sediment suspension. It is probably true that few coastal basins have sediments that are totally free of cohesive components and this very greatly complicates the determination of critical shear stress. Most cohesion is due to clay particles held together by a combination of electrostatic attraction and surface tension. Even a small fraction of clay in sediments will cause significant cohesion. Cohesion begins to be significant when the fraction by weight reaches 5% to 10%, and so modelers need to be very cautious of assigning values of critical shear stress.

One influence of grain diameter comes through the buoyancy of the particle and we define

$$\tau_d = \rho g(\rho_s/\rho - 1)d \tag{12.26}$$

where ρ_s is the density of the sediment grains. The critical shear stress is denoted by τ_c and the critical shear speed u_c^* is defined as

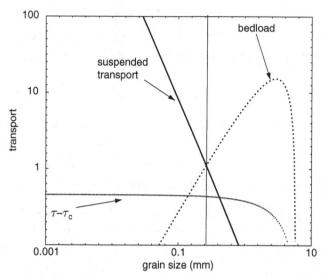

Figure 12.14 Comparative strength of suspended transport and bedload as a function of grain diameter together with the variation of excess shear stress.

$$u_c^* = \sqrt{\tau_c/\rho}. \tag{12.27}$$

The ratio of τ_c to τ_d is called the critical *Shields number S*, i.e.,

$$\tau_c = S\tau_d. \tag{12.28}$$

Although we take S to be a constant here, which varies between about 0.04 and 0.1, it is formally a function of the Reynolds number $u_c d_s / v$ for the bed. A commonly used formula for the bedload is

$$q = 8\sqrt{gd^3 \left(\frac{\rho_s}{\rho} - 1\right)\left(\frac{\tau - \tau_c}{\tau_d}\right)^3} \tag{12.29}$$

where τ is the bottom stress. We also find it convenient to use the bottom shear speed defined as

$$u^* = \sqrt{\tau/\rho}. \tag{12.30}$$

An empirical law for the rate E of suspension of sediments into the water column is

$$E = \alpha(\tau - \tau_c)^n \tag{12.31}$$

in which the exponent n is usually taken as $3/2$.

Figure 12.14 shows bedload and suspended sediment transport as functions of grain size (notice that this is a log–log plot) for a shear speed $u^* = 0.1\ \mathrm{m\,s^{-1}}$. Table 12.3

Table 12.3. *Matlab code for Figure 12.14*

```
n = 1.5; rho = 2.6; density of sediment relative to water
S = 0.1; % Shields parameter;
N = 200; number of particle diameters
inc = 4/N; %this is the logarithmic increase in d and u for each step
d0 = 10; %reference diameter
exd = – 6;
for m = 1:N;
        d(m) = 10^exd; %find d using present value of exponent exd
        uc(m) = sqrt(9.81*(rho – 1)*S*d(m));
        % this is critical shear speed
        us(m) = (d(m)/d0)^(2); % relative settling velocity
        exu = –3;
        % loop over shear speeds u(p)
        for p= 1:N;
                u(p) = 10 ^ exu; shear current
                X = u(p) ^ 2 – uc(m) ^ 2; excess shear stress
                if X > 0; s(p, m) = X; else ; s(p, m) = NaN; end;
                erode(p) = 1e – 12 * s(p, m) ^ n; % rate of erosion
                c(p) = erode(p)/us(m); % concentration of suspended sediments
                qs(p, m) = u(p) * c(p); % transport flux of suspended sediments
                q(p, m)= 8 * sqrt(9.81 *(rho – 1) * (d(m)^ 3))*s(p, m) ^ (3/2);
                exu = exu + inc;
        end;
        exd = exd + inc;
end;
mm = 100
figure; plot(log10( d(:) ), log10( qs(mm, :) ), 'r'); hold on;
plot(log10(d (:) ), log10(q(mm, :) ), 'r:'); hold on;
plot(log10( d(:) ), log10(5e – 6 * s(mm, :) ), 'b.');
```

shows the model code for Figure 12.14. Note that the depth average current $\bar{u}$ is related to u^* via the bottom drag coefficient C_d,

$$u^* = \bar{u}\sqrt{C_d}, \tag{12.32}$$

so that for a typical drag coefficient of 0.0025 (for sand) our assumed value for u^* corresponds to $\bar{u} = 2\,\mathrm{m\,s^{-1}}$. Figure 12.14 shows the variation of $\tau - \tau_c$ with grain diameter d which deceases to zero as τ_c increases with d; this can be verified from (12.28) and (12.26) (at fixed Shields number). Bedload q is derived from (12.29). The bedload in Figure 12.14 increases with grain diameter as $d^{3/2}$ at small d and then reaches a maximum and drops to zero due to the factor $\tau - \tau_c$. Physically the bottom sediments eventually reach a size at which they cannot be moved by the current. The suspended sediment transport in Figure 12.14 is determined from the rate of suspension E in (12.31) and the settling velocity w given by Stokes' law which in the steady state gives a concentration c:

$$c = E/w. \qquad (12.33)$$

Notice that the suspended transport decreases with d due to $\tau - \tau_c$ in the erosion rate (it becomes increasing difficult to suspend the sediments and also the increase in settlement rate with d [Stokes' law].)

12.7 Box model of sediment processes

Box models were presented in Chapter 3 and the method is used here as an illustration of sediment processes.

Referring to Figure 12.15, the concentration of sediment in the water entering the basin is c_0, and that of water leaving the basin is c. The speed of water, i.e., the residual current inside the basin between entry and exit of water is u. So the net flux of sediment out of the box per unit width is

$$Q = hu(c - c_0) \qquad (12.34)$$

where h is the water depth (assumed constant). If E is the rate of sediment suspension per unit area of the bed and S is the rate of sediment settlement per unit area, it follows immediately that in a steady state

$$Q = L(E - S) \qquad (12.35)$$

where L is the length of the box, i.e., the path length of the current; Q is either positive (net erosion) or negative (net settlement). If we write the mean concentration in the basin as c we can determine S from the settlement velocity, i.e.,

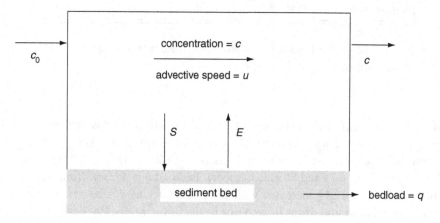

Figure 12.15 Box model of sediment processes.

$$S = cw. \tag{12.36}$$

So that substituting (12.36) in (12.35), we obtain

$$u(c - c_0) = \frac{L}{h}(E - cw). \tag{12.37}$$

One limit of (12.37) is that there is a balance between erosion and settlement and so

$$c - c_0 = \frac{E}{w}. \tag{12.38}$$

For this to be true the effects of sediment advection must be small. If we assume that c_0 and c are comparable this means that

$$\frac{L}{u} \gg \frac{h}{w}, \tag{12.39}$$

i.e., the time L/u for a particle to cross the basin is much longer than the settlement time. We can write the settlement time using Stokes' law, i.e., with typical values

$$w = c_s d^2, \quad \text{Re} \ll 1 \quad c_s \sim 4^* 10^5 \, \text{s}^{-1} \text{m}^{-1}. \tag{12.40}$$

For example, for $d = 0.1$ mm the critical value of L/hu is $100 \, \text{s m}^{-1}$. If h is 5 m and $u = 0.1 \, \text{m s}^{-1}$, this means that $L > 100$ m in order for (12.38) to hold whilst for $d = 0.01$ mm, $L > 12$ km and at $d = 0.001$ mm, $L > 1000$ km. So we can see that for coarse particles, settlement is generally complete within normal coastal basins, whereas fine sediment tends to be dominated by advection. However, if we have strong flow from rivers or tides with, for example, u approaching $1 \, \text{m s}^{-1}$, the cut-off from settlement to advection is 1 km at 0.1 mm, and 120 km for $d = 0.01$ mm. Note also that that settlement is more likely to be dominant in shallow water both because h is lower and because u tends to be reduced.

The opposite limit to (12.38) involves the concentration being dominated by advection and either suspension or settlement. In that limit, we would have either

$$\frac{c - c_0}{L} = \frac{E}{hu} \tag{12.41}$$

or

$$\frac{c - c_0}{L} = -\frac{cw_s}{hu} \tag{12.42}$$

respectively. Equation (12.41) essentially defines a characteristic concentration for erosion:

$$c_E \equiv \frac{LE}{hu} \tag{12.43}$$

Wave and sediment dynamics

and (12.42) gives a characteristic length for settlement:

$$L_s \equiv \frac{hu}{w_s}.$$
(12.44)

The analytic solution of the model when E vanishes is simply

$$c - c_0 = -c\frac{L}{L_s}$$
(12.45)

and this is called the *homogeneous* solution. With this in mind, we can solve for finite E and simply add the solutions

$$c - c_0 = -c\frac{L}{L_s} + c_E.$$
(12.46)

The time-dependent form of the model is

$$\frac{dc}{dt} = -\frac{u}{L}(c - c_0) + \frac{E - cw_s}{h}$$
(12.47)

which has the steady state solution (12.46) and can be rewritten as an explicit solution for c:

$$c = \frac{c_0 + c_E}{1 + L/L_s}.$$
(12.48)

The transport out of the box is augmented by the bedload. Both the suspension rate and the bedload are dependent on the extend to which the bottom stress exceeds the critical shear stress, i.e., on the quantity

$$\Delta\tau \equiv \tau - \tau_c$$
(12.49)

where τ_c is the critical shear stress and τ is the shear stress given by

$$\tau = \rho C_d u^2,$$
(12.50)

C_d being the bottom drag coefficient.

For coarse particles the balance between erosion and settlement is generally considered to be the dominant process that controls the sediment concentration in the water column. This has given rise to a set of models that describe the total transport, i.e., suspended sediments plus bedload, as a function of the dimensionless bottom stress, critical shear stress, settlement velocity, and some parameters of the bed. These generally must be considered more reliable than a model that attempts to derive the unsuspended concentration or details of the bed motion. Furthermore, in a three-dimensional model of suspended transport, much of the load occurs in the region of the bottom boundary, and so requires a reasonably accurate consideration of that boundary.

Table 12.4. *Classification of type of sediment transport based on Rouse number Ro and suspension parameter 1/Ro*

Type of sediment transport	Lower Rouse number (Ro)	Upper Rouse number (Ro)	Suspension parameter (1/Ro)	Type of sediment transport
Bedload	2.5	none	less than 0.4	dominantly bedload
50% suspension	1.2	2.5	0.4 to 0.8	partly suspended load
100% suspension	0.8	1.2	0.8 to 1.3	dominantly suspended load
Wash load	none	0.8	>1.3	wash load evident

There are in vogue at the time of writing a number of different transport formulations. We will here describe the Fredsøe model, developed by a research school led by Jørgen Fredsøe, a Danish pioneer of sediment transport. The suspended load transport is written as

$$q_s = 11.6u_*c_ba\left[I_1\log\left(\frac{30h}{l}\right) + I_2\right] \tag{12.51}$$

with c_b the bed concentration of suspended sediment and $a = 2d$, which is the reference level at which c_b is evaluated; I_1 and I_2 are *Einstein integrals*, h is the water depth, and l is the *Nikuradse equivalent roughness length*. The Einstein integrals are functions of a/h and the *Rouse number* defined as

$$\text{Ro} \equiv \frac{w_s}{\lambda U_f}. \tag{12.52}$$

The Rouse number dictates the mode of sediment transport. It is the ratio of particle settling velocity to the shear velocity (rate of fall versus strength of turbulence acting to suspend particles); $\lambda = 0.4$ (von Kármán constant). The inverse of the Rouse number is called the *suspension parameter*; when greater than unity, it tells us that sediment transport is dominantly by the suspended load. Table 12.4 shows a classification of sediment transport based on values of the *Rouse number Ro* and *suspension parameter 1/Ro*.

Sand is mainly transported as bed load, i.e., as a continuous distribution of sand from the bottom upwards. In contrast, finer sediments tend to settle out much more slowly and move less readily in the bed and so are mainly transported as suspended load. Their erosion is strongly affected by cohesion. So, generally when we talk about models of *sand transport* we mean a model with a continuous distribution of particles from the bed upwards, so that models which prescribe a suspension rate are most applicable to fine particles. Sand transport depends basically on models of

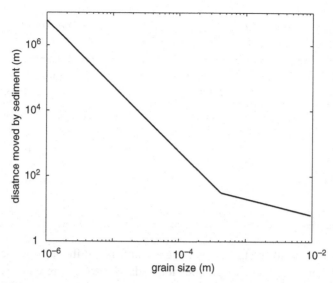

Figure 12.16 Distance moved by sediment suspended in the water column as a function of grain size for an ambient current of $0.5\,\mathrm{m\,s^{-1}}$ in a basin $5\,\mathrm{m}$ deep.

distribution of particles through the water column. This distinction is well illustrated by the point source of particles created by a dredge at work in a coastal basin. Dredging usually removes material to improve navigational access for boats and ships or for environmental management purposes. Dredges remove most of the dredge *spoil* but some sediment inevitably enters the water column. This creates a *plume*, which follows the local current and also disperses laterally. Away from the source, there is no suspension, and so particles merely settle out from the water column. If we take our expression for the settlement velocity w, we can find the time taken for particles to return to the bottom. If we then multiply by a speed u, this will give us some idea how far the particles will travel. For example, for a basin $5\,\mathrm{m}$ deep with current $u = 0.5\ \mathrm{m\,s^{-1}}$, sand grains of diameter greater than $0.1\,\mathrm{mm}$ are carried only a few hundred meters. Finer sediment is carried much further and, for example, sediment of diameter $10\,\mu\mathrm{m}$ travels $60\,\mathrm{km}$. So the plume may contain only fine sediments except near the dredge. Our model of a simple settling out of sediments is, of course, a gross oversimplification but has education value (Figure 12.16).

12.8 Turbulent mixing and settlement

If we take the vertical profile of sediment concentration as $c(z)$ and the settling velocity as W_s we can readily construct a model for the vertical profile of sediment distribution. This involves a specification for the rate of erosion E and for the coefficient of eddy viscosity in the vertical K_z. The vertical flux of sediment is

$$F_c = -K_z \frac{\partial c}{\partial z} - cw \tag{12.53}$$

where z is measured vertically upwards and F_c is positive if the flux is upward. In general K_z is a function of z. The continuity equation for sediment is

$$\frac{\partial c}{\partial t} = -\frac{\partial F}{\partial z} \tag{12.54}$$

from which we have

$$\frac{\partial c}{\partial t} = -\frac{\partial}{\partial z}\left[-k_z \frac{\partial c}{\partial z} - cw\right] = \frac{\partial}{\partial z}k_z\frac{\partial c}{\partial z} + w\frac{\partial c}{\partial z} \tag{12.55}$$

and in the steady state

$$\frac{\partial}{\partial z}k_z\frac{\partial c}{\partial z} + w\frac{\partial c}{\partial z} = 0 \tag{12.56}$$

and so integrating with respect to z,

$$k_z\frac{dc}{dz} + wc = \text{constant} \tag{12.57}$$

which is simply a statement that the steady state flux of sediment is independent of z. Since sediment cannot cross the water surface, this flux must be zero at $z = 0$ and so,

$$k_z\frac{dc}{dz} + wc = 0 \tag{12.58}$$

which is easily solved as

$$c(z) = c_0 \exp\left(-\frac{w}{k_z}z\right) \tag{12.59}$$

where c_0 is the concentration at the surface $z = 0$. Note that in the water column z is negative and so c increases exponentially towards the bottom to become

$$c(-h) = c_0 \exp\left(\frac{w}{k_z}h\right). \tag{12.60}$$

Note the characteristic length that determines the vertical decrease of c away from the bottom is K_z/w. For strong vertical mixing, this length is large and so sediment is (as we would expect) distributed over the whole water column; small w aids this effect. The other limit is high w and small K_z which tends to pack the sediment into the bottom of the water column. Equation (12.61) gives the vertical settling flux as

$$wc_0 = -wc_0 \exp\left(-\frac{w}{k_z}z\right) \tag{12.61}$$

with an equal and opposite diffusive flux. The solution is evidently dependent on the specification of a value for c_0 or $c(-h)$. Note that the flux at the bottom of the water column is necessarily zero in the steady state. We could, of course, put a source of sediment at the bottom, and a sink to allow the sediment to reenter the sediment bed. In a steady state, the difference between the source and sink would be zero. A suitable form would be

$$\text{source} = E$$
$$\sin k = wc(-h). \tag{12.62}$$

We equate the sink with the downward settling flux and E with the net diffusive flux so that the suspension flux is simply mixed upward into the water and then descends by settlement. Adding a source term to the continuity equation at $z = z_a$ is simply a matter of including a delta function, which we met in Chapter 2, at that point so that, in one dimension, (12.54) would become

$$\frac{\partial c}{\partial t} = -\frac{\partial F}{\partial z} + A\delta(z - z_a). \tag{12.63}$$

If we integrate (12.63) with respect to z from $z = -h$ to $z = 0$,

$$\frac{\partial}{\partial t} \int_{-h}^{0} c \, dz = -F(0) + F(-h) + A \tag{12.64}$$

which says that the rate of change with time of the total suspended sediment in the water column is equal to the flux outward at the surface (which is zero) minus the flux inward at the bottom plus the source of strength A. In one dimension we can equivalently write the flux as

$$F_c = -k_z \frac{\partial c}{\partial z} - cw + AU(z - z_a) \tag{12.65}$$

where U is the unit step function, which we met in Chapter 2, and defined as

$$U(z) \equiv \left\{ \begin{array}{ll} 0 & z \leq 0 \\ 1 & z > 0 \end{array} \right. . \tag{12.66}$$

The derivative of U is a unit delta function, i.e.,

$$\frac{dU(z)}{dz} = \delta(z) \tag{12.67}$$

so that the two forms (12.63) and (12.65) are equivalent. For sediments, the source and sink are situated at the bottom of the water column, i.e., $z_a = -h$ and it is natural to divide the sediment flux into two components,

$$F = F_1 + F_2 \tag{12.68}$$

$$F_1 = \begin{cases} -k_z \dfrac{\partial c}{\partial z} & z > -h \\ EU(z+h) & z = -h \end{cases} \tag{12.69}$$

$$F_2 = -wc \quad z \geq -h. \tag{12.70}$$

Hence by continuity

$$E = wc_0 \exp\left(\frac{w_s}{k_z}h\right) \tag{12.71}$$

which gives

$$c_0 = \frac{E}{w} \exp\left(-\frac{w}{k_z}h\right) \tag{12.72}$$

or equivalently

$$c(-h) = \frac{E}{w}. \tag{12.73}$$

More realistically, we define the bottom of the water column as $z = -h + a$, and consider the region $z = -h + a$ to $-h$ as sediment.

12.9 Further reading

See further reading from earlier chapters. Additional texts for this chapter are:

Baudo, R. (1990). *Sediments*. Boca Raton, FL: CRC Press.

Bennassai, G. (2006). *Introduction to Coastal Dynamics and Shoreline Protection*. Sonthampton, UK: WIT Press.

Bird, E. (2000). *Coastal Geomorphology: An Introduction*. New York: John Wiley.

Chung, T. J. (2002). *Computational Fluid Dynamics*. Cambridge, UK: Cambridge University Press.

Dean, R. G. and R. A. Dalrymple (2004). *Coastal Processes with Engineering Applications*. Cambridge, UK: Cambridge University Press.

DiToro, D. M. (2000). *Sediment Flux Modeling*. New York: Wiley-Interscience.

Dronkers, J. (2005). *Dynamics of Coastal Systems*. Singapore: World Scientific Publishing Co.

Dyer, K. R. (1986). *Coastal and Estuarine Sediment Dynamics*. New York: John Wiley.

Herbich, J. B. (2000). *Handbook of Coastal Engineering*. New York: McGraw-Hill.

Julien, P. Y. (1998). *Erosion and Sedimentation*. Cambridge, UK: Cambridge University Press.

Komar, P. D. (1997). *Beach Processes and Sedimentation*. Englewood Cliffs, NJ: Prentice Hall.

Mudroch, A. and J. Azcue (1995). *Manual of Aquatic Sediment Sampling*. Boca Raton, FL: CRC Press.

Nielsen, P. (1992). *Coastal Bottom Boundary Layers and Sediment Transport*. Singapore: World Scientific Publishing Co.

Ochi, M. K., I. Dyer, R. E. Taylor, and J. N. Newman (2005). *Ocean Waves: The Stochastic Approach*. Cambridge, UK: Cambridge University Press.

Simon, D. and F. Senturk (1992). *Sediment Transport Technology, Water and Sediment Dynamics*. Highlands Ranch, CO: Water Resources Publications.

Svendsen, I. A. (2006). *Introduction To Nearshore Hydrodynamics*. Singapore: World Scientific Publishing Co.

U.S. Army Corps of Engineers (2004). *Coastal Groins and Nearshore Breakwaters*. Honolulu, HI: University Press of the Pacific.

References

Atkinson, M. J. and R. W. Bilger (1992). Effects of water velocity on phosphate uptake on coral reefs. *Limnol. Oceanogr.* **37**, 261–272.

Atkinson, M. J., J. L. Falter, and C. J. Hearn (2001). Nutrient dynamics in the Biosphere 2 coral reef mesocosm: water velocity controls NH_4 and PO_4 uptake. *Coral Reefs* **20**, 341–346.

Baird, M. E., M. Roughan, R. W. Brander, J. H. Middleton, and G. J. Nippard (2004). Mass-transfer-limited nitrate uptake on a coral reef flat, Warraber Island, Torres Strait, Australia. *Coral Reefs* **23**, 386–396.

Bloch, S. C. (2003). *Excel for Engineers and Scientists*, 2nd edn. New York: John Wiley.

Blumberg, A. F. and G. L. Mellor (1987). A description of a three-dimensional coastal ocean circulation model. In *Three-Dimensional Coastal Ocean Models*, ed. N. Heaps. Washington, DC: American Geophysical Union, pp. 1–16.

Brander, R. W., P. S. Kench, and D. Hart (2004). Spatial and temporal variations in wave characteristics across a reef platform, Warraber Island, Torres Strait, Australia. *Marine Geol.* **207** (1–4), 169–184.

Courant, R., K. O. Friedrichs, and H. Levy (1928). Über die partiellen Differenzengleichungen der mathematischen Physik. *Math. Annalen* **100**, 32–74.

Csanady, G. T. (1982). *Circulation in the Coastal Ocean (Environmental Fluid Mechanics)*. New York: Springer.

Dauxois, T. and M. Peyrard (2006). *Physics of Solitons*. Cambridge, UK: Cambridge University, Press.

Dean, R. G. and R. A. Dalrymple (2001). *Coastal Processes with Engineering Applications*. Cambridge, UK: Cambridge University Press.

DHI (2002). *MIKE 21 Coastal Hydraulics and Oceanography, Hydrodynamic Module: Reference Manual*. Hørsholm, Denmark: DHI Water and Environment.

Ebbesmeyer, C. C. and W. J. Ingraham (1994). Shoe spilt in the North Pacific. *Eos* **73**, 361–365.

Forester, C. S. (1953). *Hornblower and the* Atropos. London: Michael Joseph.

Fredsøe, J. and R. Diehards (1992). *Mechanics of Coastal Sediment Transport*, Advanced Series in Ocean Engineering vol. 3. Singapore: World Scientific Publishing Co.

Grant, W. D. and O. W. Madsen (1979). Combined wave and current interation with a rough bottom. *J. Geophys. Res.* **84**, 1797–1808.

Hatcher, A. (1989). Variation in the components of a benthic community structure in a coastal lagoon as a function of spatial scale. *Austr. J. Marine Freshw. Res.* **40**, 79–96.

Hatcher, B. G. (1990). Coral reef primary productivity: a hierarchy of pattern and process. *Trends Ecol. Evol.* **5**, 149–155.

Hearn, C. J. (1983). *Seasonal Aspects of the Oceanography of Koombana Bay*, Environmental Dynamics Report ED–83–045. Perth, WA: The University of Western Australia.

Hearn, C. J. (1985). On the value of the mixing efficiency in the Simpson–Hunter h/u^3 criterion. *Deutsche Hydrogr. Z.* **38**(3), 133–145.

Hearn, C. J. (1998). Application of the Stommel model to shallow mediterranean estuaries and their characterisation. *J. Geophys. Res.* **103**, 10391–10404.

Hearn, C. J. (1999). Wave-breaking hydrodynamics within coral reef systems and the effect of changing relative sealevel. *J. Geophys. Res.* **104**, 30007–30020.

Hearn, C. J. (2000). The SPECIES simulation software. In *Proceedings of the United Nations Conference on Information Management and Decision Support for Marine Biodiversity Protection and Human Welfare: Coral Reefs*, pp. 87–100. Townsville, QLD: Australian Institute of Marine Science.

Hearn, C. J. (2006). The great leap forward: particle age in coastal models. *Eos Trans.* **87**(52), Fall Meeting Suppl. Abstract OS33F–01.

Hearn, C. J. and J. L. Largier (1997). The summer buoyancy dynamics of a shallow mediterranean estuary and some effects of changing bathymetry: Tomales Bay, California. *Estuar. Coast. Shelf Sci.* **45**, 497–506.

Hearn, C. J. and I. N. Parker (1988). Hydrodynamic processes on the Ningaloo Coral Reef, Western Australia. *Proc. 6th Int. Coral Reef Symp.* **2**, 497–502.

Hearn, C. J. and G. Prince (1983). *A System of Computer Programs for the Acquisition and Analysis of Drogue Data*, Environmental Dynamics Report ED-83-052. Perth, WA: The University of Western Australia.

Hearn, C. J. and B. J. Robson (2001). Inter-annual variability of bottom hypoxia in a shallow mediterranean estuary. *Estuar. Coast. Shelf Sci.* **52**, 643–657.

Hearn, C. J. and E. Schulz (in press). Hydrodynamics of flow across a coral reef. *Proc. Roy. Soc. Lond. A*.

Hearn, C. J. and H. S. Sidhu (1999). The Stommel model of shallow coastal basins. *Proc. Roy. Soc. Lond. A* **455**, 3997–4011.

Hearn, C. J. and H. S. Sidhu (2003). Stommel transitions in shallow coastal basins. *Continent. Shelf Res.* **23**, 1071–1085.

Hearn, C. J., J. R. Hunter, J. Imberger, and D. van Senden (1985). Tidally induced jet in Koombana Bay, Western Australia. *Austr. J. Marine Freshw. Res.* **36**(4), 453–479.

Hearn, C. J., M. J. Atkinson, and J. L. Falter (2001). A physical derivation of nutrient uptake rates in coral reefs: effects of roughness and waves. *Coral Reefs* **20**(4), 347–356.

Kantha, H. and C. A. Clayson (2000). *Numerical Models of Oceans and Oceanic Processes*, International Geophysics Series vol. 66. New York: Academic Press.

Kato, H. and O. M. Phillips (1969). On the penetration of a turbulent layer into a stratified fluid. *J. Fluid Mech.* **37**, 643–655.

Longuet-Higgins, M. S. and R. W. Stewart (1964). Radiation stress in water waves: a physical discussion with applications. *Deep Sea Res.* **11**, 529–562.

Luketina, D. A. and J. Imberger (1987). Characteristics of a surface buoyant jet. *J. Geophys. Res.* **92**, 5435–5447.

Mangor, K. (2004). *Shoreline Management Guidelines*. Hørshølm, Denmark: DHI Water and Environment.

Mann, K. H. and J. R. Lazier (2005). *Dynamics of Marine Ecosystems: Biological–Physical Interactions in the Oceans*. Oxford, UK: Blackwell.

Mellor, G. L. and T. Yamada (1982). Development of a turbulence closure model for geophysical fluid problems. *Rev. Geophys. Space Phys.* **20**, 851–875.

Minorsky, N. (1974). *Introduction to Nonlinear Oscillations*. Malabar, FL: Krieger.

Munk, W. H. and E. R. Anderson (1948). Notes on a theory of the thermocline. *J. Marine Res.* **3**, 276–295.

Officer, C. B. (1976). *Physical Oceanography of Estuaries and Associated Coastal Waters*. New York: John Wiley.

Paterson, J. C., P. F. Hamblin, and J. Imberger (1984). Classification and dynamic simulation of the vertical structure of lakes. *Limnol. Oceangr.* **29**, 845–861.

Pope, S. B. (2000). *Turbulent Flows*. Cambridge, UK: Cambridge University Press.

Rahmstorf, S. (1995). Bifurcations of the Atlantic thermohaline circulation in response to changes in the hydrological cycle. *Nature* **378**, 145–149.

Rahmstorf, S. and A. Ganopolski. (1999). Simple theoretical model may explain apparent climate instability. *J. Climate* **12**, 1349–1352.

Simpson, J. H. and J. R. Hunter (1974). Fronts in the Irish Sea. *Nature* **250**, 404–406.

Simpson, J. H., J. Brown, J. Matthews, and G. Allen (1990). Tidal strain, density currents and stirring in the control of estuarine stratification. *Estuaries* **13**, 125–132.

Smith, S. V., W. J. Kimmerer, E. A. Laws, R. E. Brock, and T. W. Walsh (1981). Kaneohe Bay sewage diversion experiment: perspectives on ecosystem responses to nutritional perturbation. *Pacif. Sci.* **35**, 279–395.

Smith, S. V., J. T. Hollibaugh, S. J. Dollar, and S. Vink (1991a). Tomales Bay Metabolism: C–N–P stoichiometry and ecosystem heterotrophy at the land–sea interface. *Estuar. Coast. Shelf Sci.* **33**, 223–257.

Smith, S. V., J. T. Hollibaugh, S. J. Dollar, and S. Vink (1991b). Tomales Bay, California: a case for carbon controlled nitrogen cycling. *Limnol. Oceanogr.* **34**, 37–52.

Stocker, T. F. and D. G. Wright (1991). Rapid transitions of the ocean's deep circulation induced by changes in surface water fluxes. *Nature* **351**, 729–732.

Stommel, H. (1961). Thermohaline convection with two stable regimes of flow. *Tellus* **13** (2), 224–230.

Stommel, H. (1963). Varieties of oceanographic experience: the ocean can be investigated as a hydrodynamical phenomenon as well as explored geographically. *Science* **15**, 572–576.

Tait, R. J. (1972). Wave set-up on coral reefs. *J. Geophys. Res.* **77**, 2207–2211.

Taylor, G. I. (1953). Dispersion of soluble matter in solvent flowing through a tube. *Proc. Roy. Soc. Lond. A* **219**, 186–203.

Thorpe, S. A. (2005). *The Turbulent Ocean*. Cambridge, UK: Cambridge University Press.

Thurman, H. V. and A. P. Trujillo (2003). *Introductory Oceanography*, 10th edn. New York: Prentice Hall.

Whitehead, J. A. (1995). Thermohaline ocean processes and models. *Ann. Rev. Fluid Mech.* **27**, 89–113.

Whitehead, J. A. (1998). Multiple T–S states for estuaries, shelves, and marginal seas. *Estuaries* **21**, 281–293.

Wolanski, E. and S. Spagnol (2000). Sticky waters in the Great Barrier Reef. *Estuar. Coast. Shelf Sci.* **50**, 27–32.

Zeidberg, L. D. and W. M. Hamner (2002). Distribution of squid paralarvae, *Loligo opalescens* (Cephalopoda: Myopsida), in the Southern California Bight in the three years following the 1997–1998 El Niño. *Marine Biol.* **141**, 111–122.

Index

Entries in italics refer to figures and those in bold to tables.

475